PLASMA PROTEIN TURNOVER

PLASMA PROTEIN TURNOVER

edited by

R. BIANCHI and G. MARIANI

(Second Medical Clinic and C.N.R. Clinical Physiology Laboratory at the University of Pisa, Italy), and

A. S. McFARLANE

(Clinical Research Centre, Harrow, Middlesex)

SBN 333 18475 0

ISBN 978-1-349-02646-3   ISBN 978-1-349-02644-9 (eBook)
DOI 10.1007/978-1-349-02644-9

First published 1976 by

THE MACMILLAN PRESS LTD.

London and Basingstoke
Associated companies in New York Dublin
Melbourne Johannesburg and Madras

Preface

In May 1974 the Plasma Protein Group held its sixth meeting in Torino (Italy) at the invitation of Professor Donato (Pisa). The scientific organisation of the meeting was by the Plasmaprotein Division of the Clinical Physiology Laboratory of the Italian National Research Council in Pisa, and namely by Drs R. Bianchi and G. Mariani.

The meeting was the first to be subsidised by the newly created Italian Foundation for Medical Research (F.I.R.M.) under the chairmanship of Mrs Agnelli. The symposium was held in the accommodation of the Fondazione Giovanni Agnelli in Torino, and local arrangements which were excellent were in the care of Dr. G. Mariani. The Group is particularly indebted to the Italian Foundation for Medical Research and to the Italian National Research Council (C.N.R.) without whom the meeting could not have been held.

To facilitate publication of the proceedings the services of The Macmillan Press of London and a British co-editor were sought. This book represents a factual account of the proceedings which consisted mainly of the presentation of papers. These are arranged in the order of their presentation but for the first time the programme included general lectures by four invited authors and these lectures are interspersed among the papers.

1975

R.B.
G.M.
A.S.McF.

Principal Symposium
Contributors and the Editors

R. Bianchi, Laboratorio Fisiologia Clinica C.N.R., Via Savi 8, 56100 Pisa, Italy

K. Birger Jensen, Medical Department B, Division of Gastroenterology, Københavns Amts Sygehus i Gentofte, Niels Andersens vej 65, 2900 Hellerup., Copenhagen, Denmark

S. Comoglio, Nuclear Research Centre, SORIN, Laboratorio Fisiologia Clinica C.N.R., Via Savi 8, 56100 Pisa, Italy

J. W. L. Davies, MRC Industrial Injuries and Burns Unit, Birmingham Accident Hospital, Bath Row, Birmingham, B15 1NA, UK

C. P. Dietrich, Departamento de Bioquímica e Farmacologia, Escola Paulista de Medicina, Rua Botucatu 862, Caixa Postal 20372, São Paulo, Brazil

C. O. Enwonwu, Department of Biochemistry, Faculty of Medicine, University of Nigeria, Enugu Campus, Enugu, Nigeria

J. J. Franks, Department of Medicine, University of Colorado Medical Center, 4200 East 9th Avenue, Denver, Colorado 80220, USA

A. H. Gordon, Medical Research Council, National Institute for Medical Research, The Ridgeway, Mill Hill, London, NW7 1AA, UK

W. P. T. James, University of Cambridge and Medical Research Council, MRC Dunn Nutrition Unit, Dunn Nutritional Laboratory, Milton Road, Cambridge, CB4 1XJ, UK

B. G. Johansson, Lasarettet i Lund, Department of Clinical Chemistry, University Hospital, S-221 85 Lund, Sweden

C. O. Kindmark, Department of Infectious Diseases, Akademiska Sjukhuset, S-750 14 Uppsala, Sweden

N. R. Lazarus, Diabetes Research Laboratory, The Wellcome Foundation Ltd., Temple Hill, Dartford, Kent, DA1 5AH, UK

A. S. McFarlane, Division of Clinical Investigation, Clinical Research Centre, Watford Road, Harrow, HA1 3UJ, UK

G. Mariani, Laboratorio Fisiologia Clinica C.N.R., Via Savi 8, 56100 Pisa, Italy

L. L. Miller, Department of Radiation Biology and Biophysics, School of Medicine and Dentistry, University of Rochester, Rochester, New York 14642, USA

G. Milhaud, Service d'Explorations Fonctionnelles par les Radioisotopes,

Hopital Saint-Antoine, 184, Rue du Faubourg Saint-Antoine, 75571 Paris Cédex 12, France

H. N. Munro, Department of Nutrition and Food Science, Massachusetts Institute of Technology, Cambridge, Massachusetts 02139, USA

R. Quast, Hygiene-Institut und Medizinal-Untersuchungsamt der Universität Marburg, Pilgrimstein 2, S-355 Marburg/Lahn, West Germany

E. B. Reeve, Division of Laboratory Medicine, University of Colorado Medical Center, 4200 East 9th Avenue, Denver, Colorado 80220, USA

E. Regoeczi, Department of Pathology, Room 4N55, McMaster University Health Sciences Center, 1200 Main Street West, Hamilton, Ontario, L8S 4J9, Canada

J. Robbins, Department of Health, Education, and Welfare, Clinical Endocrinology Branch, National Institute of Arthritis, Metabolism, and Digestive Diseases, N.I.H., Building 10, Room 8N 315, Bethesda, Maryland 20014, USA

U. Rosa, Laboratorio Fisiologia Clinica C.N.R., Via Savi 8, 56100 Pisa, Italy

V. M. Rosenoer, G. I. Research Unit, Lahey Clinic Foundation, 605 Commonwealth Avenue, Boston, Massachusetts 02215, USA

N. Rossing, Department of Clinical Physiology, Finseninstitutet, 2100 København ø, Strandboulevarden 49, Copenhagen, Denmark

M. A. Rothschild, Department of Nuclear Medicine, Veterans Administration Hospital, First Avenue at East 24th Street, New York 10010, USA

I. H. Scheinberg, Albert Einstein College of Medicine of Yeshiva University, 1300 Morris Park Avenue, Bronx, New York 10461, USA

E. Shafrir, The Hebrew University, Hadassah Medical School, Jerusalem, Israel

T. P. Stein, The Graduate Hospital, University of Pennsylvania, 19th and Lombard Streets, Philadelphia, Pennsylvania 19146, USA

Y. Takeda, University of Colorado Medical Center, 4200 East 9th Avenue, Denver, Colorado 80220, USA

A. S. Tavill, Division of Clinical Investigation, M.R.C. Clinical Research Centre, Watford Road, Harrow, Middlesex, HA1 3UJ, UK

A. Vermeulen, Department of Endocrinology and Metabolism, Akademisch Ziekenhuis, De Pintelaan 135, B-9000 Gent, Belgium

F. Vítek, Institute of Biophysics, Faculty of General Medicine, Charles University, Salmovská 3, 120 00 Prague 2, Czechoslovakia

T. A. Waldmann, Metabolism Branch, National Cancer Institute, National Institutes of Health, Building 10, Room 4N 117, Bethesda, Maryland 20014, USA

Contents

PART 1

Tracer Preparation

1
Preparative isoelectric focusing of plasma proteins

R. QUAST

Even though the purpose of this paper is to review techniques of preparative isoelectric focusing with special regard to the separation of plasma proteins, it should be made clear from the start that it will be limited to the discussion of isoelectric focusing in pH gradients based on ampholytes, so-called 'natural pH gradients'[1]. It will not deal with transient pH gradients to be observed in electrolysis experiments with different buffer systems, nor with stable pH gradients produced by temperature gradients in a buffering solution as described by Luner and Kolin[2]. Furthermore, special attention will be focused on the amount of protein which can be separated, or which actually has been separated in a single experiment.

If we agree that 'preparative' means a sample of 100 mg or more, the vast literature on isoelectric focusing is substantially reduced, and there remain only a few examples which describe the separation of such an amount of protein in a single experiment, although the theoretical limits of present day techniques are much higher.

The theoretical approach to the presently most used technique for isoelectric focusing has been founded by H. Svensson[1,3,4], while the practical realisation rests on the production of suitable ampholytes[5-7]. These are polyamino-polycarboxylic acids resembling in many properties natural amino acids. A comparison between two physical separation techniques used in molecular biology may elucidate the principles underlying the unsurpassed resolving power of isoelectric focusing. In isoelectric focusing the concentration of macromolecular solute in a single zone is given by $C = C_0 \exp(-x^2/2\sigma^2)$ whereas in equilibrium density gradient centrifugation it is given by $m = m_{r_0} \exp(-(r-r_0)^2/2\sigma^2)$ where σ, the standard deviation of the Gaussian function, is defined for each technique as follows:

$$\sigma = \pm \sqrt{\frac{1}{fN(\mathrm{d}u/\mathrm{d}(\mathrm{pH}))} \times \frac{1}{E} \times \frac{1}{(\mathrm{d}(\mathrm{pH})/\mathrm{d}x)} \times \mathrm{R}\,T}$$

(isoelectric focusing)

$$\sigma = \pm \sqrt{\frac{\rho_0}{M_s} \times \frac{1}{\omega r} \times \frac{1}{(d\rho/dr)_{eff}}} \times RT$$

(equilibrium density gradient
centrifugation)

The formal similarity of the equations, rewritten from those of Svensson[1] and Trautman[8], is clearly visible. In both techniques, isoelectric focusing as well as equilibrium density gradient centrifugation, a steady state is reached in which the resolving power depends largely on the slope of the respective gradient—on the pH gradient, or on the density gradient—and the actual resolution depends additionally on the field strength—electric field, or gravity field—and on molecular parameters of the macromolecular solute—diffusion coefficient and slope of the titration curve in the vicinity of the isoelectric point, or density and molecular weight. As is shown by the standard equation, the mass concentration of a focussed solute through a zone in the steady state follows the law of Gaussian distribution. Diffusional forces tend to move the molecules away from the concentration maximum in the middle of the zone, while forces exerted by the field and by the influence of the gradient on the molecules collect them again. The 'natural' pH gradient in isoelectric focusing is established by a series of amphoteric substances of a molecular weight in the range 300–800 units. Some readily available laboratory chemicals, mostly amino acids, have been used by Svensson in his first experiments[4]. However, the number of ampholytes in those experiments was rather small compared with ampholyte mixtures used nowadays, and so the separation power was also small. Today, under the trade name Ampholine, ampholytes are marketed in mixtures which are ideally suited for isoelectric focusing. Similar substances were recently synthesised which according to the authors prove to be comparable to the commercially available product[9]. The Ampholine materials span a range from pH 2.5–11, and may be obtained in several subranges. Successful separations of very acidic proteins have been reported recently[10].

Although preparative and analytical isoelectric focusing have many aspects in common, both differ very much in the question of resolution of zones. When resolution is defined as a clearly detectable difference between two adjacent zones, based on the theoretical considerations of Svensson[11], this may suffice as a guideline for successful analytical experiments. In preparative isoelectric focusing, however, complete separation between two neighbouring zones is expected, so that both components may be recovered in a pure state. Thus, requirements for preparative isoelectric focusing are much more stringent. At the same time, the quantitative differences between both applications must be considered. In typical analytical experiments with the 110 ml column of LKB, a maximum of 1–4 mg of protein per zone width of 1–2 mm may be applied, whereas preparative isoelectric focusing should yield purified protein in the order of more than 200 mg. Fortunately, good resolving power and high yield are not contradictory in density gradient stabilised columns, as has been pointed out by Rilbe and Petterson[12]: 'The capacity of a density gradient stabilised column rises with the square of the resolving power with which it operates for one and the same protein system and for otherwise unchanged experimental conditions', and further, 'Extremely sharp bands—as found in analytical runs—indicate low resolving power' (which in other words, indicate steep pH gradients), 'whereas a fine resolution within a narrow pH range is characterised by comparatively broad zones'. These authors calculated the maximum load for a 110 ml density gradient stabilised column to be 1.074 g for a single protein, and achieved 74 per cent of this value in experiments with a newly designed vertically cooled short column.

Some other problems which are accentuated in preparative experiments relate to the chemical properties of the ampholytes and of the density gradient substances. After a successful separation, one often wishes to separate the ampholytes from the proteins, because they interfere with protein determination and protein staining techniques. Three different methods for this purpose have been reported: dialysis, gel-filtration[13], and ammonium sulphate precipitation of the protein[14]. In some cases, one can use ion-exchange chromatography instead[15]. Although the ease with which the ampholytes may be separated from the proteins by dialysis or by gel-filtration implies that firm complexes between both are not formed[13], artefacts have been observed in analytical gel electrofocusing which must be interpreted as changes of the isoelectric points of proteins by complex formation with ampholytes[16]. Another type of artefact produced by the ampholytes has been reported to be a reduction of ferric myoglobin in gel electrofocusing[17]. Both these unwanted influences most probably do not occur in preparative runs, but one should be on guard against such surprises. The beneficial property of ampholytes to form relatively stable complexes with metal ions, thus preventing them from interacting with free thiol-groups of proteins[18], may lead to some embarrassment, when the complexed metal ions are necessary for the activity of an enzyme[19]. A similar ambiguity of influences may be observed with thiol-reagents, such as β-mercaptoethanol or dithiothreitol. These substances have a protective effect on the activity of, for instance, adenylate kinase when added to the density gradient solutions[20]. On the other hand, a number of unexpected zones were found in experiments with subunits of fumarase which were attributed to mixed disulphide formation with thiol-reagents[21]. In our laboratory, it was observed that Cleland's reagent (dithiothreitol) can be detected by its smell only in the fractions collected between pH 3.7–7.5 after electrofocusing a 10 mM solution in a pH 3–10 gradient. Whether that means a destruction in the pH regions outside that range, or a slow focusing has not been further investigated. Other additives often used in protein chemistry include urea in high concentrations. Urea may be used in isoelectric focusing experiments if one takes care not to introduce cyanate which may react with

R. Quast

proteins, and if one is aware of the pH-shift caused by the urea[22]. A protein stabilising effect has been reported for 6 M ethyleneglycol on J^{131}-gamma-globulin[23]. A similar effect is attributed to the sucrose in density gradient stabilised columns. When using sucrose, one should be informed of the occurrence of impurities[24] which can be only partially removed by membrane filtration of all solutions directly before filling the column[25]. Sucrose, moreover, is not suitable for the stabilisation of pH-gradients in the far alkaline region, because it dissociates and acquires electrical charge[26]. It can be replaced in such applications by glycerol or ethyleneglycol.

In all experiments with isoelectric focusing, the zones must be stabilised against convectional mixing brought about by gravitational or thermal instabilities. While it is not difficult to conduct an experiment so that the fluid in which the separation takes place is always in thermal equilibrium, the focusing of solutes in small zones inevitably leads to density differences and hence to gravitational instabilities. This effect is more pronounced in preparative work where higher mass concentrations are to be expected. In order to solve this problem, several stabilisation methods have been explored (*see* table 1.1). At present, the most promising approaches to preparative isoelectric focusing seem to be the large density gradient stabilised column, like that produced by LKB, the thick layer of granulated gel, the apparatus for which is sold by DESAGA, and the principle of zone convection. However, for the latter, no suitable apparatus is on the market. Examples of experiments done employing each of the three principles of stabilisation shall be demonstrated.

Table 1.1 Preparative isoelectric focusing in natural pH gradients.

Stabilisation of zones against gravitational remixing	Apparatus	Manufacturer	Literature
1. Density gradient	vertical column	LKB-Producer	Haglund (1967)[27]
	short column	experimental	Rilbe (1973)[28]
2. Gel layer	horizontal layer of granulated gel	DESAGA GmbH	Radola (1971)[29]
3. Semipermeable membranes	closed chambers	experimental	Rilbe (1969)[30]
4. Zone convection	horizontal chamber	experimental	Valmet (1969)[31,32]
	sea-serpent tube	experimental	Rilbe (1969), Rose & Harboe (1969)[33]
	coiled tube	experimental	this paper
5. Others	zero gravity field	not yet tried	Space-Lab

The most thoroughly investigated of these three techniques is that using a density gradient to stabilise the liquid[4]. With this technique, excellent results have been obtained. Although considerable loads have been separated successfully, the beginner is warned not to overload the density gradient, otherwise droplet sedimentation will occur which will spoil the whole experiment. This can be observed in electrofocusing a high load of a coloured protein, such as haemoglobin, in a steep pH gradient and at a high field strength.

An experiment done with a mixture of different haemolysates may serve to point out some details of interest for a successful application of a density gradient stabilised column in preparative isoelectric focusing (figure 1.1). First, although acceptable focusing has been achieved, the different haemoglobins were not separated, because a steep pH gradient was used (Ampholine 3–10). The same reason applies to the submaximal load of only 40.6 mg haemoglobin determined by spectrophotometry of the reduced pyridine haemochromogen[34,35]. Forcing separation in such an experiment by increasing the field strength of approximately 31 V cm^{-1} may lead to droplet sedimentation, because the local concentration of haemoglobin may become higher than can be carried by the density gradient. It is much better in such a situation to collect the fractions of interest and to repeat the run in a more shallow pH gradient.

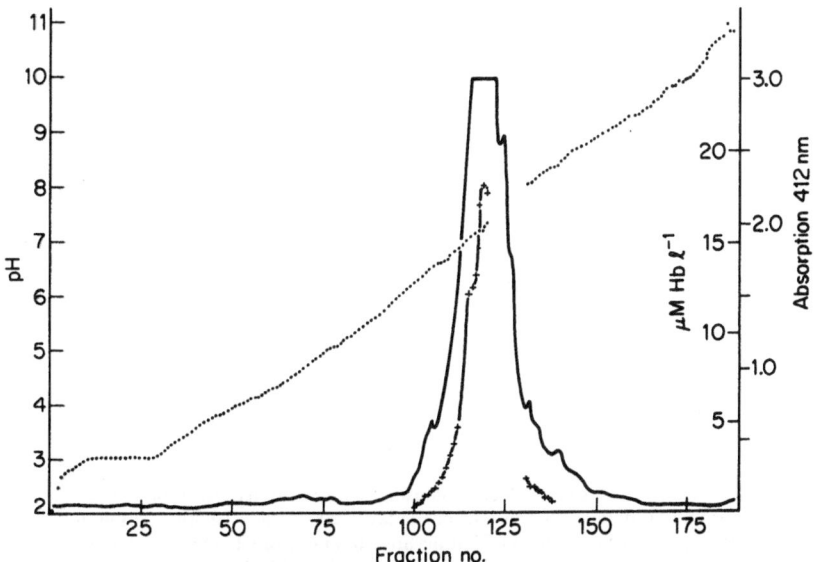

Figure 1.1. Isoelectric focusing in a preparative, density gradient stabilised column. Experimental details: total volume 550 ml; Ampholine 3–10, 1%; density gradient 0–50% w/v sucrose; sample 40.6 mg haemoglobin (haemolysates of human cord blood, human blood, horse blood, turkey blood, and crystalline bovine haemoglobin); voltage 300–1500 V; maximum power 12.1 W; temperature 4°C; duration of the run 28 hours.

The preparative variant of isoelectric focusing in thin layers of granulated gel has been introduced by Radola[29]. The equipment for this technique may be obtained from DESAGA, Heidelberg. We did a pilot experiment with 225 mg of serum proteins, following the manual, and found the technique very

R. Quast

useful. It seems to be important to have a real stiff gel, which before reducing
the water content by evaporation from the poured layer must be thoroughly
evacuated. The electrodes were placed in our experiment directly on top of
filter paper strips wetted with the respective electrode solutions and laid across
the ends of the gel layer, one long side of the strips folded and stuck between
gel and the wall of the gel trough. This arrangement resembles that of the
analytical Multiphor equipment of LKB. Considering that a wide range pH
gradient has been used, an acceptable separation was obtained (figure 1.2).

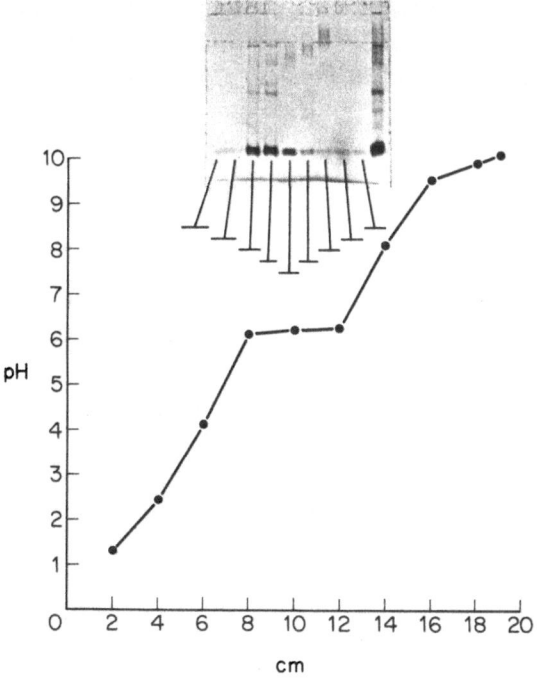

Figure 1.2. Preparative isoelectric focusing in a granulated gel stabilised layer.
Experimental details: dimensions of the layer 200 × 200 × 10 mm; Ampholine
3–10, 1%; Gel: Sephadex G-75 superfine; temperature 12–13°C (tapwater);
sample: 5 ml human serum, dialysed against 10 mM glycin, precipitates removed by ultracentrifugation, total protein 225 mg; voltage 200–400 V;
maximum power 40 W; duration of the run 4 hours.

The pH gradient as measured with a flat-membrane electrode directly on the
gel was not straight, which may be attributed to the relatively short time of the
run, or to the glycin content of the sample and the low buffering capacity of
the Ampholine 3–10 in the neutral region. If one really intends to use a wide
range (steep) pH gradient for preparative work, the pH gradient should be
fortified by adding ampholytes isoelectric in the neutral range—e.g. 5–7 or
6–8—as is done in analytical applications[36]. Although with this technique

no electroendosmotic flow has been reported up to now, one should be on the alert, for in all gel electrofocusing techniques precipitated protein may give rise to fixed charges and hence to electroendosmotic flow[30] which is directly proportional to the field strength applied. Another point which is critical in such experiments is the danger of local overheating, especially with gel layers of 10 mm cooled only from below. Since the greatest voltage drop occurs in zones of lowest conductivity, which typically are found in the neutral region of the pH gradient[6], and since the heat production is proportional to the electric power, local overheating is first to be expected in the neutral region.

The next experiment demonstrates an inexpensive approach to the technical realisation of the zone convection principle[30-33]. A comparable arrangement was already used for analytical purpose, independently[37,38]. A simple coil of glass tube, internal diameter about 6 mm, with 35 turns served as separation chamber, each turn being one compartment. During the run, the proteins condense in zones and tend to sediment to the lower part of the respective turn in which they are focused, thus creating a density gradient by themselves in each ascending section of a turn. In order to illustrate this technique, the coil is shown after such an experiment in figure 1.3. Collecting the fractions after the run presented considerable difficulties, and much of the separation was blurred by diffusion and by convectional mixing, especially by precipitates sticking to the walls of the tube. However, the absorption peaks spaced at nearly equal intervals (figure 1.4) indicate the successful application of the zone convection principle. The difficulties mentioned above may be over-

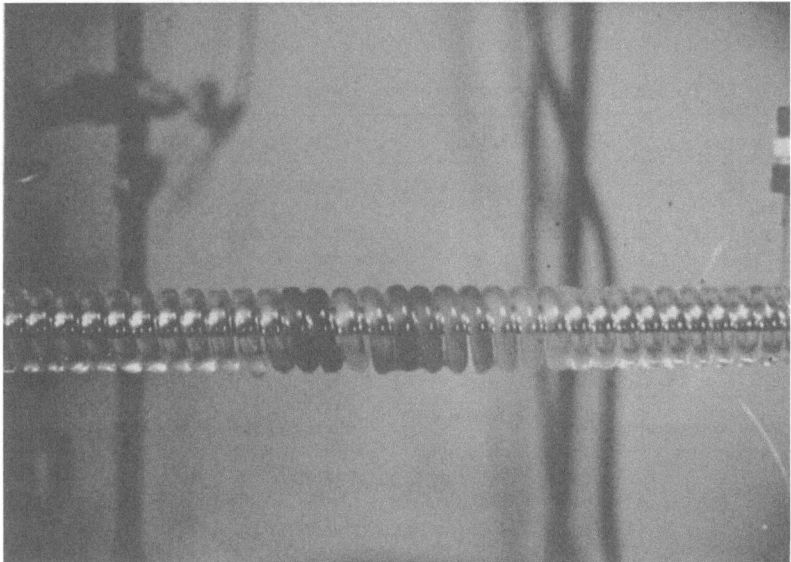

Figure 1.3. Preparative isoelectric focusing by zone convection in a coiled glass tube. Focusing of haemoglobin clearly visible. For experimental details see fig. 1.4.

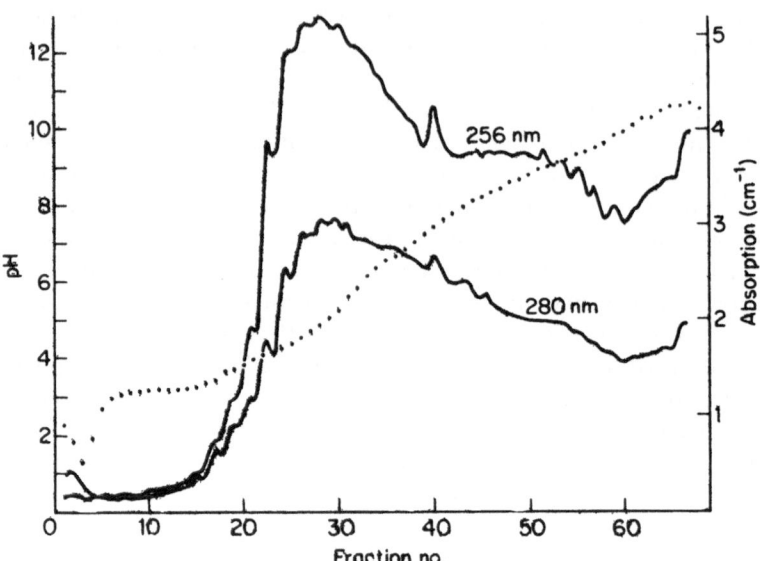

Figure 1.4. Preparative isoelectric focusing by zone convection in a coiled glass tube. Experimental details: tube length 295 cm; volume 88 ml; coil of 35 turns; Ampholine 3–10, 2% in 10% sucrose and 5% ethyleneglycol; sample: 10 ml human serum (see fig. 1.2), total protein 450 mg; voltage 2820 V throughout the run; maximum power 8.5 W; temperature 10°C; duration of the run: 138 hours. The absorption peaks between fraction No. 15 and No. 30 indicate condensed protein zones from the lower part of turns.

Table 1.2 A comparison of three different techniques for preparative isoelectric focusing

Instrument	vertical column	flat bed	coiled tube
Stabilisation	density gradient	granulated gel	zone convection
Availability	LKB	DESAGA	laboratory made
Volume	440 ml	400 ml or 800 ml	100 ml to 300 ml
Protein load	up to 1 g	up to 0.8 or 1.6 g	minimum 0.45 g; upper limit not yet known
Resolving power	high	medium	for a given pH gradient fixed by the number of turns
Ampholine consumption	high	high	low
Additional reagents	Sucrose, or other density gradient substance	Sephadex G 75 superfine, or Bio-Gel P-60 minus 400 mesh	none
Drawbacks	droplet sedimentation, when overloading density gradient	electroendosmotic flow, when precipitates become fixed in gel	high voltage (over 3 kV)

come by a new model of the glass coil in which a small straight tube is fused to the top of each turn. By this means the uppermost portion of the liquid in all turns may be removed at the same time with a battery of syringes, leaving the rest of the liquid in the middle and lower part of the turns as separated fractions. Subsequently, each single turn of the coil can be emptied in succession without the danger of contamination by adjacent fractions. The long time of the run—about 6 days—can only be reduced by applying a higher field strength, more than the 9.7 V cm^{-1} of the experiment. Therefore, high tension in the order of 10–30 kV d.c. appears necessary. Although this technique is at present by no means perfect, it is promising and will be explored further in this laboratory.

Another highly interesting technique is at present under development in the laboratories of Prof. Rilbe in Gothenburg. He employs very short columns in which a strong density gradient stabilises the liquid[12,28]. Field strengths of up to 75 V cm^{-1} can be applied and the time required for the separation is drastically reduced to half an hour. Very high resolutions were obtained with the newest models of the short density gradient stabilised columns. Once such

Table 1.3 The protein load in some experiments on preparative isoelectric focusing.

Protein sample	Load (mg)	Ampholine	Author
1. 110 ml *density gradient column, LKB*			
Liver prep., alkaline phosphatase	20	3–10	Latner et al.[19]
Casein in 7 M urea	50–80	3–6	Josephson et al.[22]
Specific antibodies	5–25		Freedman[39]
Specific antibodies	5–20		Freedman & Painter[40]
Albumin, J^{131}-Albumin	20	3–10, 4–6	Reis & Wetter[41]
J^{131}-Gamma-Globulin in 6 M ethyleneglycol	20	5–8	Jones et al.[23]
Albumin	50	4–6	Valmet[42]
Cohn Fract. IV	60	3–6	Li & Li[43]
Enzyme prep. from beef-kidney mitochondria	20	7–10	Scandurra et al.[44]
2. 440 ml *density gradient column, LKB*			
Plasminogen	60	3–10, 5.8–7	Wallén & Wiman[15]
Guinea pig serum protein	~225	3–10	Stewart-Tull & Arbuthnott[45]
Liver cytosol	30–40	3–10	Criss et al.[20]
Specific antibody	50–250	3–10, 5–8, 7–10	Freedman[39]
Specific antibody	40–200	3–10, 5–8, 6.2–7.4	Freedmann & Painter
Cohn Fract. II	250	5–9	Howard & Virella[46]
3. *Zone convection apparatus, not available*			
Albumin	250	5.7–6.3	Valmet[42]
Serum protein in 25% Glycerol	750	4–8	Rose & Harboe[33]
4. *Granulated gel-layer, DESAGA*			
Sperm whale myoglobin	1000	7–9	Radola[29]

apparatus will be commercially available—its domain appears to be the separation of labile proteins in very short runs.

A comparison of the main features of the three techniques demonstrated in experiments above may give an impression of the capabilities, the advantages, and the disadvantages one must take into account when deciding to use one of them for preparative work (see table 1.2).

It cannot be the aim of this paper to give a detailed treatment of the various plasma proteins and derivatives thereof, and of their behaviour in isoelectric focusing experiments. Nevertheless, some examples of applications of preparative isoelectric focusing found in the literature will be given (table 1.3). It is obvious that the maximum load has not been introduced in any one of these experiments. On the other hand, it may be seen that quite different proteins may be submitted to preparative isoelectric focusing. Furthermore, high loads are found in separations employing density gradient stabilised columns either in experiments with narrow-range ampholytes, or with very heterogeneous protein samples, thus demonstrating the important relationship between resolving power and sample load capacity already mentioned above.

I am afraid, that much confusion will arise from uncritical employment of the unexcelled resolving power of isoelectric focusing, especially when multiple molecular forms of proteins[26] are separated after chemical or enzymatical conversion to tracer proteins. In any case, it seems wise to use the corresponding analytical technique in pilot separations as a control and to avoid unnecessary losses of valuable material.

Acknowledgement

I am greatly indebted to Professor H. Rilbe for the information concerning the developments in his laboratory. The skilful technical assistance of Mrs. B. Snaidero in the experiments cited above is thankfully acknowledged. Most of the analytical instrumentation used in this work is a gift of the Deutsche Forschungsgemeinschaft.

References

1. Svensson, H. *Acta Chem. Scand.*, **15** (1961), 325–341
2. Luner, S. J. and Kolin, A. *Proc. Natl. Acad. Sci. U.S.*, **66** (1970), 898–903
3. Svensson, H. *Acta Chem. Scand.*, **16** (1962), 456–466
4. Svensson, H. *Arch. Biochem. Biophys.*, **Suppl. 1**, (1962), 132–138
5. Vesterberg, O. *Brit. Patent*, No. 1106818 (1968)
6. Davies, H. *Protides of Biological Fluids*, Proc. 17th coll., Brugge, Pergamon Press, Oxford, 1970, pp. 389–396

7. Haglund, H. *Methods of Biochem. Anal.*, **19** (1970), 1–104
8. Trautman, R. Ultracentrifugation. In: *Instrumental methods of experimental biology* (Newman, D. W., ed.) MacMillan Company, New York, 1967
9. Vinogradov, S. N., Lowenkron, S., Andonian, M. R., Bagshaw, J., Felgenhauer, K. and Pak, S. J. *Biochem. Biophys. Res. Comm.*, **54** (1973), 501–506
10. Vesterberg, O. *Acta Chem. Scand.*, **27** (1973), 2415–2420
11. Svensson, H. *J. Chromatog.*, **35** (1966), 266–273
12. Rilbe, H. and Petterson, S. Preparative isoelectric focusing in short density gradient columns with vertical cooling. In *Isoelectric Focusing*, J. P. Arbuthnot and J. A. Beeley, eds., Butterworths, London, 1975
13. Vesterberg, O. *Science Tools*, **16** (1969), 24–27
14. Nilsson, P., Wadström, T. and Vesterberg, O. *Biochim. Biophys. Acta*, **221** (1970), 146–148
15. Wallén, P. and Wiman, B. *Biochim. Biophys. Acta*, **257** (1972), 122–134
16. Frater, R. *J. Chromatog.*, **50** (1970), 469–474
17. Quinn, J. R. *J. Chromatog.*, **76** (1973), 520–522
18. Vesterberg, O. *Protides of Biological Fluids*, Proc. 17th coll., Brugge, Pergamon Press, Oxford, 1970, pp. 383–387
19. Latner, A. L., Parsons, M. E. and Skillen, A. W. *Biochem. J.*, **118** (1970), 299–302
20. Criss, W. E., Litwack, G., Morris, H. P. and Weinhouse, S. *Cancer Res.*, **30** (1970), 370–375
21. Penner, P. E. and Cohen, L. H. *J. Biol. Chem.*, **246** (1971), 4261–4265
22. Josephson, R. V., Maheswaran, S. K., Morr, C. V., Jenness, R. and Lindorfer, R. K. *Anal. Biochem.*, **40** (1971), 476–493
23. Jones, R. E., Hemmings, W. A. and Faulk, W. P. *Immunochemistry*, **8** (1971), 299–301
24. Earland, C. and Ramsden, D. B. *J. Chromatog.*, **35** (1968), 575–576
25. Delmotte, P. *Science Tools*, **17** (1970), 51
26. Vesterberg, O. VII *Symposium on Chromatography & Electrophoresis*, Publ. Press Acad. Europeennes, Brussels, 1973, pp. 81–98
27. Haglund, H. *Science Tools*, **14** (1967), 17–23
28. Rilbe, H. *Ann. N.Y. Acad. Sci.*, **209** (1973), 80–93
29. Radola, B. J. *Protides of Biological Fluids*, Proc. 18th coll., Brugge, Pergamon Press, Oxford, 1971, pp. 487–491
30. Rilbe, H. *Protides of Biological Fluids*, Proc. 17th coll., Brugge, Pergamon Press, Oxford, 1970, pp. 369–382
31. Valmet, E. *Science Tools*, **15** (1969), 8–13
32. Valmet, E. *Protides of Biological Fluids*, Proc. 17th coll., Brugge, Pergamon Press, Oxford, 1970, pp. 401–407
33. Rose, C. and Harboe, N. M. G. *Protides of Biological Fluids*, Proc. 17th coll., Brugge, Pergamon Press, Oxford, 1970, pp. 397–400

34. Paul, K. G., Theorell, H. and Åkeson, Å. *Acta Chem. Scand.*, **7** (1953), 1284
35. Falk, J. E. *Porphyrins and metalloporphyrins. Their general, physical and co-ordination chemistry, and laboratory methods.* BBA Library Vol. 2, Elsevier, Amsterdam, 1964
36. Söderholm, J., Allestam, P. and Wadström, T. *FEBS Letters*, **24** (1972), 89–92
37. Macko, V. and Stegemann, H. *Anal. Biochem.*, **37** (1970), 186–190
38. Bours, J. *Exp. Eye Res.*, **16** (1973), 501–515
39. Freedman, M. H. *J. Immunol. Methods*, **1** (1972), 177–198
40. Freedman, M. H. and Painter, R. H. *J. Biol. Chem.*, **246** (1971), 4340–4349
41. Reis, H. E. and Wetter, O. *Klin. Wschr.*, **47** (1969), 426–430
42. Valmet, E. *Protides of Biological Fluids*, Proc. 17th coll., Brugge, Pergamon Press, Oxford, 1970, pp. 443–448
43. Li, Yu-Teh and Li, Su-Chen *Protides of Biological Fluids*, Proc. 17th coll., Brugge, Pergamon Press, Oxford, 1970, pp. 455–463
44. Scandurra, R., Cannella, C. and Elli, R. *Science Tools*, **16** (1969), 17–19
45. Stewart-Tull, D. E. S. and Arbuthnott, J. P. *Science Tools*, **18** (1971), 17–21
46. Howard, A. and Virella, G. *Protides of Biological Fluids*, Proc. 17th coll. Brugge, Pergamon Press, Oxford, 1970, pp. 449–453

Discussion

MILLER

I have a question about the technique with the coil. It occurred to me and it is probably obvious to a lot of people that you might conceivably use a plastic coil that should have the same properties as glass or better, perhaps, in terms of thermal conductivity, and yet after your separation could simply be clamped off with haemostat-like clamps, and cut out without worrying about preserving the coil. Have you tried this kind of thing?

QUAST

We did not try it, but V. Macko and H. Stegemann[1] as well as J. Bours[2] have already done such experiments using quick freezing, for instance in liquid nitrogen, and just cut the coil. We did not use plastic coils, because plastic does not have adequate heat conducting properties, since we intend to use very high voltage in the range of 10 to 30 kV dc in order to shorten the time of the run which was six days in the reported case.

MUNRO

Have you experienced problems with multiple forms of protein which are in fact artefactual, in other words albumin carries tryptophan on 10% of the series of bands on isoelectric focusing, and you have the problem of proving molecules and also variable amounts of free fatty acids. This may give rise to a that you are dealing with separate species. Have you any experience or comments on this kind of problem associated with isoelectric focusing?

QUAST

I have no experience with multiple molecular forms myself, but it is well known from the literature that adsorption of fatty acids to albumin actually gives rise to a broad spectrum of albumins to be found between pH 4.8 and 5.8; that has been reported first by Erkki Valmet[3]. Then you have to think of the multiple forms of the immunoglobulins; maybe you have seen the fantastic resolution obtained in analytical gel electrofocusing where you get about 10 to 20 different bands of immunoglobulins. I think this is the real problem with this technique: that you will not see the forest, because there are too many trees.

GORDON

I would like to say two things about isoelectric focusing. I think it is the best technique for the investigation of microheterogeneity of proteins. If you use proper controls, then artefactual microheterogeneity due to ampholytes can be eliminated. My second point is that I feel that microheterogeneity of proteins is another subject and one on which the group could easily spend a good deal of time in the future.

References

1. Macko, V. and Stegemann, H. Free electrofocusing in a coil of polyethylene tubing. *Anal. Biochem.*, **37** (1970), 186
2. Bours, J. Free isoelectric focusing of bovine lens gamma-crystallins. *Exp. Eye Res.*, **16** (1973), 501
3. Valmet, E. The heterogeneity of human serum albumin. In *Prot. Biol. Fluids*, Vol., 17 (Peeters, H. ed.) Pergamon Press Ltd., Oxford, 1970, p. 443

2

Continuous fractionation of proteins by simultaneous gel filtration and electrophoresis*

C. P. DIETRICH

Electrophoresis up to the present is one of the best methods for identification and characterisation of proteins. We can even say that the development of electrophoresis has permitted the discovery of a wide variety of non-enzyme proteins.

Several improvements of the electrophoretic method, using different supports besides paper have been developed in the last years. For instance, refined separations of mixtures of proteins using polyacrylamide and starch gels as supports have been exhaustively reported (for a review, *see* reference 1).

Nevertheless, attempts to obtain large amounts of pure protein fractions using those electrophoretic methods have been to a large extent relatively unsuccessful. Large sample loads decrease proportionally the resolution of the protein fractions. Also, recovery of the fractions from the gels is usually poor and contamination of the sample with the support is usually a drawback[1].

Theoretically the continuous loading electrophoresis principle[2] could eliminate the two major drawbacks of the single load systems mentioned above. The sample to be fractionated would show at any specific time a relatively small concentration into the system and could be completely recovered.

At least three models of electrophoresis apparatus using this principle are commercially available. Nevertheless not many reports using these types of apparatus for standard large-scale fractionation of proteins have appeared, which is in a sense surprising. The main criticism that has been made regarding the commercially available continuous electrophoresis systems is variation of the elution profile during the continuous run.

We have designed an electrophoresis chamber in which these variations are minimised[3]. The chamber also can be used with any given support that permits flow of liquid such as cellulose acetate, Sephadex, etc. With the use of Sephadex, molecular sieving was introduced in the electrophoretic fractionation.

*Aided by grants from FAPESP (Fundaçao de Amparo à Pesquisa do Estado de São Paulo), CNPq (Conselho Nacional de Pesquisas), and FINEP (Financiadora de Projetos)

Design

The apparatus, shown in figure 2.1, consists of a $12 \times 15 \times 0.2$ cm inner chamber (electrophoresis chamber) and two $10 \times 15 \times 0.2$ cm outer (refrigeration) chambers. The entire system is constructed of lucite. The two main lucite sections have been designed purposely to allow easy removal of one of the refrigeration chambers as shown in figure 2.1. This improved design permits

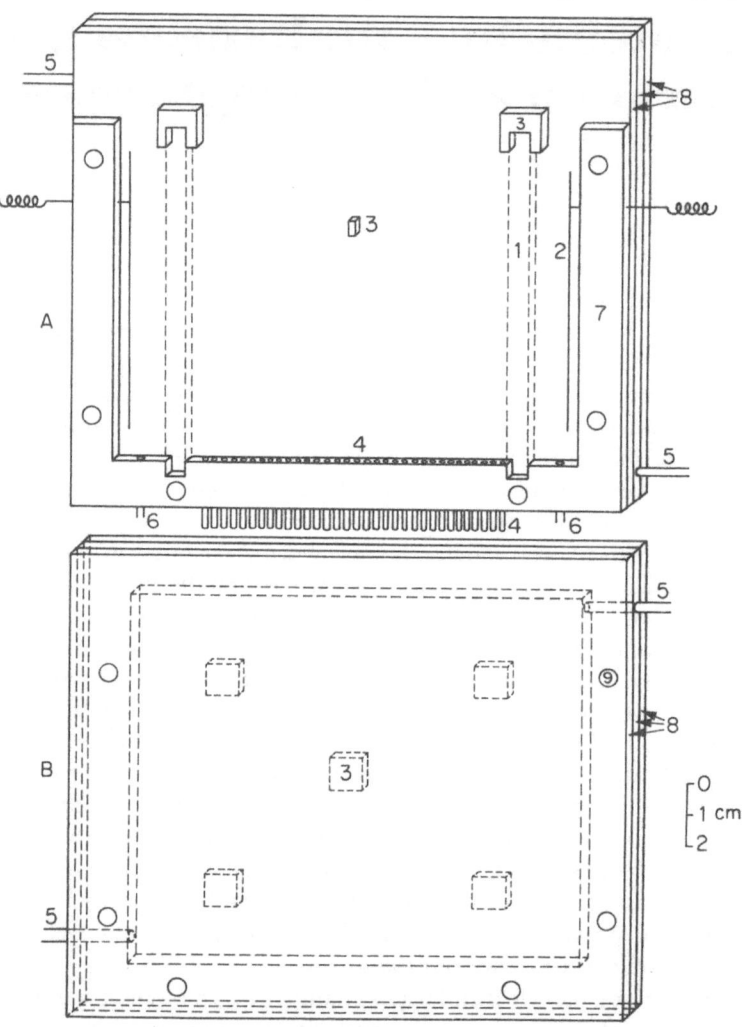

Figure 2.1. Scheme of the disassembled continuous loading electrophoresis apparatus. A, refrigeration chamber with the components of the electrophoresis chamber. B, refrigeration chamber. 1, filter paper pads; 2, platinum electrode; 3, lucite blocks; 4, eluate outlets; 5, cold water inlets and outlets; 6, buffer inlet; 7, wall of the electrophoresis chamber; 8, wall of the refrigeration chambers; 9, holes for the bolts.

the manipulation of the electrophoresis chamber. During operation the two sections are held together by six strategically placed bolts and wing nuts (figure 2.2). The inner chamber is lubricated with silicon grease to avoid leakages of buffer and Sephadex.

The electrophoresis chamber is divided into a central chamber and two lateral compartments by two paper pads consisting of six strips of Whatman 3 MM paper (1, figure 2.1) tightly packed together and placed vertically at 2 cm from each end of the chamber. The pads permit the passage of the electric current through the electrophoresis chamber and confine the Sephadex to this chamber. Lucite spacers (3, figure 2.1) are placed strategically within the refrigeration and electrophoresis chambers to avoid binding or buckling of the lucite walls and to hold the paper pads, as shown in figure 2.2.

Figure 2.2. A frontal view of the continuous loading electrophoresis system in operation.

Polyethylene tubing connections are used for the circulation of water to the refrigeration chambers. The diameter of the water inlet tubes is one-third the diameter of the outlet lines to prevent pressure build up.

The bottom of the electrophoresis chamber has 40 outlets connected to 0.1 cm diameter polyethylene tubes. The lower extremities of these tubes are loosely packed with cotton, which is held in position by other polyethylene tubes of small diameter. The latter are mounted in a lucite sheet ($0.2 \times 25 \times$ 10 cm) which is put on top of a rack containing the collecting test tubes. The cotton filling permits the flow of liquid and holds the gel inside the electrophoresis chamber.

The buffer solution is continuously introduced by a peristaltic pump at the bottom of the electrode compartments (6, figure 2.1) at the rate of 10–30 ml min^{-1}. The buffer flows out from the top of these chambers along the external walls and is drained through cotton wicks attached to these walls.

A third buffer outlet is situated on top of the spacer that holds the paper pads (figure 2.2). This outlet is placed parallel to the surface of the gel and in the direction of the electrophoresis chamber. In this position the buffer that flows from this outlet does not disturb the surface of the gel. A difference of level (0.5 cm) between these spacers and the walls of the electrophoresis chamber precludes the flow of buffer from the electrodes to the middle compartment.

Operation

With the apparatus in vertical position enough buffered Sephadex.G-25 (coarse) is introduced into the electrophoresis chamber to fill the 40 polyethylene outlets at the bottom of the chamber. Subsequently buffered gel of the chosen grades is introduced up to the lucite blocks that hold the paper pads. (It was found necessary to coat the paper pads with hot agarose (1% in buffer) just prior to the introduction of the Sephadex.) As soon as the gel settles, the refrigeration and the buffer injection pumps are started and the tube outlets introduced into the test tubes. The sample to be fractionated is continuously injected into the electrophoresis chamber through a capillary tube which is dipped into the gel to a depth of 1 cm. The continuous sample injector is turned on at a rate of $0.25–1$ ml h^{-1}. At the same time the power supply is switched on to the desired potential. At the end of the injection the whole system is kept in operation for about 2 hours to ensure complete elution of the sample.

Fractionation of proteins

The fractionation of human serum proteins with Sephadex G-75 (superfine) is shown in figure 2.3. A potential of 500 volts was applied for the fractionation. Serum was processed at a rate of 1 ml h^{-1} during this run. The analytical method employed to detect the proteins was agarose gel electrophoresis in 0.05 M barbital buffer pH 8.6[4]. A complete separation of γ, β_1 and β_2 as well as α_2-globulins was obtained using the Sephadex G-75 as support for the electrophoresis and Tris-HCl, 0.04 M, pH 8.6, as the eluent buffer. Albumin and α_1-globulin were not separated from each other in this fractionation. In another group of experiments using plasma instead of serum, fibrinogen was also obtained free of the other protein fractions.

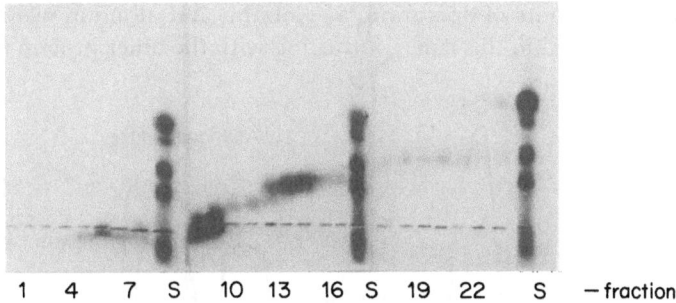

1   4   7   S   10 13 16 S   19 22   S   — fraction

Figure 2.3. Agarose gel electrophoresis of aliquots of the protein fractions obtained by the continuous loading gel electrophoresis in Sephadex G-75 (superfine). S, 2 μl of human serum.

Among the Sephadex gels the G-75 (superfine) seemed to be the one that has the highest load capacity. When Sephadex G-25 was used (figure 2.4), the plasma proteins were fractionated in groups; γ-globulin together with fibrinogen were separated from the β-globulins which in turn were separated from the α-globulins contaminated with albumin. The different pattern of fractionation obtained by the use of these two types of gel stresses the importance of the sieving in the fractionation.

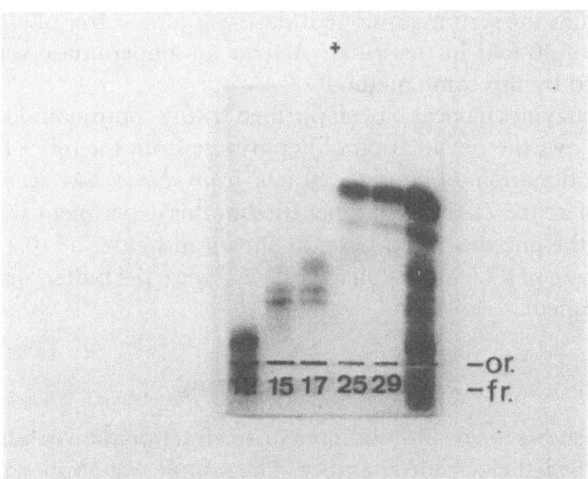

Figure 2.4. Experiment as described in figure 2.3 except that Sephadex G-25 was used as support for the fractionation.

Figure 2.5 shows the fractionation of three enzymatic activities, one aminopeptidase and two kininases (carboxypeptidases), in Sephadex G-200*. The load of this run was 0.25 ml of serum per hour. The three enzymes were completely separated from each other. 18 ml of serum was processed in some of the runs in 72 hours of operation. α_1-globulin and albumin were still not completely separated in this run, contrasting with the other protein fractions.

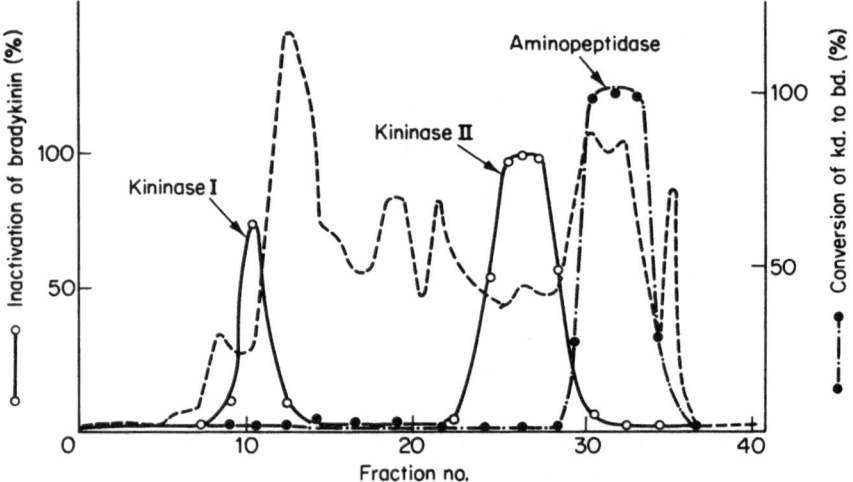

Figure 2.5. Fractionation pattern of human serum and two kininases and an aminopeptidase by continuous gel electrophoresis in Sephadex G-200.

The fractionation of α_1-globulin and albumin was achieved using 0.01 M Tris-maleate buffer pH 6.0 in Sephadex G-25, as shown in figure 2.6. Under these conditions the serum aminopeptidase was almost free of albumin with a purification of 40-fold in this run*. A liver aminopeptidase was also extensively purified by this same method[6].

Bacterial enzymes have also been purified by the continuous loading system. Figure 2.7 shows the purification of heparinase from the other bacterial proteins by fractionation of crude extracts from *Flavobacterium heparinum* (for details, *see* reference 7). The chamber used in this experiment was three times longer than the one described here, as shown in figure 2.8. 0.1 M ethylenediamine acetate pH 7.0 and Sephadex G-200 were the buffer and the gel used in this experiment.

Trouble-shooting

Figure 2.9 shows the results obtained in one fractionation in which there was a leakage of the left electrode chamber. The sample was applied in the electro-

*The enzymatic assay of the plasma proteins was performed by Dr. J. A. Guimarães whom I thank for permission to reproduce some of his data in this paper. For details see reference 5.

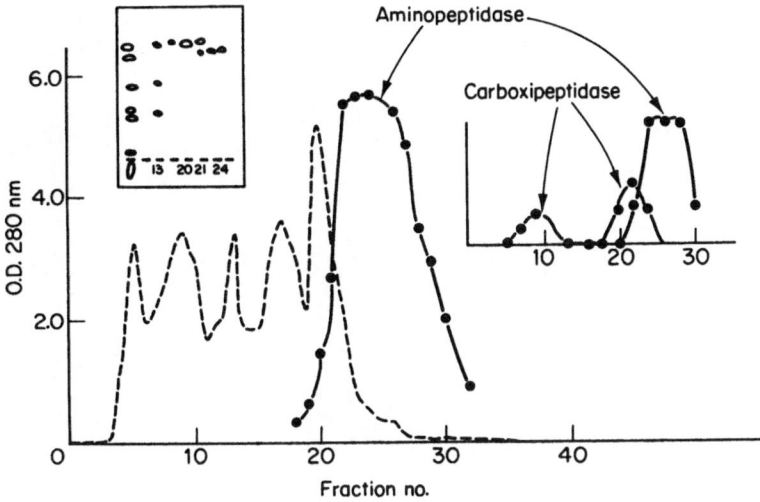

Figure 2.6. Fractionation pattern of the aminopeptidase by the continuous gel electrophoresis in Sephadex G-25 at pH 6.0.

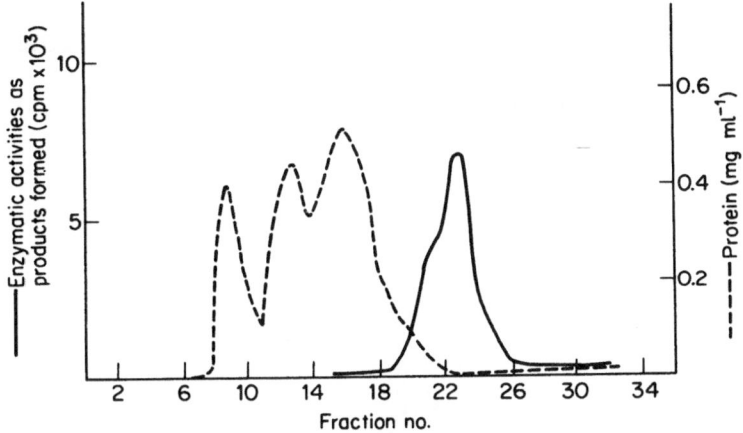

Figure 2.7. Purification of the heparinase from *F. heparinum* by continuous gel electrophoresis in Sephadex G-200.

phoresis chamber at 3 cm from the left side. The protein fractions came out from the chamber at the right side occupying only the last tubes. Figure 2.10 shows the pattern of fractionation when the injection of the sample was faster than the flow of the curtain (overload). Under this condition a cross contamination of most of the fractions was obtained.

The proper use of the continuous electrophoresis method here described might be useful for the preparation in large scale of a wide variety of proteins as well as other charged compounds.

C. P. Dietrich

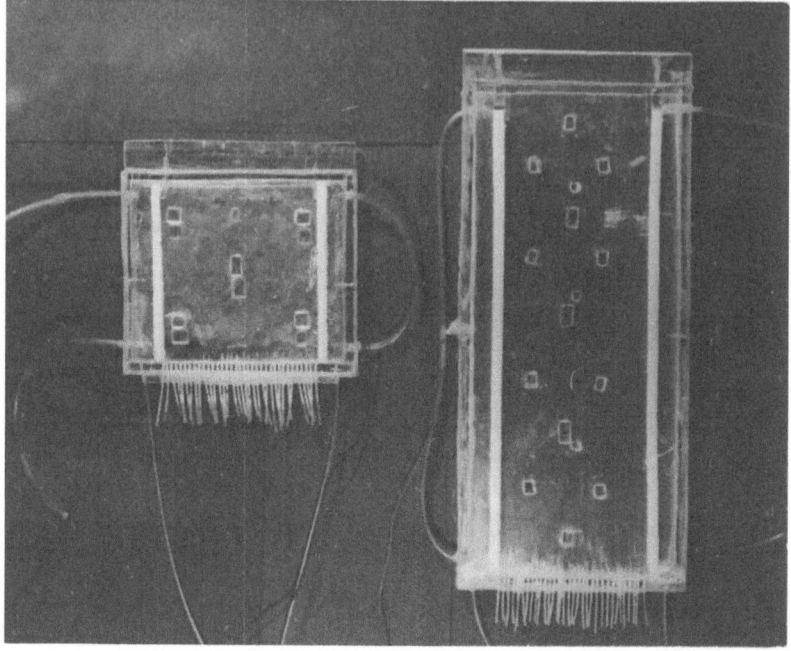

Figure 2.8. Chamber used for the experiment shown in figure 2.7 (right) compared to the conventional chamber (left).

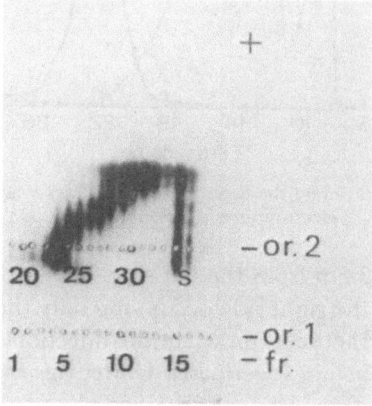

Figure 2.9. Aliquots of the continuous fractionation applied in agarose gel electrophoresis. Note that the samples were applied in two different rows in the gel.

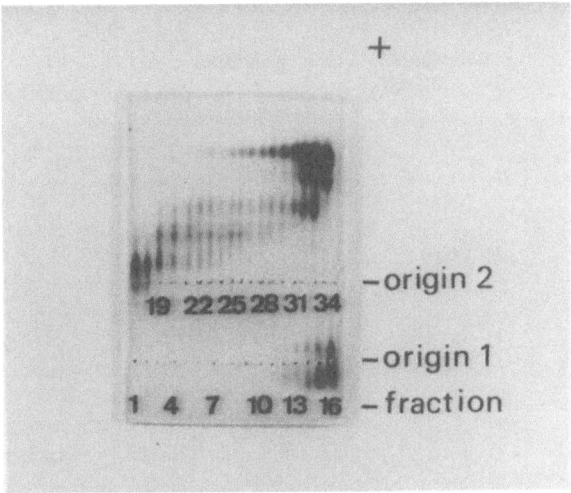

Figure 2.10. Sample overload of the continuous electrophoresis chamber.

References

1. Gordon, A. H. *Electrophoresis of proteins in polyacrylamide and starch gels.* (1969). North-Holland Publishing Co., Amsterdam, London
2. Svensson, H. and Brattsten, I. *Arkiv Kemi,* **1** (1949), 401
3. Dietrich, C. P. *Anal. Biochem.,* **51** (1973), 345
4. Dietrich, C. P. and Dietrich, S. M. C. *Anal. Biochem.,* **46** (1972), 209
5. Guimarães, J. A., Borges, D. R., Prado, E. S. and Prado, J. L. *Biochem. Pharmacol.,* **22** (1973), 3157
6. Borges, D. R., Prado, J. L. and Guimarães, J. A. Naunyn-Schmiedeberg's *Arch. Pharmacol.,* **281** (1974), 403
7. Dietrich, C. P., Silva, M. E. and Michelacci, Y. M. *J. Biol. Chem.,* **248** (1973), 6408

Discussion

ROSA

What is the maximum amount of protein that can be fractionated by the continuous loading method?

DIETRICH

Theoretically, there is no limit, if the conditions of the run, e.g., voltage and flow of the buffer, remain constant during the operation. We have maintained the curtain operating uninterruptedly for 72 hours; during this period there was no appreciable change of the variables, and we were able to fractionate 1.8 g of serum proteins.

3

Labelling of polypeptide hormones at high specific activities

G. MILHAUD

Radioimmunoassays and membrane binding studies require the availability of polypeptide hormones labelled at high specific activities and retaining respectively their immunological and their biological activity.

Several methods are used for the radioactive labelling. Iodine 125 is widely used, especially if tyrosine or histidine are present in the polypeptide. The chloramine-T method of Hunter, Greenwood and Glover[1] is most commonly used. This procedure has the advantage of being simple and yielding preparations of high specific activity. It can induce oxidative changes, particularly if the polypeptide contains methionine or disulphide linkage, the consequence being losses in both immunological and biological activity. MacFarlane[2] has advocated the use of iodine monochloride to reduce the damage, which is indeed effective, but at the cost of a diminished specific activity due to the addition of carrier iodine. Rosa, et al.[3] have used electrolytic iodination and found no appreciable changes in the protein during the labelling process. We found the use of this technique difficult in the presence of minute volumes of reagents. More recently, Bolton and Hunter[4] have proposed a method of labelling the protein by conjugating it with a radioiodinated acylating agent. 3-(4-hydroxyphenyl) propionic acid hydroxysuccinimide ester is reacted with the protein to be labelled, preventing damage to the protein from impurities present in the radioiodine solution. According to this procedure, human growth hormone, human luteinizing hormone and human thyroid stimulating hormone have been obtained.

Enzymatic iodination[5] is a very attractive tool: the general idea is that the specificity of the enzymatic oxidation of iodide to 'active iodine' is not likely to induce non-specific oxidative changes in the polypeptide hormone to be labelled. Lactoperoxidase radioiodination of insulin, glucagon and proteins was performed by Thorell and Johansson[6]. We have been using this procedure to label human and salmon calcitonin (figure 3.1), which are 32-aminoacid polypeptides; bovine parathyroid hormone, an 84-aminoacid polypeptide; and the active fragment of human parathyroid hormone, a 34-aminoacid residue (figure 3.2), which we have synthesised with the solid phase technique[7,8].

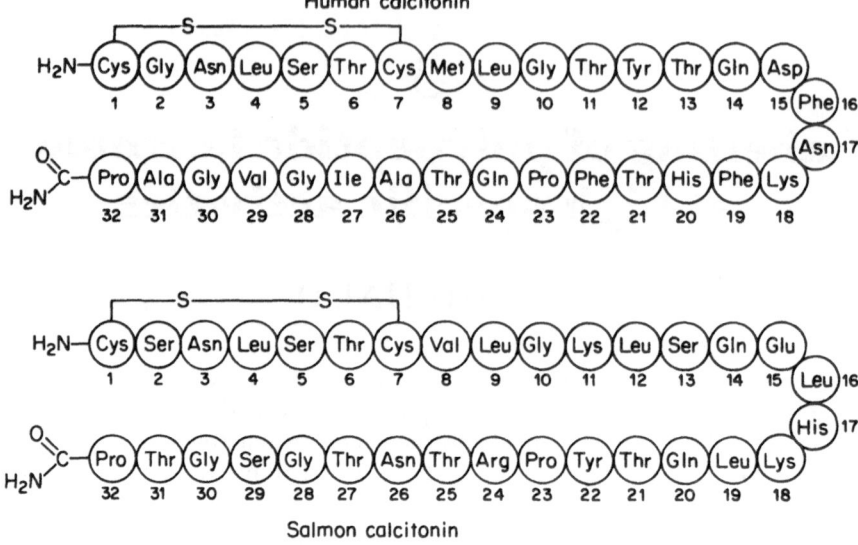

Figure 3.1. Structure of human and salmon calcitonin.

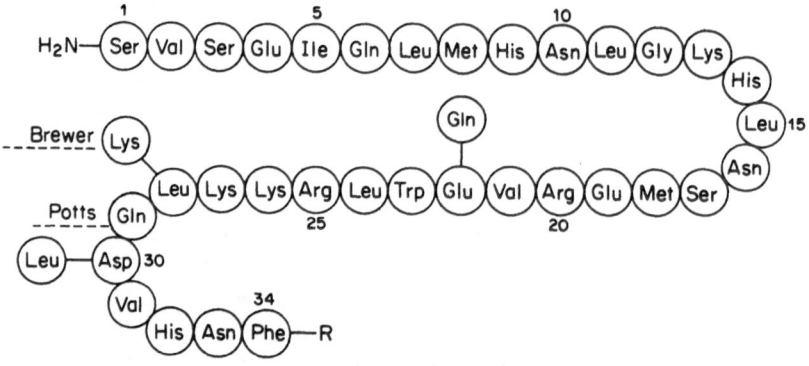

Figure 3.2. Structure of the active fragment of human parathyroid hormone.

Lactoperoxidase was prepared from milk and partially purified using adsorption on GM Sephadex 50 ion-exchange resin and ammonium sulphate precipitation. Carrier free I^{125} without reducing agents (Radiochemical Centre, Amersham) was used in the presence of a minute amount of hydrogen peroxide. Phosphate buffer pH 7.5, 0.05 M was the solvent for reagents and polypeptides. The reaction was allowed to proceed for lengths of time varying from 2 s to 2 min, and then stopped. The purification of the iodinated polypeptide was performed using QuSo adsorption and Sephadex G50 column presaturated with 1 ml of 1% serum albumine.

The specific activity of the preparation amounted to 200 μC μg^{-1}, corresponding to a substitution of 1 g-atom of iodine per mole polypeptide. The yield of the iodination procedure varied from 80 to 90%

Results

1. Human calcitonin labelled with the aid of lactoperoxidase is perfectly suitable for radioimmunoassay and the detection limit is around 7 pg per assay.

The biological activity assayed using the hypocalcemic response in the rat after weaning is at least 80% of the starting material. Kidney membrane binding was not assessed as the affinity of this polypeptide to the membrane is low.

2. Salmon calcitonin labelled by the same procedure keeps practically all its biological activity in the rat assay, binds strongly to kidney membranes and can be displaced reversibly by cold salmon calcitonin but not by the human hormone.

3. Bovine parathormone radioiodinated by the same technique retains the immunological activity and is currently used in our laboratory for radioimmunoassay of the circulating human hormone using a heterologous system. Labelled parathormone is biologically active and can be used for membrane receptor studies and assays.

4. The active 1-34 fragment of human parathormone does not contain tyrosine and is therefore not a good candidate for enzymatic labelling. We synthesised two analogues of the hormone, one with tyrosine in the position 1[8] and another one with 35 residues, adding in the position minus one tyrosine[8]. The iodination proceeds satisfactorily and we are using the tyrosine analogues to establish an assay of the circulating 1-34 fragment, which is present next to the 1-84 hormone. It is likely that radioimmunoassay of the whole hormone as well as of the fragment will be of major clinical interest.

References

1. Greenwood, F. C., Hunter, W. M. and Glover, J. S. The preparation of ^{131}I labelled human growth hormone of high specific activity. *Biochem. J.* **89**, (1963), 114

2. McFarlane, A. S., Efficient trace-labelling of proteins with iodine. *Nature*, **182** (1958), 53

3. Rosa, V., Pennisi, F., Bianchi, R., Federighi, G. and Donato, L. Chemical and biological effects of iodination on human albumin, *Biochim. Biophys. Acta*, **133** (1967), 486

4. Bolton, A. E. and Hunter, W. M. A new method for labelling protein hormones with radioiodine for use in the radioimmunoassay. *J. Endocrin.*, **55** (1972)

5. Fawcett, D. M. and Kirkwood, S. Tyrosine iodinese. *J. Biol. Chem.*, **209** (1954), 249

6. Thorell, J. I. and Johansson, B. G. Enzymatic iodination of polypeptides

with ^{125}I to high specific activity. *Biochim. Biophys. Acta.*, **251** (1971), 363

7. Milhaud, G., Rivaille, P. and J. F. Staub. Parathyroid-hormone fragment. *Lancet*, **ii** (1973), 440

8. Milhaud, G., Rivaille, P., Staub, J. F. and Jullienne, A. Synthèse et activité biologique d'une séquence 1–34 de la parathormone humaine. *C.R. Acad. Sci. Paris*, Série D, **279** (1974), 1015

Discussion

HOFFENBERG

Did I understand correctly that in the amino acid substituted fragments there is still biological activity and particularly in the last, the acylation procedure? Does the fragment retain biological activity?

MILHAUD

For the acylation procedure I do not know; I just reported the proposal of Dr Hunter, but what I can tell you is that if you introduce tyrosine in position 1 or − 1, you retain immunological activity. We have not tested this for biological activity, but we will be testing it on membrane binding, due to the fact that you need rather large amounts (250 to 300 μg of the substance) to test for biological activity. So, before you decide this step you must have enough material to use for this. What I can tell you is that in the radioimmunoassay system the amino acid substituted material reacts quite well with antiserum, and also it can be displaced by the straight fragment 1 to 34, so it behaves in this respect to our satisfaction.

ROSA

Did you experience with lactoperoxidase the strong variability, depending on the quality of the radioiodine, which is characteristic of the chloramine-T procedure?

MILHAUD

Yes, we found the same degree of variability.

WALDMANN

My question to the Chairman, Dr Rosa, is: did you develop a system for electrolytic iodination to a very small size sample, and if you have done so what is the smallest size of sample you can handle, and what type of specific radioactivity can you get on smaller molecules like 32, 50 or 100-polypeptide hormones?

ROSA

We developed 3 or 4 years ago a micro-electrolytic cell, specially designed for the labelling of human growth hormone and TSH. The total volume of the cell was 0.8–0.9 ml, and the minimal protein concentration, in terms of proteins like insulin for instance, was of the order of 30–35 μg. As for the specific activity, you know that electrolysis requires a carrier of radioiodine, and the maximum specific activity attainable in this condition was of the order of 60–70 up to 100 mCi mg^{-1}.

4

Advances in protein labelling

U. ROSA

Introduction

Recent progress in protein labelling ·has been obtained from two related areas, metabolic and RIA (radioimmunoassay) studies. In the former case the labelled derivative should keep the properties of the unlabelled molecule. A great deal of work on iodination has resulted. It was found that, in order to keep the metabolic properties of the unlabelled compound, the iodination of a labelled protein must be such that the monoiodinated derivative is the main product[1].

In the latter case the requirements are not so strict. However, in order to keep the maximum sensitivity of an antiserum, the tracer must not lose too much of its antigenicity. This may be achieved by minimising the modifications due to iodination. On the other hand, RIA requires labelled antigens with a relatively high specific activity, so labelling must be done with carrier-free radioactive iodine. In these conditions, the ID (iodination degree) cannot be kept lower than 1 i.a.m. (iodine atoms per molecule) while figures of 0.1–0.2 i.a.m. are currently quoted for labelled proteins used in metabolic research. The ID is difficult to compute, since the exact quantity of iodine used for labelling cannot be accurately evaluated. Furthermore, the labelling conditions adopted for the preparation of tracers for RIA are much more drastic than in metabolic studies. The occurrence of degradation processes in the protein is then more favoured. Studies on purification procedures that could separate the less damaged form of the protein have been the obvious outcome of this line of work.

Cross-fertilisation between the two approaches has led to an improvement in labelling techniques and to the definition of the elements of a labelling strategy for each protein. The more relevant aspects of this problem will be discussed in the present paper.

Iodination studies

The ideal models to study the effect of the iodination are small polypeptides, like angiotensin II. This compound has a single tyrosyl group, which is mono-

U. Rosa

and diiodinated when the labelling is done under controlled conditions and in the absence of strong oxidising agents. In figure 4.1 is shown the evolution of the two analogues with increasing iodine content. Appreciable quantities of DIA (diiodinated angiotensin) are formed at the beginning of the iodination. Since the two analogues can be separated[2] their biological activity was measured with an acceptable degree of precision. It was found that MIA (monoiodinated angiotensin) retains more than 70% of the original biological activity, while DIA is almost entirely inactive. In order to prepare a tracer for metabolic studies with high specific activity, either the formation of DIA has to be inhibited or MIA and DIA must be separated. Both approaches can be adopted in the case of angiotensin.

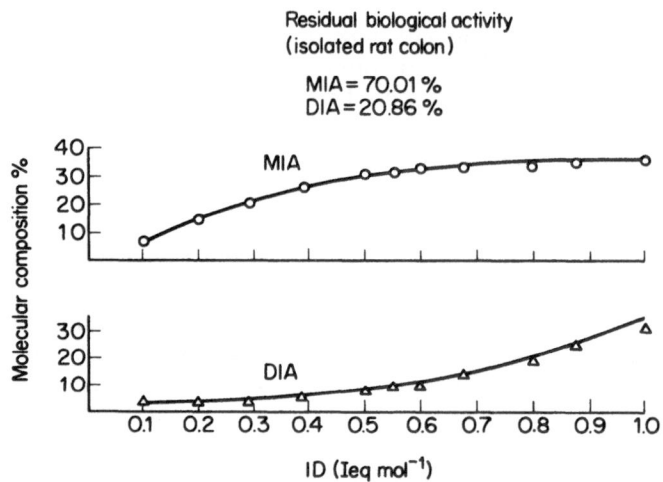

Figure 4.1. Per cent composition of iodinated angiotensin with increasing degree of iodination: (○), monoiodinated angiotensin; (△), di-iodinated angiotensin[2].

Insulin, with four tyrosyl groups per monomer unit, can originate 80 derivatives by iodination. These will differ from each other both in regard to the intramolecular location and in the net content iodine atoms. It is difficult if not impossible to fractionate into homogeneous classes such a complex mixture of compounds. Conditions must be used to ensure that only the analogues retaining biological activity originate from the labelling procedure.

In figure 4.2[4] is summarised the relationship between the level of iodination and biological activity. With a single exception, there is a sharp decrease of biological activity at an average ID between 1 and 2 iodine atoms per molecule. This result depends on the bioassay technique and, probably, also on the iodination method. The data show that one or more of the insulin analogues with a low iodine content were obtained by keeping the ID lower than 1–2 i.a.m. while retaining the biological activity. In complex proteins like insulin,

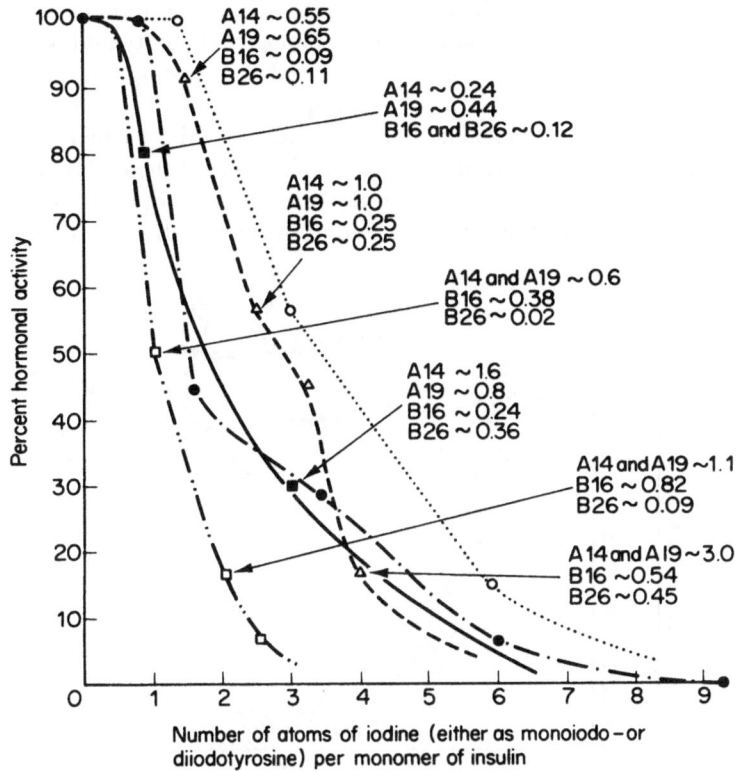

Figure 4.2. The relation between degree of iodination of tyrosines and potency. Key to biological assays: (□) fat-cell[5]; (○) mouse convulsion[6]; (●) rat diaphragm[6]; (△) fat-pad[7]; (■) blood-sugar depression[8].

to maximise the probability that monoiodinated analogues predominate in the labelled products, the ID should be kept low (0.1–0.2 i.a.m.).

However, unless monoiodinated insulin is separated and tested, experiments like these preclude any categorical statement about the biological activity of the iodinated analogues. Thus, the central problem of protein iodination can be stated: viz. how to control the process in order to increase the probability of obtaining an adequate tracer. A first empirical approach is to avoid the formation of diiodinated tyrosyl groups, and to check their absence in a tracer. In fact, their formation in a protein should be avoided because they are a source of important conformational damage. There is now enough evidence to support this point.

As an example, some years ago, we showed[7,9] that diiodination of tyrosine A19 in insulin produces a major conformational change. One of the disulphide bridges, probably A20–B19, becomes unreactive towards sodium sulphite or glutathione–insulin transhydrogenase. The splitting of the molecule into separated chains, a possible step in insulin catabolism, is inhibited. The effects of iodination on the biological activity and on the reactivity of the interchain S—S bonds measured by titrating the thiol groups formed by sulphitolysis are compared in figure 4.3. Such an effect is closely related to the loss of biological activity. This was explained when the electron density distribution of the molecule was found. One of the ortho positions in the A19 tyrosine is less accessible than the other in any degree of aggregation. When the A19 ring is diiodinated, the conformation is disrupted.

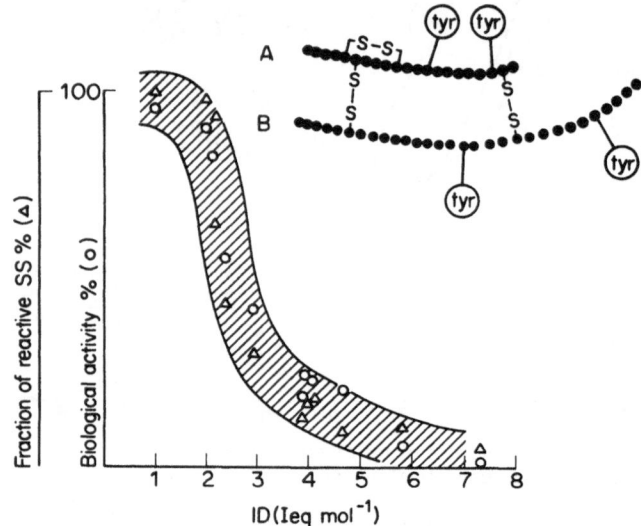

Figure 4.3 Extent of the reaction between iodoinsulins and Na_2SO_3 at pH 7 ($\triangle$)[7]; the SH groups formed per mole of insulin were measured and the results were expressed as a percentage with respect to the SH groups formed when native insulin reacts with Na_2SO_3 under the same conditions ($\bigcirc$), biological activity being measured by CO_2 production on rat epididymal fat-pad.

In more recent years, evidence has accumulated that diiodination leads to a loss of biological activity and almost invariably to a large loss of immunoreactivity, not only in polypeptides like angiotensin II[3] or in relatively small proteins like insulin, but also in large protein hormones[1].

The only way to avoid the damage is by keeping the ID low. However this is not always the case with several proteins. The iodination of one tyrosine may have cooperative effects on further iodination of the same group, thus favouring disubstitution. It is only when several tyrosyl groups with the same accessibility are available in a given protein that the formation of DIT can be reduced by keeping low the average substitution degree. However in HGH with 8 tyrosyl groups, only 2 or 3 of these groups can freely react with iodine. Chromatographic analysis of an hydrolysate of GH iodinated at ID 3–4 has shown[1] that almost 80% of the radioactivity is in the form of DIT, despite the fact that only 20–25% of the potential capacity of the protein to bind iodine on the tyrosyl group has been saturated.

Labelling of insulin with small amounts of iodine results in substitution mainly on the two A-chain tyrosyl residues over those of the B-chain[10]. This can be clearly seen from figure 4.4. Also, this effect is shown by experiments like those reported in figure 4.5, showing the distribution of the iodine between A and B chains of insulin, with increasing iodine content (these results are expressed as % iodination of each chain over the total iodine incorporated).

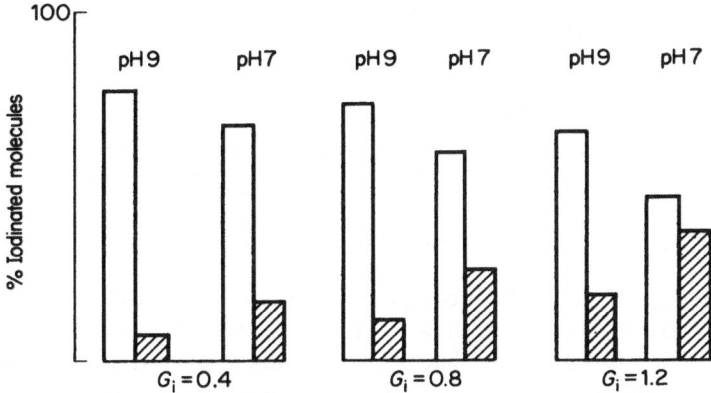

Figure 4.4. Effect of the labelling pH on the relative concentration of mono-iodinated angiotensin (MIA, open columns) and diiodinated angiotensin (DIA, hatched columns) in preparation of iodinated angiotensin. Each pair of preparations had the same iodine content. G_i = iodine atoms per molecule and differed only for the pH at which the labelling had been carried out. Iodinated angiotensins (labelled with ^{125}I) were submitted to enzymic hydrolysis and the radioactivity in form of MIT (monoiodotyrosine) and of DIT (diiodotyrosine) was measured after chromatographic analysis of the hydrolysate. The concentrations of MIA and DIA were calculated from these data, assuming that all the iodine contained in the polypeptide was bound to the tyrosyl group[2].

There is a reasonable possibility also that the majority of large proteins iodinate as selectively as insulin. Diiodination might also occur at low levels of substitution. In some cases, however, this can be avoided or reduced by varying the pH or the composition of the reaction medium. With angiotensin II, for instance, or with small polypeptides, the formation of the monoiodinated derivative is pH dependent. The highest yields are obtained at a basic pH. Figure 4.4 shows that the % of MIA (monoiodinated angiotensin) found in preparations of angiotensin II, labelled at the same ID, under the same conditions except of pH, varies with the pH of the reaction medium[2].

The use of basic pH conditions to obtain monoiodinated molecules cannot be applied to highly structured proteins. The change from a physiological to a basic pH involves the risk of iodinating tyrosyl groups which normally are inaccessible. In insulin, when changing from pH 7–7.5 to basic values, the

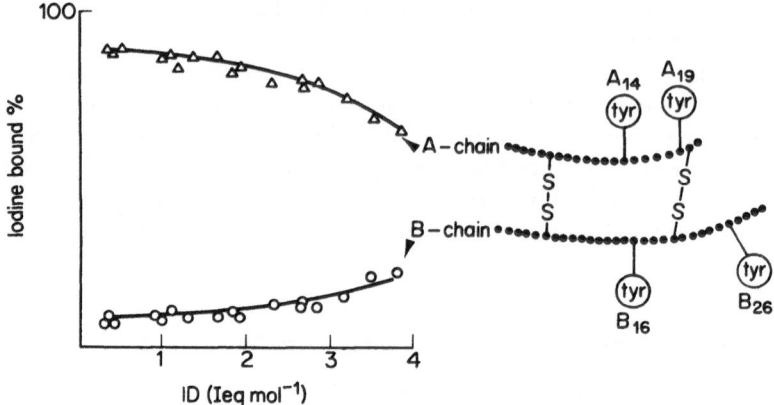

Figure 4.5. Effect of incorporation in insulin of increasing amounts of iodine on the distribution among the tyrosyl groups of the A-chain ($\triangle$) and of the B-chain ($\bigcirc$)[9]. ^{125}I-labelled iodoinsulins were split into chains by sulphytolysis in presence of 8 M urea; the S-sulphonated chains were separated by electrophoresis and their radioactivity was counted[10].

tyrosyl groups of the B-chain compete with those of the A-chain for iodine[10]. A clear illustration of this is shown in figure 4.6 where the ratio of the iodine content of the B-chain over that of the A-chain is plotted against the pH of iodination. The degree of iodination in the A-chain decreases with increasing

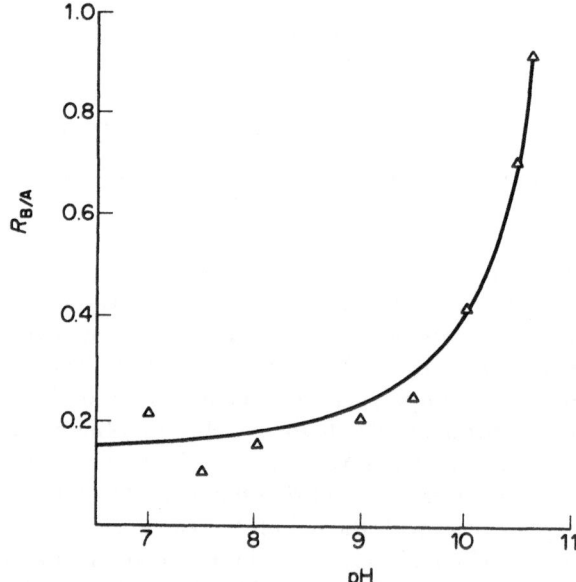

Figure 4.6. Effect of the pH at which the iodination of insulin is carried out on the distribution of iodine between the A- and B-chains. $R_{B/A} = $ [125]I radioactivity on the B-chain divided by the ^{125}I radioactivity on the A-chain[10].

pH. However, A- and B-chains are labelled in the same manner at a pH above 9.

Strategy of labelling

Iodination studies have thus provided a substantial body of information on the behaviour of iodinated proteins. They suggest that, when planning a labelling experiment, a strategy should be defined. The main elements to be accounted for are the properties of the protein itself.

The concentration of the protein in the reaction medium should be kept low. There is a lower risk of labelling aggregates or polymers, thus avoiding di-iodination of the more accessible molecules. This is true especially for glyco-proteins like growth hormone, thyroid-stimulating hormone and human placental lactogen. However, the reaction is less controlled at low concentrations. Further, there may be adsorption on the reaction vessel and some proteins, e.g. insulin, are more damaged as monomers. This problem can be overcome by using a low iodine-to-protein ratio.

If iodine carrier is used, this requires a relatively large amount of protein, favouring monoiodination and the control of the ID, but leading to a low specific activity, a situation often incompatible with the need to inject a small mass of the tracer to avoid modification of the protein pool. With carrier-free radioiodine there is the risk of damage from excess of oxidising agents and from the contaminants of radioiodine, whose degradative effect on glycoproteins for example, is very high.

Varying the composition of the supporting medium, salts or solvent, or the labelling pH, can, as has previously been shown for insulin, change the intra-molecular location of the iodine; with highly structured proteins one can iodinate tyrosyl groups which are unexposed when the protein is at physio-logical conditions (remember the case of the tyrosines of the B-chain in insulin). On the other hand, with some polypeptides, one can favour monoiodination by changing the pH (remember the case of angiotensin).

With certain procedures (e.g. electrolysis) one can vary the rate of iodine formation. At a low rate of formation of molecular iodine, one can more easily control the process and minimise the risk of overiodination. On the other hand, electrolysis which takes a much longer time than other procedures because of the slow rate of formation of the iodine, gives bad results with fibrinogen due to the inherent instability of the protein.

Viewed against this background, almost all the iodination methods currently available have advantages and disadvantages depending on the protein under study.

Iodination methods

Chloramine-T[11]

While this procedure has no competitor for labelling antigens in RIA, it has

certain disadvantages for the preparation of metabolic tracers:
1. There is no control of the reaction.
2. The ID is uncertain.
3. The damage from the oxidising agent and especially from radioiodine contaminants is very high.

In addition, preliminary iodination studies, which are essential to define the properties of the iodoprotein, are usually carried out with procedures requiring higher protein concentration and a carrier of radioiodine. When passing to Chloramine-T for routine labelling it is difficult to extrapolate the results of these studies to the preparation of the tracer.

Electrolysis at constant current[12]

The main advantages are the absence of oxidising agents and the strict control of the iodination process, since the rate of formation of the iodine depends on the electrolysis current. Electrolysis is an excellent tool for iodination studies since it permits, in a single run, the preparation of samples of different iodine content under the same experimental conditions. However, also available as a semimicrotechnique[13], it requires a minimal concentration of the protein that is 10 times larger than that required for chloramine-T labelling.

Iodine monochloride[14]

This technique has a great versatility and, in skilled hands, can yield the same results as electrolysis with less technical complications.

Lactoperoxidase[15]

This technique shares with electrolysis the important advantage that the protein is not exposed to strong oxidising agents; in addition it can be applied with carrier-free radioiodine. Advantages and disadvantages are specifically discussed by Milhaud in the previous paper.

It is worth mentioning here that an entirely different approach in labelling has recently been developed by Hunter et al.[16]. By this technique the protein is treated with a ^{125}I-labelled acylating agent which reacts with the free amino groups in the protein forming highly stable amide bonds. The reaction is very fast, the yield is high and no reagent other than the labelled group comes in contact with the protein. To my knowledge it has never been applied to the preparation of labelled proteins for metabolic studies.

Purification methods

In the preparation of labelled proteins for metabolic studies, the purification has been traditionally considered as the last step in the labelling sequence, by which the tracer is separated from an excess of iodine and from other reagents. Studies of protein labelling for RIA, however, have shown that the existence of very small differences in charge and probably in shape between uniodinated and iodinated molecules can be exploited, also with large proteins, to enrich the tracer in the monoiodinated form with respect to the uniodinated and to the overiodinated forms.

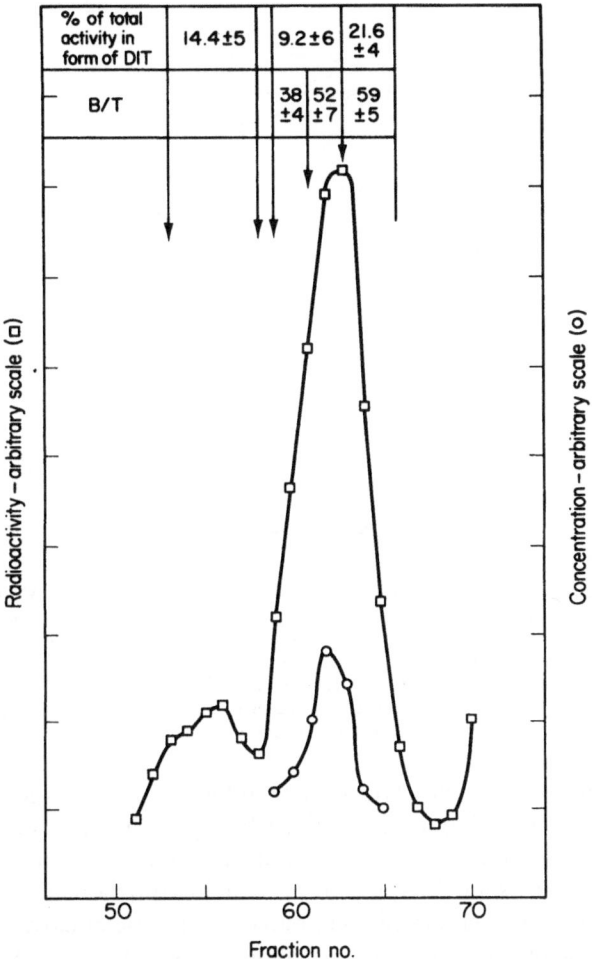

Fig. 4.7. Elution pattern of ^{125}I-labelled-HGH ($\square$) (135 mCi mg^{-1}) and unlabelled HGH (O) from a G-200 Sephadex column equilibrated for 48 h (1.5×85 cm; 4°C; buffer: 0.1 M veronal-Na pH 8.6 containing 0.1% BSA; sample: 0.8 ml corresponding to about 10 μg of HGH).

For some, pooled fractions are indicated: the % of radioactivity found in the form of DIT (diiodotyrosine) after enzymic hydrolysis and the binding ability (B/T) measured under comparable conditions using the same dilution of the antiserum and an equal amount of radioactivity are given. The elution pattern of unlabelled HGH was obtained on the same column after exhaustive washing; the HGH concentration was measured by RIA[1].

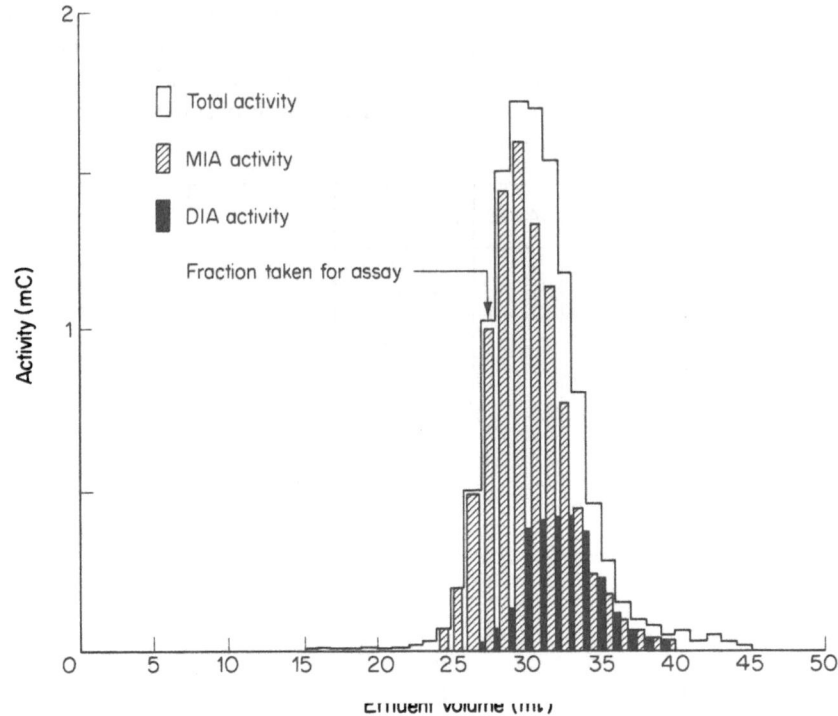

Figure 4.8. Elution pattern of iodinated angiotensin, composed of 80% of monoiodinated and 20% of diiodinated angiotensin (indicated as MIA and DIA, respectively) on a Sephadex G-25 column (9 × 520 mm) equilibrated with 0.01 M Tris-buffer pH 7.4. The continuous profile is the total radioactivity of each fraction. The hatched area and the black area are proportional to the radioactivity in the form of MIA or DIA measured in each fraction. The arrow indicates the fraction which is used as a tracer in metabolic studies or in radio-immunoassay[2,3].

In several cases one can label the protein at low ID and low specific activity, then isolate a tracer at high specific activity by separating most of the uniodinated molecules[3]. A merit of such a technique, now currently employed for the preparation of tracers for RIA, is that of reducing the damaging potential of procedures like that of chloramine-T, based on the use of carrier-free radio-iodine. The risk of damage when labelling a protein, at 20–30 mCi mg^{-1} instead of 200–300 mCi mg^{-1}, is substantially reduced, since with chloramine-T the damage due to contaminants and to the extent of the substitution is directly proportional to the total activity used for labelling (i.e. to the specific activity one has to attain).

As an example in figure 4.7 are shown the peaks obtained when eluting I^{125}-labelled human growth hormone from a G-200 Sephadex column. An analysis of the fractions taken from the apparently homogeneous central peak shows that the elution of the iodinated form is delayed. The effect is too small to allow a sharp separation into peaks but one can see that the relative DIT content in the labelled protein increases when passing from the head to the tail of the peak. At the same time there is an enrichment effect of the iodinated over the cold protein. The B/T values can be taken as an index directly proportional to the specific activity of the hormone.

Obviously the effect is more pronounced with small polypeptides like angiotensin II. The homogeneous peak (see figure 4.8) obtained by eluting iodinated angiotensin II from a G-25 Sephadex column contains fractions enriched in MIA with respect to DIA and vice versa. Still better results can be obtained using electrophoresis on a column of polyacrylamide gel. By this method the specific activity of a small peptide like angiotensin II is increased from 30–50 to 100–1200 mCi mg^{-1}, and almost pure monoiodinated angiotensin can be obtained[1]. This technique can be advantageously applied to larger protein molecules (TSH, LH, FSH, etc.).

Conclusions

The last time that problems of protein iodination were specifically discussed in this group was in 1967 in Saltsjöbaden. Also on that occasion I had the honour to be asked to give a lecture on this subject and I used the results of an iodination study of HSA[12] to discuss the general problems connected with the preparation of tracers for metabolic research. The present lecture, on the same general subject, points to the results of iodination studies that have been carried out on protein and polypeptide hormones. I have tried to put forward evidence on some points that, in my opinion, can help to a better understanding of the mechanism of protein iodination and to a better design of a successful labelling experiment. It is hoped this may prove more valuable than a catalogue of the achievements in this field during the last seven years.

References

1. Rosa, U. and Malvano, R. *Advances Nuclear Energy in Biology*. Proceedings of the IAEA/WHO Symp., Copenhagen, 1973, p. 91
2. Kurcbart, H., Vancheri, L., Coli, A. and Rosa, U. *Biochim. Biophys. Acta*, **230** (1971), 160
3. Gandolfi, C., Malvano, R. and Rosa, U. *Biochim., Biophys., Acta*, **251** (1971), 254
4. Blundell, T., Dodson, G., Hodgkin, D. and Mercola, D., *Advances in Protein Chemistry*, vol. XXVI, Academic Press, N.Y. 1972, p. 279
5. Garratt, G. J., Harrison, D. M. and Wicks, M. *Biochem. J.*, **23** (1972), 312
6. Izzo, J. L., Roncone, A., Izzo, M. J. and Bale, W. F. *J. Biol. Chem.*, **239** (1964), 3749

7. Rosa, U., Massaglia, A., Pennisi, F., Cozzani, I. and Rossi, C. A. *Biochem. J.*, **108** (1967), 407

8. De Zoeten, L. W., Havinga, E. and Everse, J. *Rec. Trav. Chim. Pays-Bas*, **80** (1961), 917

9. Rosa, U., Rossi, C. A. and Donato, L. *Pharmac. of Hormon. and Polyp. Proteins*, Plenum Press, 1968, p. 336

10. Massaglia, A., Rosa, U., Rialdi, G. and Rossi, C. A. *Biochem. J.*, **115** (1969) 11

11. Hunter, W. M. and Greenwood, F. C. *Nature* (London), **194** (1962), 495

12. Rosa, U., Pennisi, F., Bianchi, R., Federighi, G. and Donato, L. *Bioch., Biophys. Acta*, **133** (1967), 486

13. Pennisi, F. and Rosa, U. *J. Nucl. Biol. Med.*, **13** (1969), 64

14. McFarlane, A. S. *Nature*, **182** (1958), 53

15. Thorell, J. L. and Johansson, B. G. *Biochim. Biophys.*, *Acta*, **251** (1971), 363

16. Bolton, A. E. and Hunter, W. M. *Biochem. J.*, **133** (1973), 529

Discussion

REGOECZI

Dr. Rosa, I understand that you are able to label some proteins like glucagon, thrombin etc. by substitution not only with iodotyrosine, but with iodohistidine as well. I would like to know whether you can foresee or devise simple tests to measure the amount of free histidine substituted residues and, additionally, whether the iodo-label behaves differently at that site. Do you think that iodohistidine residues would be released more readily than mono- or diiodotyrosine? I wonder if you have some information on that.

ROSA

Above all, it must be said that the iodination of the histidine residues can be minimised or avoided through the control of the labelling pH. In fact, the reaction is pH dependent, and it occurs with high yield at a pH greater than 8.5–9. This has been shown by Covelli and Wolff in iodination studies of insulin and lysozyme. They give a procedure to measure the amount of the iodohistidine in the hydrolysate of an iodinated protein but, because of the instability of these derivatives, the results must be regarded as semiquantitative. For the same reason, if iodohistidine is formed, this results in a continuous release of iodine from the labelled protein.

ROBBINS

I do not know if Wolff and Covelli used this method, but a straightforward way to identify iodotyrosine would be by spectrophotometric titration, and this would differentiate iodotyrosine from iodohistidine.

ROSA

In general, spectral titration cannot find practical application in iodination studies like those I have reported, since the amount of protein used is too small. In addition, the amount of iodinated tyrosine formed in preparations at very low iodination degree is very small.

Moreover, if one wants to study the distribution of iodine among the tyrosil groups of a protein, one has to split the protein into fragments, as I reported in the case of insulin. In this case one has to do with such a small amount of material at the end of fragmentation that the advantage of the technique based on the measurement of radioactivity is very high in terms of sensitivity.

MILLER

Having achieved very high efficiency in iodination, it is obvious that the maximum utility of a preparation depends on the shelf-life, if you will. I wonder whether you have anything more to add to the usual precaution of including some kind of protective protein immediately after you have purified your iodinated material to prolong the shelf-life.

ROSA

We prefer to use freshly labelled protein for metabolic studies. The practice of adding to the labelled hormone a protective protein, like HSA, is very popular in radioimmunoassay, and we have adopted the same procedure to store labelled hormones at high specific activity. HSA acts also as a protector against the wall adsorption, which is one of the factors responsible for the degradation during storage. Storage in the dark, at low temperature, but not lower than 4–5°, is a common practice. I wish to stress, however, that metabolic studies should be systematically carried out with freshly labelled tracers.

MCFARLANE

Dr Rosa, I think you may well be right that if we could avoid the formation of diiodotyrosine in any labelling procedure we would avoid the biggest damage effects. What is puzzling me is how can we do that with most plasma proteins. Although you showed in a figure—and I think it was angiotensin—that as you become more alkaline than 8–8.5 you get a preferential formation of mono-iodotyrosine, I find that with most plasma proteins I must iodinate at least at pH 7. They will not iodinate below 7, as you know, and in that case we get more diiodotyrosine than monoiodotyrosine. I have the impression that you are aware of this difficulty.

ROSA

I agree with you, Dr McFarlane, that a plasma protein must be iodinated at pH around the neutrality. However, in general terms, when modifying the iodination pH the relative concentration of the available tyrosine can either decrease or increase; the first condition favours monoiodination. In addition, as I showed in the case of insulin, the intramolecular distribution of the iodine can be different.

My point with angiotensin was that pH can also affect the probability that a given tyrosil group is mono- or diiodinated. The reason for this is that the concentration of the phenolate form, required for the iodination, depends on the pH of the reaction mixture. In any case, any labelled protein should be checked for its diiodotyrosine content, as an elementary criterion to judge on its adequacy.

MILLER

Dr Rosa, you mentioned fibrinogen and I would like to know whether some-one in the audience, like Dr Regoeczi for instance, who has a lot of exper-ience with fibrinogen, could say something about the problems with it. I would like to suggest that anybody interested in fibrinogen could try to use the technique of Bolton and Hunter which was mentioned by Dr Milhaud and myself, because the potential capability of this technique for labelling proteins has not yet been completely exploited. This method is a true alternative sub-stitution procedure to iodine, because the reagent group on the proteins is the amino group of lysine. There is the possibility with pH to increase or decrease the relative concentration of an amino group which is available.

According to Quatrecasas, for instance, it should be possible to favour iodination of 'alpha' amino groups instead of 'epsilon' amino groups, so that it is a technique which should be tried in the case of complicated proteins.

ROSA

As I mentioned, I do not have personal experience on the use of this technique, which looks very promising with the 'difficult' proteins like fibrinogen.

5

Albumin and IgG turnover in ulcerative colitis and Crohn's disease*

K. BIRGER JENSEN and S. JARNUM

A decreased serum albumin is almost invariably present in chronic inflammatory bowel disease. Although an abnormal intestinal protein loss is well documented it is not known to what extent it contributes to the hypoalbuminaemia, since no reports published so far have compared the size of protein loss with albumin synthesis.

By contrast with albumin, IgG is usually present in normal concentration in serum. This is due to an increased synthesis of IgG.

The present study was undertaken primarily to establish the relation between intestinal protein leakage and, on one hand, the degradation rate of albumin and IgG, a relationship which is to be expected, and on the other hand, the synthetic rate of these two proteins. A comparison has also been made between the turnover data and the severity of the disease and biochemical findings in general.

Methods

Patient population

In all, 98 patients were studied of which 23 suffered from ulcerative colitis and 75 from Crohn's disease. Nine of the 23 patients with ulcerative colitis were later operated on. Of the 75 patients with Crohn's disease, 45 were unoperated and 30 had previously been operated on and, following the study, 43 were subjected to intestinal resection. Four patients were studied both before and after surgery when the presence of recurrent Crohn's disease had been established.

The diagnosis of ulcerative colitis and Crohn's disease was based on characteristic clinical, radiological, and histological findings as well as the exclusion of tuberculosis and parasitic and neoplastic disease.

*This work was supported by grants from The Danish State Research Fund and Christian d. X's Fund.

The clinical activity states of ulcerative colitis and Crohn's disease at the time of investigation were defined according to criteria summarised in tables 5.1 and 5.2, respectively.

Table 5.1 Criteria used for classification of patients with ulcerative colitis into three groups according to clinical and routine laboratory findings.

Inactive or mild	: 1–4 bowel movements per day with occasional or no visible blood
	: Normal or almost normal laboratory findings
	: Normal working capacity
Moderate	: 2–10 bowel movements per day with visible blood
	: Abnormal laboratory findings, 3 of 4: haemoglobin, E.S.R., serum albumin, serum orosomucoid
	: +/− weight loss
	: Normal or reduced working capacity
Severe	: 6 or more bowel movements per day with visible blood
	: Abnormal laboratory findings, 3 of 4: haemoglobin, E.S.R., serum albumin, serum orosomucoid
	: Poor general condition with weight loss
	: No or severely reduced working capacity
	: +/− parenteral nutrition

Table 5.2. Criteria used for classification of patients with Crohn's disease into three groups according to clinical and routine laboratory findings*.

Inactive or mild	: 1–5 bowel movements per day with occasional or no visible blood
	: Normal or almost normal laboratory findings
	: Abdominal pains 0–4 times monthly
	: Normal or almost normal working capacity
Moderate	: 1–10 bowel movements per day +/− visible blood
	: Abnormal laboratory findings, 3 of 4: haemoglobin, E.S.R., serum albumin, serum orosomucoid
	: Frequent bouts of abdominal pain, once or twice weekly.
	: Reduced working capacity
Severe	: 1–10 or more bowel movements per day +/− blood
	: Abnormal laboratory findings, 3 of 4: haemoglobin, E.S.R., serum albumin, serum orosomucoid
	: Daily abdominal pains or attacks of subileus several times a month
	: Poor general condition with weight loss +/− parenteral nutrition
	: No or severely reduced working capacity

*This classification is independent of former operations, pathoanatomical localisation and radiographical extension of the disease.

Isotope studies

Gastrointestinal protein loss was quantified by means of ^{59}Fe-labelled iron dextran (The Radiochemical Centre, Amersham, England). After intravenous administration of 0.04 μCi of ^{59}Fe-labelled iron dextran per kg body wt stools were collected until they became red following administration of 1 g carmine 96 hours after injection. The faecal excretion of ^{59}Fe (Q_F) was expressed as a percentage of the injected dose, and the gastrointestinal clearance of ^{59}Fe-labelled iron dextran was calculated as the percentage of intravascular ^{59}Fe excreted in the stools per day. ^{59}Fe-iron dextran studies were made in all 23 patients with ulcerative colitis and in 67 of 75 patients with Crohn's disease. The ^{59}Fe-iron dextran study was made simultaneously with the albumin and IgG turnover studies.

In four patients with Crohn's disease, [51]CrCl$_3$ (0.3 μCi per kg body wt) was used for estimation of gastrointestinal protein loss[1]. The [51]CrCl$_3$ study was made 5–7 days before the albumin and IgG turnover studies.

Turnover studies

[131]I-labelled human albumin (Institutt for Atomenergi, Kjeller, Norway, Code MISN) and metabolically homogeneous human IgG, isolated from normal human serum by means of DEAE-cellulose chromatography[2] and labelled with ^{125}I according to the ICl-method of McFarlane[3], were used for turnover studies.

About 0.4 μCi ^{131}I and 0.15 μCi ^{125}I per kg body wt were administered intravenously. By means of plasma radioactivity analysis the turnover data were calculated according to the method of Nosslin[4]. (For details, *see* reference 5).

Statistical analysis

Differences between mean values were analysed by a two-tailed Mann–Whitney rank sum test. Correlation estimations were carried out by Spearman's test, or, when the number of observations exceeded 48, the Student *t*-test.

Results

Fractional catabolic rates of albumin and IgG and intestinal protein loss

Ulcerative colitis

Hypoalbuminaemia was present in practically all moderate and severe cases of ulcerative colitis, whereas the majority of cases maintained a normal serum IgG concentration irrespective of the clinical severity. This was not due to preferential catabolism of albumin, since a fairly close and positive correlation existed between the catabolic rates of the two proteins (figure 5.1).

A simple explanation of this correlation would be that both proteins shared equally an intestinal protein loss. The size of intestinal protein loss was estimated by means of ^{59}Fe-iron dextran clearance. It turned out to be highly correlated with the fractional catabolic rate of both albumin (figure 5.2) and IgG (figure 5.3).

We conclude that the major factor determining the fractional catabolism of albumin and IgG in ulcerative colitis is the size of intestinal protein loss.

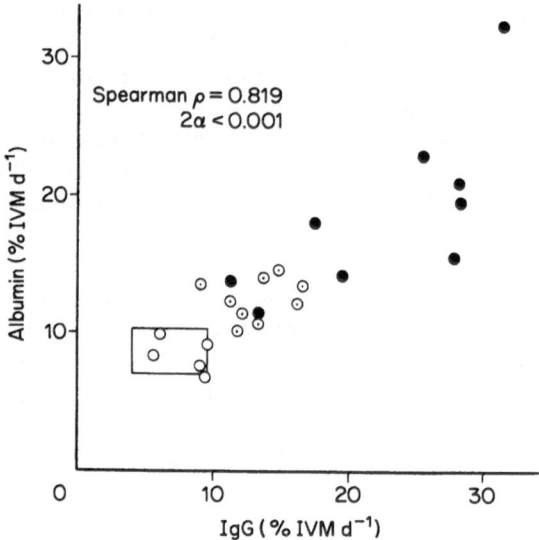

Figure 5.1. Ulcerative colitis. Fractional catabolic rates (% intravascular mass, IVM, degraded per day) in 23 patients. The box indicates normal range. ● = severe, ☉ = moderate and, ○ = inactive or mild disease activity

Crohn's disease

The fractional catabolic rates of albumin and IgG were also positively correlated in Crohn's disease, although not to the same high level as in ulcerative colitis. A high correlation was present between the size of intestinal

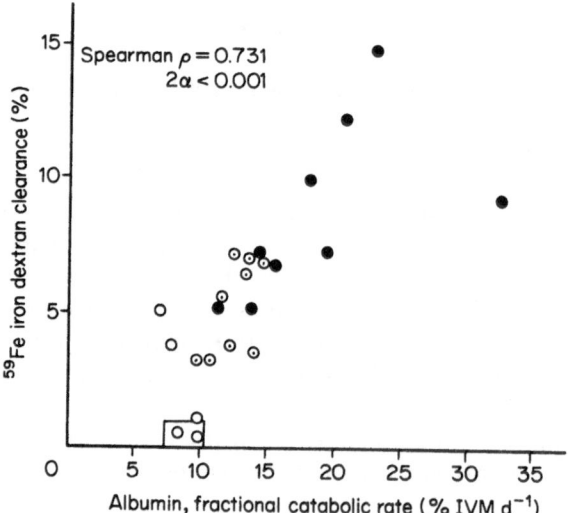

Figure 5.2. Ulcerative colitis. ⁵⁹Fe-iron dextran clearance and fractional catabolic rate of albumin in 23 patients. Symbols as in figure 5.1.

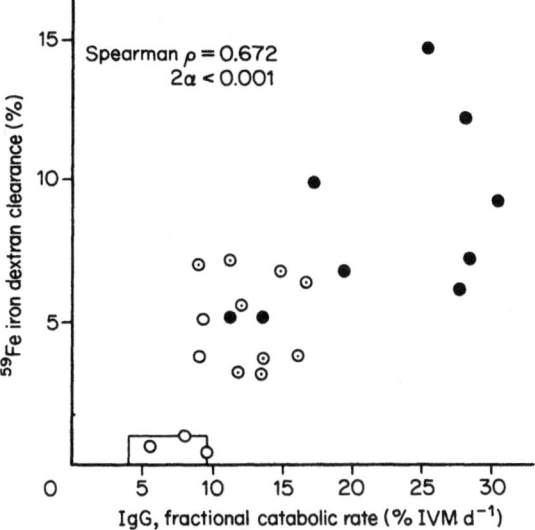

Figure 5.3. Ulcerative colitis. ^{59}Fe-iron dextran clearance and fractional catabolic rate of IgG in 23 patients. Symbols as in figure 5.1.

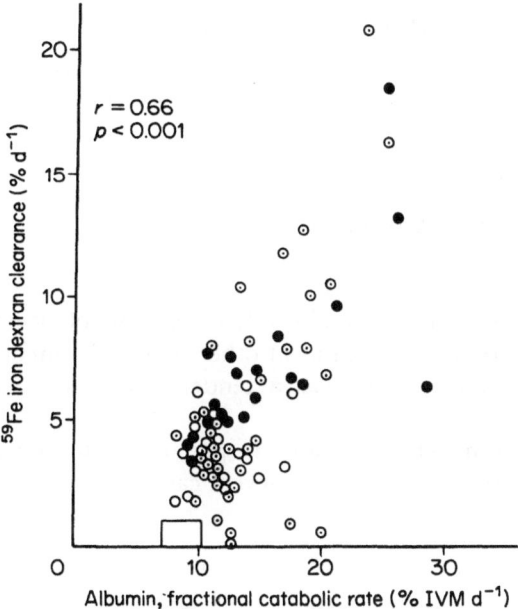

Figure 5.4. Crohn's disease. ^{59}Fe-iron dextran clearance and fractional catabolic rate of albumin in 76 patients. Symbols as in figure 5.1.

protein loss (faecal ^{59}Fe-iron dextran clearance) and albumin catabolism (figure 5.4) and, to a lesser degree, between protein loss and IgG catabolism (figure 5.5).

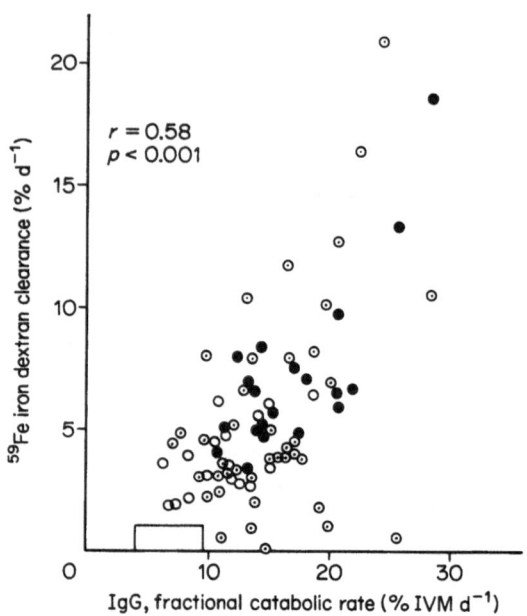

Figure 5.5. Crohn's disease. ^{59}Fe-iron dextran clearance and fractional catabolic rate of IgG in 76 patients. Symbols as in figure 5.1.

Synthetic rate of albumin and IgG

Ulcerative colitis

The synthetic rate of albumin was normal in most cases of ulcerative colitis. No compensatory increase was seen as in other, non-inflammatory protein-losing gastroenteropathies. There was a tendency to decreased synthetic rates in severe cases.

IgG synthesis was somewhat increased in patients with moderate disease activity and often markedly increased in severe cases.

Crohn's disease

The same pattern of synthetic rates of albumin and IgG was found in Crohn's disease, only there were more abnormal values (low albumin synthesis and high IgG synthesis in severe cases).

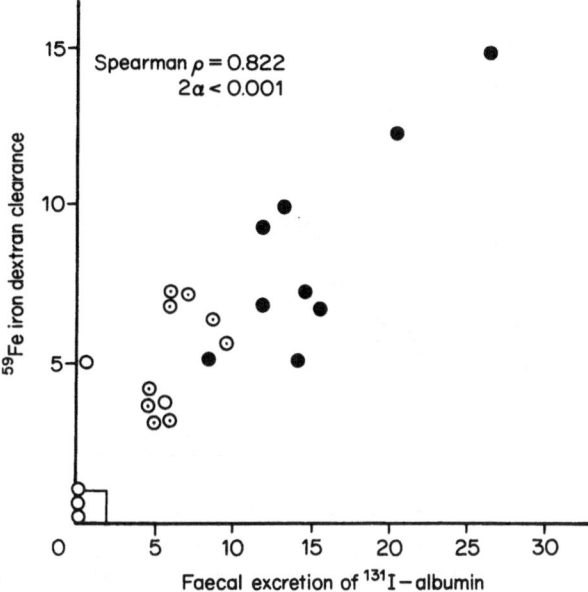

Figure 5.6. Ulcerative Colitis. ^{59}Fe-iron dextran clearance and faecal excretion of ^{131}I from albumin (% of injected dose in five days) in 23 patients. Symbols as in figure 5.1.

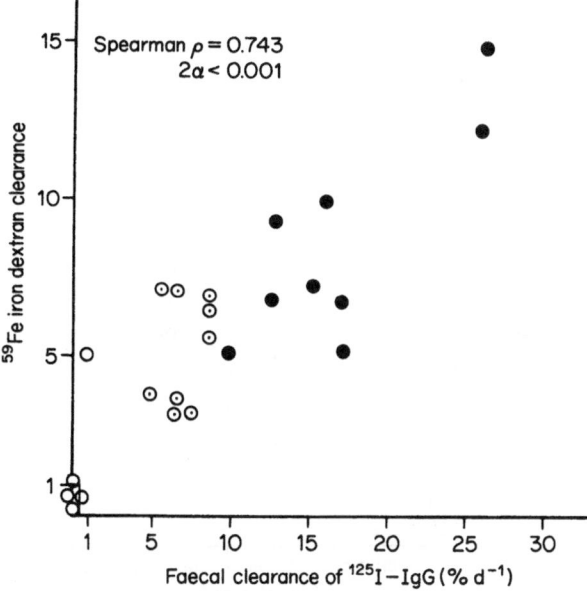

Figure 5.7. Ulcerative colitis. ^{59}Fe-iron dextran clearance and faecal excretion of ^{125}I from IgG (% of injected dose excreted in five days) in 23 patients. Symbols as in figure 5.1.

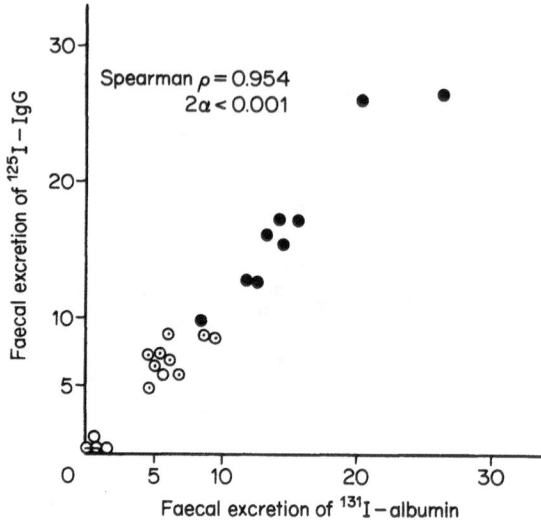

Figure 5.8. Ulcerative colitis. Faecal excretion of radioiodine from IgG and albumin (% of injected dose in five days) in 23 patients. Symbols as in figure 5.1.

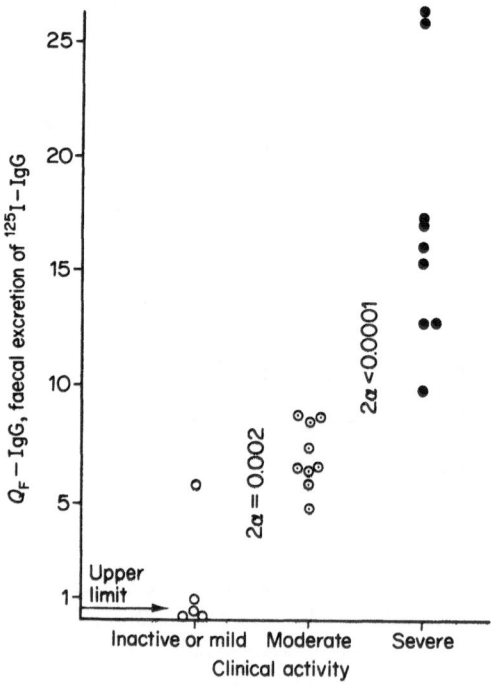

Figure 5.9. Ulcerative colitis. Faecal excretion of ^{125}I from IgG and disease activity in 23 patients. Symbols as in figure 5.1.

Faecal radioiodine excretion

Ulcerative colitis

Intestinal protein loss is likely to determine the amount of radioiodine excreted in the stools in ulcerative colitis, since a rapid colonic transit leaves little possibility of radioiodine from albumin and IgG being reabsorbed. In fact, a very good correlation was present between intestinal protein loss (^{59}Fe-clearance) and stool excretion of radioiodine from both albumin (figure 5.6) and IgG (figure 5.7).

Also the correlation between faecal radioiodine from albumin and IgG was extremely close (figure 5.8). The faecal radioiodine (^{125}I) excretion from ^{125}I-IgG turned out to be highly dependent on the clinical activity (figure 5.9). Faecal radioiodine (^{131}I) from ^{131}I-albumin had a similar discriminatory value (figure 5.10).

The same held true, although to a lesser degree, for stool weight versus the degree of clinical activity (figure 5.11). A daily stool weight of 350 g or more was almost invariably associated with moderate or severe disease activity.

Crohn's disease

Contrary to ulcerative colitis no correlation was present between ^{59}Fe-clearance and stool excretion of radioiodine from albumin or IgG in Crohn's

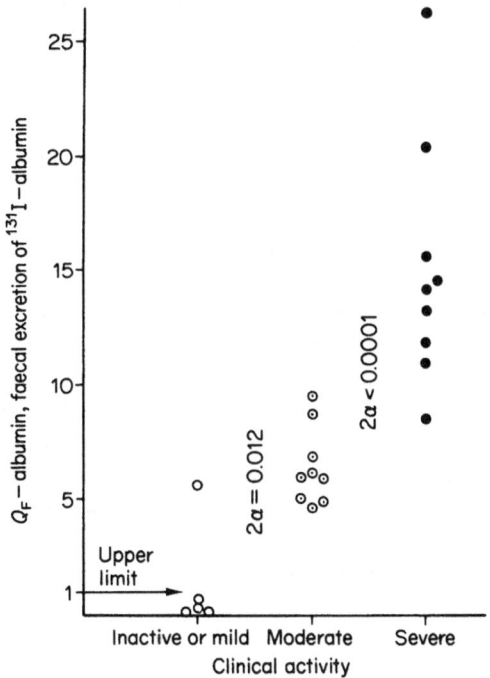

Figure 5.10. Ulcerative colitis. Faecal excretion of ^{131}I from albumin and disease activity in 23 patients. Symbols as in figure 5.1.

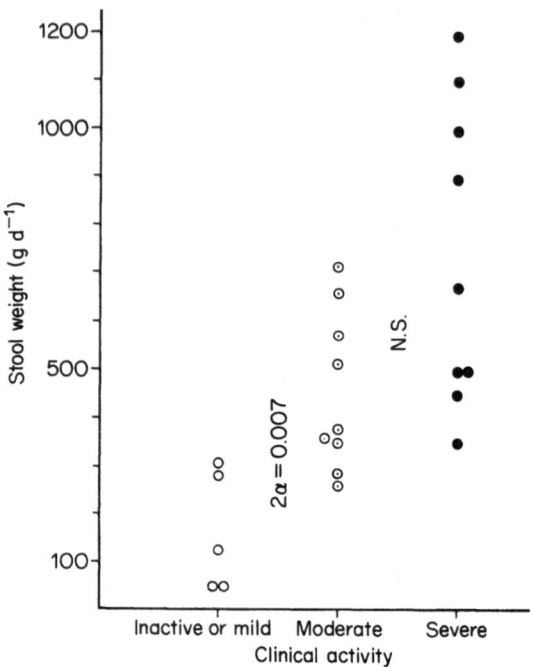

Figure 5.11. Ulcerative colitis. Average stool weights and disease activity in 23 patients. Symbols as in Figure 5.1.

disease. However, it should be pointed out that this is valid only for the gross overall correlation between intestinal protein loss (^{59}Fe-clearance) and faecal radioiodine excretion, and that no attempt has been made, in this report, to distinguish between various groups of Crohn's disease, as for instance granulomatous colitis or pure small bowel lesion.

No correlation existed between faecal radioiodine excretion from albumin and IgG in Crohn's disease, and there was only a statistically insignificant trend towards a correlation between the degree of clinical severity and faecal radioiodine excretion from albumin (figure 5.12) and IgG (figure 5.13).

There was no statistically significant correlation between stool weight and clinical severity (figure 5.14), although stool weights above 1050 g per day occurred only in moderate and severe cases. Part of the explanation of this lack of correlation was that all three groups of clinical activity included patients with ileostomy and/or ileal resection.

Biochemical findings

Ulcerative colitis

Since haemoglobin, E.S.R., serum albumin and orosomucoid were used

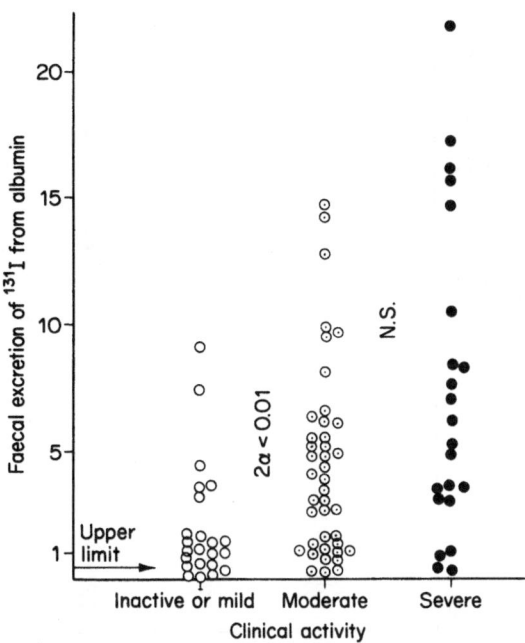

Figure 5.12. Crohn's disease. Faecal excretion of ^{131}I from albumin (% of injected dose in five days) and disease activity in 86 patients. Symbols as in figure 5.1.

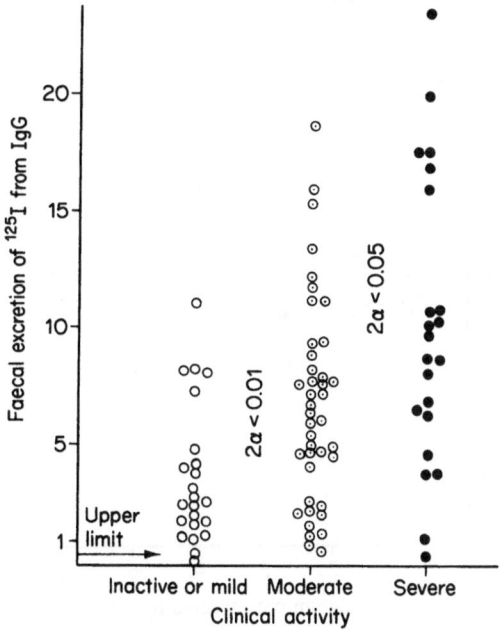

Figure 5.13. Crohn's disease. Faecal excretion of ^{125}I from IgG (% of injected dose in five days) and disease activity in 86 patients. Symbols as in figure 5.1.

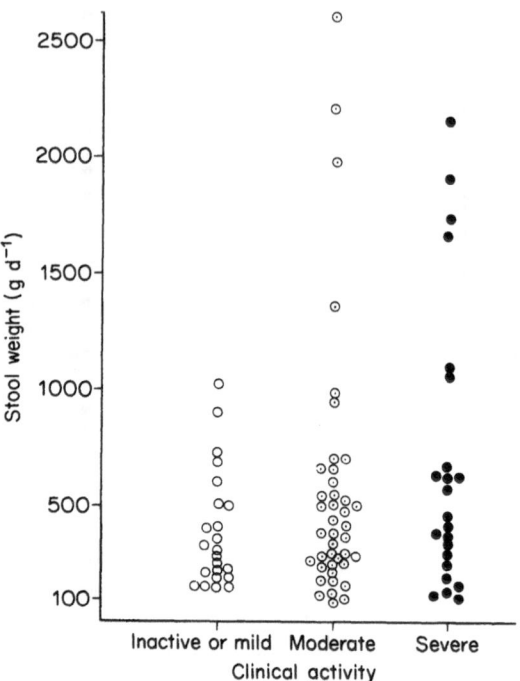

Figure 5.14. Crohn's disease. Average stool weights and clinical disease activity in 83 patients. Symbols as in figure 5.1.

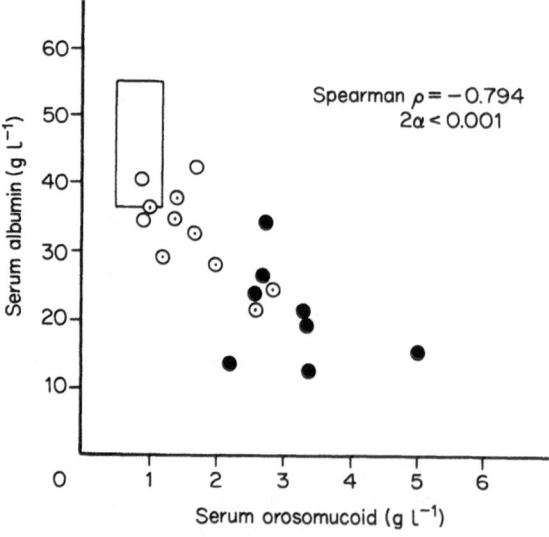

Figure 5.15. Ulcerative colitis. Serum albumin and orosomucoid concentrations in 19 patients. Symbols as in figure 5.1.

for definition of various degrees of clinical activity, a comparison was made between these variables.

Serum albumin and orosomucoid were closely and inversely correlated (figure 5.15), whereas no correlation existed between serum albumin and orosomucoid on the one hand, and haemoglobin and E.S.R. on the other hand (figure 5.16). Among other laboratory findings serum transferrin and thrombocyte count turned out to be highly correlated to both orosomucoid and albumin (figures 5.17 and 5.18).

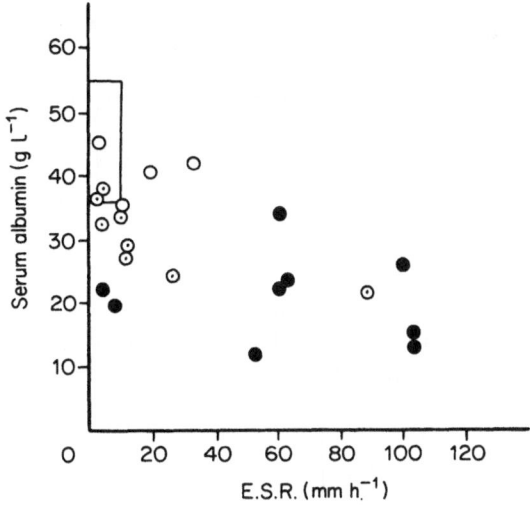

Figure 5.16. Ulcerative colitis. Serum albumin concentrations and E.S.R. in 21 patients. Symbols as in figure 5.1.

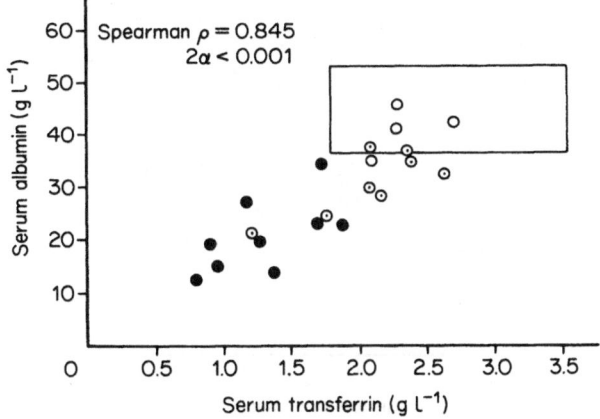

Figure 5.17. Ulcerative colitis. Serum albumin and transferrin concentrations in 19 patients. Symbols as in figure 5.1.

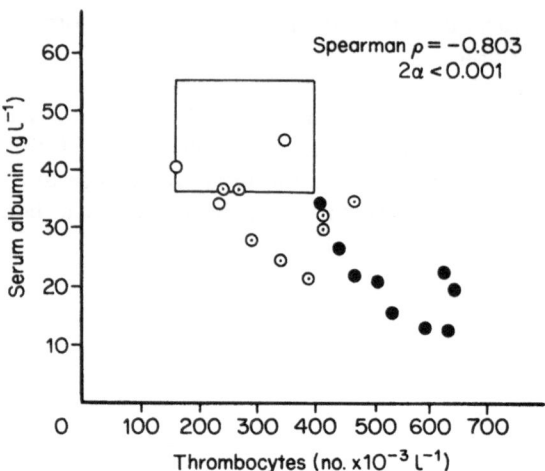

Figure 5.18. Ulcerative colitis. Serum albumin concentrations and thrombocyte counts in 20 patients. Symbols as in figure 5.1.

Crohn's disease

In Crohn's disease exactly the same correlations were present, although, statistically, to a less significant level. There was a fairly good inverse relationship between serum albumin and orosomucoid (figure 5.19). There was no

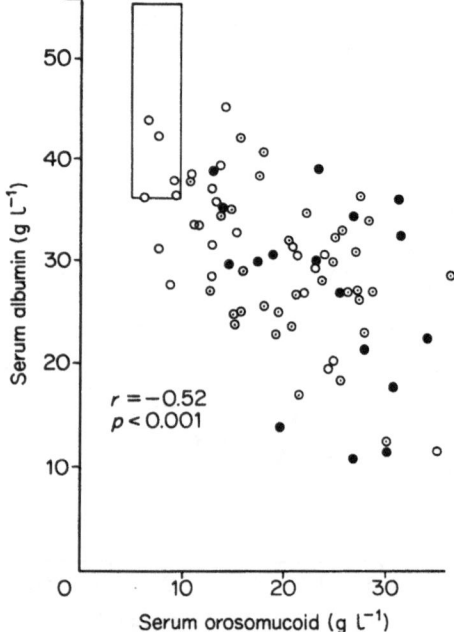

Figure 5.19. Crohn's disease. Serum albumin and orosomucoid concentrations in 81 patients. Symbols as in figure 5.1.

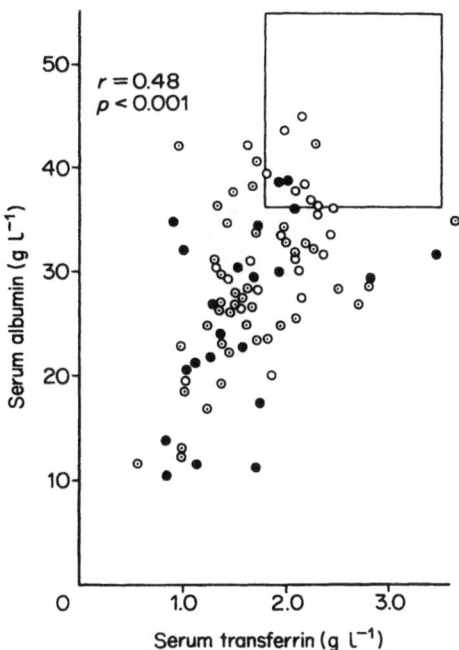

Figure 5.20. Crohn's disease. Serum albumin and transferrin concentrations in 87 patients. Symbols as in figure 5.1.

correlation between these two proteins and haemoglobin and E.S.R., but a statistically significant relation between serum orosomucoid and albumin on the one hand, and serum transferrin and thrombocyte count on the other hand (figures 5.20 and 5.21).

Conclusions

Our studies of 98 patients with ulcerative colitis or Crohn's disease have shown that the increased fractional catabolism of albumin and IgG in these diseases is primarily due to abnormal intestinal protein loss. A compensatory increase of albumin synthesis, as seen in other protein-losing gastroentero-pathies, is not present in chronic inflammatory bowel disease. A low synthetic rate was often present in severe cases of both ulcerative colitis and Crohn's disease. In contrast, IgG synthesis was normal or, in severe cases, often markedly increased.

Faecal radioiodine excretion following intravenous injection of ^{131}I-albumin and ^{125}I-IgG bore a highly significant positive relation to intestinal protein loss in ulcerative colitis. It also, with high accuracy, reflected the degree of clinical activity.

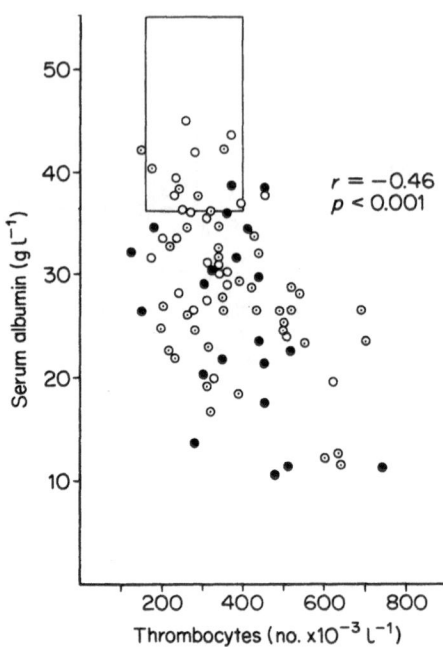

Figure 5.21. Crohn's disease. Serum albumin concentrations and thrombocyte counts in 84 patients. Symbols as in figure 5.1.

In Crohn's disease, however, faecal radioiodine excretion was not correlated to intestinal protein loss, but there was a significantly higher faecal radioiodine excretion in moderate and severe cases than in mild and inactive cases. No distinction was made, in this report, between groups of patients with different pathoanatomical site and extension of the lesion.

In ulcerative colitis, severe disease activity was almost invariably associated with a daily stool weight of 350 g or more, serum albumin less than 25 g l^{-1}, serum orosomucoid above 2.5 g l^{-1}, decreased serum transferrin and increased thrombocyte count. In Crohn's disease there was no correlation between clinical severity and stool weight, serum albumin and serum orosomucoid, serum transferrin and thrombocyte count. In both ulcerative colitis and Crohn's disease E.S.R. was of no discriminatory value in the distinction between various degrees of clinical severity.

References

1. Jarnum, S., Westergaard, H., Yssing, M. and Jensen, H. Quantitation of gastrointestinal protein loss by means of Fe[59]-labelled iron dextran. *Gastroenterology*, **55** (1968), 229–241

2. Petersen, E. A. and Sober, H. A. *J. Am. Chem. Soc.*, **78** (1956), 751
3. McFarlane, A. S. Efficient trace labelling of proteins with iodine. *Nature* (Lond), **182** (1958), 53
4. Nosslin, B. Applications of tracer theory to protein turnover studies. *J. Nucl. Biol. Med.*, **9** (1966), 18–19
5. Jarnum, S. and Jensen, K. B. Plasma protein turnover (albumin, trans-ferrin, IgG, IgM) in Ménétrier's disease (giant hypertrophic gastritis): Evidence of non-selective protein loss. *Gut*, **13** (1972), 128–137

Discussion

VESIN

Just a comment on this very interesting paper. We made studies of this type before, but with a smaller number of patients than you did. They brought us to the same finding, and I think the interpretation of the protein leakage must be correlated to the activity or to the extension of the lesions in the colon, and in the bowel as in Crohn's enteritis. We followed several patients at various stages, especially in ulcerative colitis, and protein leakage disappeared when the patient became in a quiet state.

I think that one of the main interests of these studies for the clinician is to show that low albumin figures in these patients are reliable indexes of the stage of disease, and whether the medical treatment can reduce the extent of the lesions. You will see albumin rise very rapidly as well as protein leakage, as demonstrated by these methods, has disappeared.

MUNRO

I calculate that you have far more than 30 g of protein leaking into the gut from albumin and other plasma proteins, without adding additional protein from haemoglobin and desquamated cells.

The question is: How much reabsorption of the products occurs in ulcerative colitis? What proportion of the stuff that is secreted into the gut undergoes sufficient reabsorption to be reutilised?

BIRGER-JENSEN

I cannot answer your question with exact figures by means of the present data. If the fractional catabolic rate of albumin is increased, e.g. 5 per cent per day due to colonic protein loss, this corresponds to an albumin loss of 5 g or less. The IgG loss can be estimated in a similar way to about 1 g per day.

REGOECZI

I would like to ask you what advantage do you see in using ^{59}Fe-dextran as compared to other tracers. Second question: Do you need some special precautions to get a good iron dextran for clinical use?

BIRGER-JENSEN

[59]Fe-dextran is a very good tracer in protein losing gastroenteropathies, since the iron could not go into the urine and the tissues. Because of its high energy,

[59]Fe can be used simultaneously with ^{131}I and ^{125}I, in contrast with ^{51}Cr which has the same energy as [131]I. [51]CrCl$_3$ is usable for determining GI protein loss, but it prolongs the study by 5 to 10 days. Unfortunately, the Radiochemical Centre in Amersham stopped the production of ^{59}Fe-iron dextran a month ago. On the other hand, I must say we have seen some mild anaphylactic reactions with ^{59}Fe-dextran which we have never seen with ^{51}Cr-albumin.

ANDERSEN

Have you ever determined the albumin or IgG content of faeces radio-immunologically?

BIRGER-JENSEN

No, I have not. It is very difficult, because to my knowledge you cannot do it immunochemically, since you are handling split products and so on, which will give you an erroneously high concentration, by binding relatively much specific antibody. It is a hard problem to tackle.

ROTHSCHILD

Since the protein loss by the stool contributes to your catabolic rate, therefore we can suppose that the endogenous catabolism was significantly reduced in your studies.

BIRGER-JENSEN

This is highly probable. We were unable to show it by our techniques. However, Waldmann's group convincingly demonstrated that endogenous catabolism of [67]Cu-labelled ceruloplasmin was diminished in the presence of abnormal protein loss.

JAMES

We have measured the nitrogen output in the stools of our patients and we found fantastic losses of protein or nitrogen, very much higher than the figure you mentioned. Indeed, we have had losses equivalent to 120 g of protein a day in patients with ulcerative colitis. So, did you in fact measure protein output in the stool in association with your studies? Of course, this might be relevant to your low finding. Furthermore, I am interested in whether you controlled the diet prior to actually doing the turnover study.

BIRGER-JENSEN

Taking the second question first, we did not control the dietary nitrogen input. The patients were on a normal diet and in a steady-state condition. We used the inert macromolecular ^{59}Fe-dextran to prove that there was a loss which accounted for the increased catabolic rate. We can calculate only indirectly the amount of protein lost, assuming that the increase of the fractional catabolic rate was due to the loss.

TAVILL

Could I ask you another question about ^{59}Fe-dextran? Iron deficiency is a common accompaniment of Crohn's disease and ulcerative colitis. In that sort of situation would you not expect the iron stores and the bone marrow to be extremely avid for iron? Knowing that you give only a trace dose, is it not possible that an artefactual result could occur because of increased uptake of ^{59}Fe in these sites?

BIRGER-JENSEN

I see what you mean, but since we use a plasma clearance calculation it is a matter of no concern to which extent ^{59}Fe is taken up by iron storage sites.

HOFFENBERG

If you can make allowance for the amount of protein broken down in the gut and for the reabsorption of iodine from that, should not it be possible to measure endogenous breakdown by measuring the urine radioiodine excretion and expressing that as a function of mean plasma specific activity during the day?

BIRGER-JENSEN

Unfortunately, it is not possible to do so. Part of the iodine excreted in the urine is iodine from radioiodine labelled protein exuded into the gut and split by gut enzymes with ensuing reabsorption of the iodine label, and we cannot determine by the present methods how much comes from endogenous catabolism and how much of the label comes from reabsorption of exuded labelled proteins.

6

Preparative agarose gel electrophoresis

B. G. JOHANSSON

Abstract

Simple and rapid techniques for preparation of proteins are often in demand in studies of protein metabolism. Various preparative gel electrophoretic techniques employing polyacrylamide have been described, but their use is sometimes hampered by the fact that they are laborious and time-consuming.

A method employing agarose gel which makes it possible to fractionate 50–100 mg proteins in a few hours will be described in this communication. The method is a modification of an analytical agarose gel electrophoretic procedure developed for use in clinical laboratory work and was described in detail earlier[1].

A 3 mm thick layer of 1% (w/v) agarose dissolved in 0.075 M barbital buffer, pH 8.6, is allowed to gel between two glass plates (11.0 × 20.5 × 0.1 cm) separated by a 3 mm thick ∪-formed plastic frame. After gelling, one of the plates is removed, and a slit of dimensions 0.2–0.5 × 16 cm is cut in the gel. The sample which should consist of 0.5–1 ml of a solution containing at most 100 mg protein is carefully mixed with one-quarter its volume of 3% (w/v) agarose solution which has been held at 45–50°C in a water bath. The mixture is immediately transferred to the slit. The plate is placed on the cooled support in the electrophoresis apparatus[1] and contact with the electrode reservoirs is secured by several layers of lint or soft cellulose tissue. The electrophoretic separation is performed with a potential gradient of 10 V cm^{-1}. When a satisfactory separation is anticipated, usually after 2–3 hours, the positions of the proteins are indicated by making a paper replica of the separation serving as a guide for the further procedure. Agarose strips containing the protein fraction(s) wanted are cut out of the agarose and frozen. On subsequent thawing more than half the liquid leaves the gel and can easily be collected. The rest of the liquid can be collected by centrifuging the gel in specially designed cups with a porous bottom.

An alternative method for elution of a *single* fraction is to cut a slit (0.3 ×

16 cm) just in front of the desired fraction, with the paper replica as a guide. This slit is filled with buffer and the electrophoresis continued with repeated changes of buffer in the slit at intervals of 2–3 minutes. This alternative procedure is suitable for proteins with relatively high migration rate and which are sensitive to freezing.

The procedure can also be used for proteins which are difficult to separate in ordinary agarose gels, if agarose with very low content of charged groups is used instead of the qualities commercially available[2].

The electrophoretic technique described has been widely used in our laboratory in the preparation of plasma proteins for analytical purposes and for use as antigens in antibody production against various plasma proteins.

References

1. Johansson, B. G. Agarose gel electrophoresis. *Scand J. clin. Lab. Invest.*, suppl. 124, **29** (1972), 7
2. Johansson, B. G. and Hjertén, S. Electrophoresis, crossed immunoelectrophoresis and isoelectric focusing in agarose gels with reduced electroendosmotic flow. *Anal. Biochem.*, **59** (1974), 200

Discussion

MCFARLANE

I am absolutely convinced that preparative agarose gel electrophoresis is the method of choice for preparing pure proteins.

As far as I can judge from the crossed immunoelectrophoresis you showed, the quality of albumin you can get by this method is far beyond anything I have seen anywhere else. Do you use the Uppsala's apparatus?

JOHANSSON

No, we use a simple apparatus (see ref. 1 in the abstract) developed in Professor Laurell's laboratory in Malmö. The most important feature is effective cooling. The elution is performed manually.

GORDON

I want to ask first of all how far the pre-treatment which you mentioned by ion exchange is responsible for the excellent band sharpness obtained by your technique in agarose. My second question concerns the residues which are obtained from all gels—acrylamide or agarose—when you elute from them. In fact, this is a problem that we always have, and its importance depends on what you want to do with the recovered materials, e.g. analytical studies, turnover studies, and so on.

JOHANSSON

The pre-treatment of the agarose does not increase band sharpness but is used for other purposes. As to the second question, we have had no troubles with gas chromatographic analysis of sugar after elution from agarose electrophoresis in studies of carbohydrate composition of glycoproteins. So, there cannot be too much matrix substances eluted.

GORDON

Is it your opinion that such low levels of material from the matrix is the result of the pre-treatment of the agarose with anion exchanger?

JOHANSSON

No, I do not think so.

GORDON

Then, the only advantage of the pre-treatment is to obtain increased band sharpness?

JOHANSSON

No. The advantage is that we are able to separate proteins which adsorb on ordinary agarose gels, e.g. cationic proteins like certain complement factors, C-reactive protein and lysozyme.

SCHEINBERG

Which proteins did you find do not stand freezing, so that when you try to elute them you have to use the centrifugation methods?

JOHANSSON

α_2-Macroglobulin and lipoproteins belong to this group of proteins.

ROSENOER

Can you comment upon the quantities of pure albumin you can achieve by this technique?

JOHANSSON

The upper limit for protein application is around 100 mg, but it is easy to perform several runs.

TAVILL

You mentioned that you could perform preparative gel electrophoresis only with proteins of high migration rates. And yet on the list of proteins you have prepared by this method you have pretty well covered the whole spectrum.

JOHANSSON

I am sorry that there was a misunderstanding: the elution with the *ditch* technique is most suitable for proteins with high migration rates, but it is always possible to cut out agarose strips and elute the protein by freezing and thawing.

PART 2

Kinetic Models

7

Albumin distribution from short-term tracer studies in man*

R. BIANCHI, G. MARIANI, A. PILO
and M. G. TONI

Introduction

The subject of the present paper is an application of the deconvolution approach to the study of extravascular (EV) albumin space. The method developed enables us to estimate not only the rate of transfer outside the vascular space and the total EV mass of the protein, but also the mass of EV albumin exchanging with the intravascular protein in a given time interval. This last parameter can be obtained using data from short-term tracer experiments (48–24 hours or less), as its computation does not imply extrapolation.

The proposed approach relies on the assumption of steady-state conditions throughout the study; its main advantage, however, is that meaningful information can be obtained also from short-term studies, during which steady-state conditions are more probable.

Results obtained in normal subjects by this approach are presented, with the aim of using such information for interpreting the mechanisms of trans-capillary passage of the albumin molecules.

Theoretical basis of the method

Definition of the model and its parameters

The model schematically represented in figure 7.1 can be used for the study of the kinetics of albumin and of other compounds whose renewal takes place within or close to the intravascular compartment. This model relies on a minimum of compartmental assumptions; in fact, while intravascular protein is assumed to behave as a single mixing pool, no compartmental hypothesis is made on the extravascular space.

*Partly supported by Public Health Service U.S.A., Contract No. PH-43-68-707.

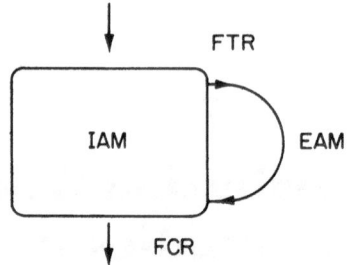

IAM : intravascular mass

EAM : extravascular mass

TAM = IAM + EAM : total mass

FCR : rate of catabolism as % of IAM/time unit

FTR : rate of intra/extra transfer as % of IAM/time unit

Figure 7.1. Model used for the study of albumin kinetics. Note that, while intravascular albumin is assumed to behave as a single compartment, no compartmental hypothesis is made on extravascular albumin (represented by a ring arrow).

The model can be characterised by the following four parameters: the total mass, subdivided into its intravascular (IAM) and extravascular (EAM) fractions; the turnover rate of the whole system (FCR), generally expressed as a percentage of the intravascular mass per time unit; and the transfer rate of the protein from the intra- to the EV space (FTR), also expressed as a percentage of the IAM per unit time.

Evaluation of the parameters by standard methods

The formulae currently employed to estimate the parameters IAM, EAM, FCR and FTR are summarised below.

$P(t)$ plasma concentration of the labelled protein *A; $p(t)$ plasma activity of *A as % of injected dose.

IAM = dose/$P(0)$ × albuminaemia

FCR   1. From $p(t)$ only: FCR = $1/\int_0^\infty p(t)\,\mathrm{d}t$

        2. From $p(t)$ + activity curve of labelled iodide released from *A breakdown in urine or in plasma.

FTR   From initial slope of $p(t)$:

$$FTR + FCR = \mathrm{d}p(t)/\mathrm{d}t|_{t=0}$$

EAM 1. As TAM − IAM

       where TAM = FCR × IAM × $\bar{t}$

       and $\bar{t} = \dfrac{\int_0^\infty t p(t)\,\mathrm{d}t}{\int_0^\infty p(t)\,\mathrm{d}t}$

2. By equilibrium time method:

$$\text{EAM/IAM} = e(t)/p(t)|_{t=t_E}$$

It is assumed that, after instantaneous i.v. injection of the labelled protein *A, its plasma disappearance curve $p(t)$ has been measured. First of all, intravascular albumin mass is obtained by the dilution principle.

Fractional catabolic rate (FCR) can be estimated as the reciprocal of the area under the activity curve $p(t)$ according to the Stewart–Hamilton principle. This approach requires experimental determination of the plasma curve only; however, it implies extrapolation of $p(t)$ to infinity and hence the performance of long-term experiments (2 weeks or more).

If we can measure, in addition to the plasma curve, the appearance curve of the iodide released from the labelled protein catabolism, either in the urines or in plasma, the experimental interval required to determine FCR can be shortened to about 1 day[1-3].

The estimate of intra/extra-vascular transfer rate (FTR) is obtained starting from the initial slope of the plasma disappearance curve; it is worthwhile to note that the measurement of this slope implies very accurate measurement of the initial part of the plasma disappearance curve.

Total body protein mass (TAM) can be obtained as the product of turnover rate (IAM × FCR) by the mean transit time ($\bar{t}$) of the whole system. The computation of $\bar{t}$ requires the extrapolation of $tp(t)$, and hence a very good determination of the final slope of $p(t)$ and long term experiments (at least 2 weeks). Extravascular mass, EAM, can then be obtained as the difference between TAM and IAM.

An alternative approach is the method of the equilibrium time, which works also in shorter experiments (about 1 week). This last approach, which applies in theory only to the case of a monocompartmental extravascular space, in practice gives results in good agreement with those obtained by the previously cited more general method.

Evaluation of the EV transfer rate and mass (FTR, EAM) by the deconvolution approach[4,5]

Up to this point, only two parameters of the extravascular albumin space have been defined and calculated, namely its mass (EAM) and its flow (FTR × IAM). Actually, much more information on extravascular space can be obtained starting from the plasma disappearance curve of the labelled protein, once FCR has been evaluated by the previously described methods. In fact, the whole transit time distribution function through the EV space can be obtained using the formulae given above.

The first equation makes it possible to compute the activity present at each time in the EV space, $e(t)$, as the difference between the injected dose, the tracer present in the vascular compartment, $p(t)$, and the tracer lost from the system up to that time by metabolic degradation, FCR $\int_0^t p(t)\,dt$.

$$e(t) = 1 - p(t) - \text{FCR} \int_0^t p(t)\,dt$$

$$\text{FTR} \times p(t) \longrightarrow \boxed{e(t)} \longrightarrow$$

$h(t)$ = transit time distribution function
of extravascular space

$$\text{FTR} \times p(t) * [1 - \int_0^t h(t)\,dt] = e(t)$$

and $p(t) * G(t) = e(t)$

where: $G(t) = \text{FTR} [1 - \int_0^t h(t)\,dt]$

$G(0) = \text{FTR}$ and

$$\int_0^\infty G(t)\,dt = \text{EAM/IAM}$$

Figure 7.2. Formulas employed to compute FTR and EAM/IAM by deconvolution approach.

Now, we can consider the EV space as an isolated system, and we know its input, i.e. the product FTR $p(t)$, and the activity present in it at each time, $e(t)$. The following relation then holds between these two functions and the transit time distribution function, $h(t)$, of the EV space†

$$p(t) * G(t) = e(t)$$

where

$$G(t) = \text{FTR}[1 - \textstyle\int_0^t h(t)\,dt].$$

As $p(t)$ and $e(t)$ are directly derivable from the experimental data, the equation can be used to evaluate the function $G(t)$, by deconvolution analysis.

Once $G(t)$ has been evaluated, both the flow and the mass of the EV albumin space can be obtained by means of the formulae which are given above; note that the evaluation of EAM requires the extrapolation of the function $G(t)$.

Application of the deconvolution approach to the analysis of short term experiments[7]

The deconvolution approach is also suitable for the analysis of short-term experiments; in fact, the function $G(t)$ can be determined by deconvolution in all cases, obviously for times not longer than the end of the experiment itself.

Now we can define new parameters of EV albumin which can be computed even with only a knowledge of $G(t)$ up to the time T (end of the experiment) without the need for any extrapolation. If T is the end of our experiment, we can consider the extravascular albumin mass as if it were divided into two

†The symbol $*$ is used to denote a convolution product operation.

subsystems: the first one constituted by all albumin molecules exchanging with intravascular albumin pool in times shorter than T, the second one by all the other albumin molecules.

The mass of the first subsystem, EM_T, can be computed starting from the function G, according to the formulae illustrated in figure 7.3 (for mathematical demonstration, *see* Appendix).

While $G(0) = FTR$ represents the overall flow of albumin into the extravascular space, $[G(0) - G(T)]$ is the fraction of flow (FTR_1) associated with the mass of extravascular albumin exchanged during the experimental interval (EM_T); furthermore, $G(T)$ is the flow directed towards the remaining extravascular mass (FTR_2), characterised by exchanging times longer than T.

From the lower formula shown in figure 7.3 we can also compute RM_T, which represents the mass of extravascular albumin renewed up to time T, and it is the sum of EM_T plus the product $G(T) \times T$ (*see* Appendix).

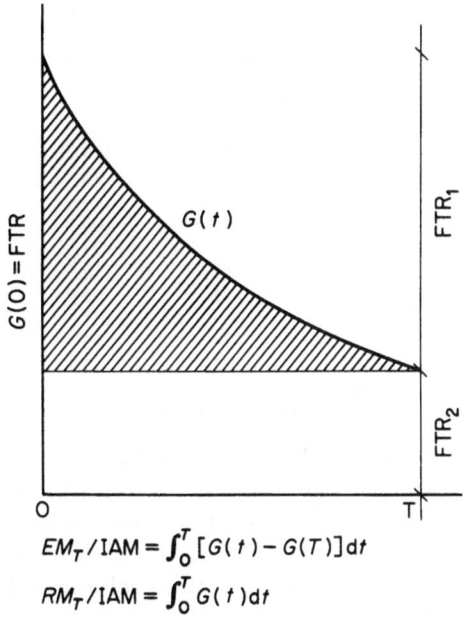

$$EM_T / IAM = \int_0^T [G(t) - G(T)]\,dt$$

$$RM_T / IAM = \int_0^T G(t)\,dt$$

Figure 7.3. Formulae used to compute extravascular albumin mass (EM_T) exchangeable in the time interval [0, T] and extravascular albumin mass renewed during the same interval (RM_T).

In conclusion, the following parameters can be obtained from the knowledge of $G(t)$ up to time T:

1. The total FTR.
2. The flow and mass of extravascular albumin completely exchangeable with the intravascular pool within the interval from 0 time to T (FTR_1, EM_T).

3. The flow associated with the remaining extravascular mass (FTR_2).
4. The mass of extravascular albumin renewed during the experimental interval (RM_T).

Results and discussion

The distribution results obtained by using this approach in 14 normal subjects are reported in Table 7.1. We found very large variations of the intra/extravascular transfer rate, which ranges in this group from 52 to 184% of IAM per day.

Table 7.1 Results obtained in a group of 14 normal subjects.

Case	FTR (% IAM/day)	Duration of the study (hours)	EAM (% IAM) eq. time	EAM (% IAM) $\int_0^T G(t)\,dt$	RM_{24} (% IAM)	EM_{24} (% IAM)
217	100	150	133	130	59	26
222	96	48	—	90	61	22
224	72	168	158	146	52	16
233	120	168	123	119	64	36
254	48	72	—	70	37	12
255	52	168	—	114 .	41	12
257	104	120	102	100	53	30
258	80	48	—	77	49	18
259	80	144	111	112	55	21
260	52	120	115	103	45	9
264	184	144	133	137	77	48
268	56	168	—	115	42	14
269	104	168	107	112	56	28
239	124	48	—	100	66	30

In the fourth column of the table we report the mass values of EV albumin computed by the integral of G till the end of the experiment (whose duration is reported in the second column), hence without extrapolation. There is a close agreement between these estimates of EAM and those obtained by the equilibrium time method (reported in the third column of the table), when the duration of the study was 6–7 days.

An interesting characteristic of EV albumin that appears in normals from these data is that its mass is quite stable, ranging from about 100 to 150% of IAM, and not proportional to the intra/extravascular transfer rate, which varies as much as 200% in the same group (figure 7.5). This fact implies that variations of transfer rate induce modifications in the distribution of albumin within the EV space.

A visual reflection of this fact is given by figure 7.4, where we report four G functions corresponding to subjects with very different intra/extravascular transfer rates; the functions are plotted for the first 3 days. It can be seen that

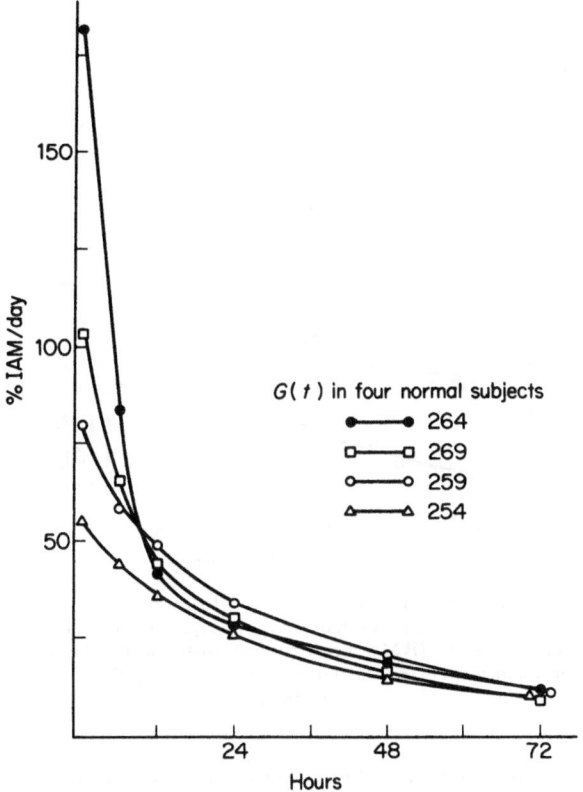

Figure 7.4. Comparison between 4 G functions obtained in 4 normal subjects
with very different fractional transfer rates (FTR = $G(0)$).

the zero intercepts, i.e. fractional transfer rates, are quite different. However,
from about the 24th hour onward, the four functions are fairly similar. In
terms of flow and mass this means that the intra/extravascular flow which
concerns albumin molecules which remain in the extravascular pool for more
than 24 hours is quite similar in the four subjects.

The increment of transfer rate corresponds to the increment only of the
rapidly exchangeable albumin (EM_{24}), which is a relatively small fraction of
the total extravascular protein.

A more quantitative proof of the variations of extravascular albumin
distribution in relation to transfer rate is presented in figure 7.5. Extravascular
mass renewed during one day and extravascular mass exchangeable during
one day are plotted against the fractional transfer rate. Renewed mass and
exchangeable mass both appear to be positively correlated with the overall
transfer rate, and these two parameters decrease in an almost parallel fashion.
As a direct consequence, no correlation exists between transfer rates and the

R. Bianchi, G. Mariani, A. Pilo and M. G. Toni

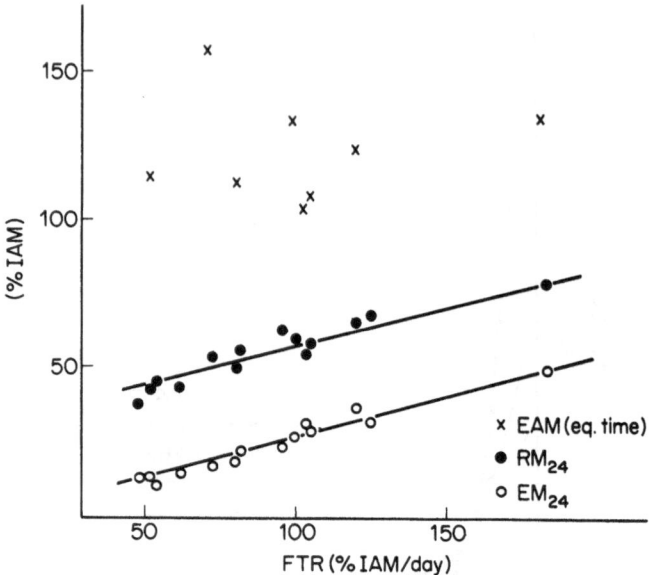

Figure 7.5. Correlation between FTR, total extravascular albumin mass (EAM), extravascular mass exchangeable (EM_{24}, closed circles) and extravascular mass renewed (RM_{24}, open circles) during a one day interval. The crosses represent total extravascular mass computed by the equilibrium time method.

difference between renewed and exchangeable masses, i.e. between fractional transfer rates and the flow associated with extravascular albumin exchangeable in times longer than one day.

These findings suggest that variations of transfer rates induce modification of the distribution of EV albumin; in fact, variations of fractional transfer rates appear to be associated with changes of rapidly exchangeable albumin, while the flow affecting the more slowly exchangeable albumin mass tends to be fairly constant.

Appendix

As the input to the extravascular space is given by the product of FTR and the intravascular activity, $p(t)$, the extravascular activity, $e(t)$, can be written as

$$e(t) = \text{FTR}\, p(t) \ast [1 - H(t)] \tag{1}$$

where $H(t) = \int_0^t h(t)\, dt$ is the integral of the transit time distribution function of the EV space, and symbol $\ast$ means a convolution product operation.

Defining

$$G(t) = \text{FTR}[1 - H(t)] \tag{2}$$

we can write equation (1) as

$$e(t) = p(t) * G(t). \tag{3}$$

As $H(0) = 0$, then we have

$$G(0) = \text{FTR}. \tag{4}$$

Moreover, since $\int_0^\infty [1 - H(t)]\, dt$ is the mean transit time of EV albumin (figure 7.5), and FTR × IAM is its turnover rate, we can write EAM as the product of flow by the mean transit time, according to

$$\text{EAM} = \text{FTR} \times \text{IAM} \int_0^\infty [1 - H(t)]\, dt \tag{5}$$

or, in other terms

$$\frac{\text{EAM}}{\text{IAM}} = \int_0^\infty G(t)\, dt. \tag{6}$$

Let us consider that for a given time T the extravascular mass is subdivided into 2 subsystems: the first one constituted by all albumin molecules which exchange with the intravascular pool in times shorter than T (the end of the experiment), the second one by all other albumin molecules.

Accordingly, the transit time distribution function (TTDF) of the first subsystem can be defined as:

$$h_1 = \begin{cases} h(t)/\mathcal{Z} & t \leqslant T \\ 0 & t > T \end{cases} \tag{7}$$

where $\mathcal{Z} = \int_0^T h(t)\, dt$ is introduced for normalisation purposes.

Similarly, the TTDF of the second subsystem will be

$$h_2 = \begin{cases} 0 & t \leqslant T \\ h(t)/(1 - \mathcal{Z}) & t > T \end{cases} \tag{8}$$

Hence $h(t)$ can be written as

$$h(t) = \mathcal{Z} h_1(t) + (1 - \mathcal{Z}) h_2(t) \tag{9}$$

and $G(t)$ as

$$G(t) = \text{FTR}[1 - \mathcal{Z} H_1(t) - (1 - \mathcal{Z}) H_2(t)] \tag{10}$$

where $H_1(t) = \int_0^t h_1(t)\, dt$ and $H_2(t) = \int_0^t h_2(t)\, dt$. As $H_2(t) = 0$ for $t \leqslant T$, we have

$$G(t) = \text{FTR}[1 - \mathcal{Z} H_1(t)] \qquad t \leqslant T. \tag{11}$$

Equation (11) can be rewritten as

$$G(t) = \text{FTR}_1[1 - H_1(t)] + \text{FTR}_2 \tag{12}$$

where $\text{FTR}_1 = \text{FTR} \times \mathcal{Z}$ represents the flow of albumin through the first subsystem (as a percentage of IAM), while $FTR_2 = FTR - FTR_1$ represents the flow through the second one.

It is evident that, since $H_1(T) = 1$, then $FTR_2 = G(T)$ and $FTR_1 = G(0) - G(T)$, as can be seen in figure 7.3.

Integrating equation (12) we obtain

$$\int_0^T [G(t) - G(T)]\, dt = FTR_1 \int_0^T [1 - H_1(t)]\, dt \qquad (13)$$

As the right hand side of this equation is the product of flow (as a percentage of IAM) multiplied by the mean transit time of the first subsystem, we obtain

$$\int_0^T [G(t) - G(T)]\, dt = EM_T/IAM \qquad (14)$$

where EM_T represents the mass of the first subsystem.

The product $G(T) \times T$ represents the mass (as a percentage of IAM) of the second subsystem, renewed during the interval $[0, T]$, as this subsystem cannot exchange in time shorter than T, while the total area $\int_0^T G(t)\, dt$ is the mass of extravascular albumin renewed till time T, i.e.

$$\frac{RM_T}{IAM} = \int_0^T G(t)\, dt \qquad (15)$$

where RM_T indicates the mass of extravascular albumin renewed in a time equal to the experimental interval.

References

1. Bianchi, R., Mariani, G. and Pilo, A. (1972) Short-term measurement of plasma protein catabolism in man. In *Radioaktive Isotope in Klinik und Forschung*, 10ed., K. Fellinger and R. Höfer, eds., München, p. 94
2. Bianchi, R., Mariani, G., Pilo, A., Toni, M. G. and Donato, L. (1973) Short-term determination of plasma protein turnover by a two tracer technique using plasma only or plasma and urine data. In *Protein Turnover* ASP, Amsterdam, p. 47
3. Bianchi, R., Mariani, G., Pilo, A. and Toni, M. G. Short-term determination of serum albumin catabolism in man from plasma data only. *J. Nucl. Biol. Med.*, **17** (1973), 117
4. Vitek, F., Bianchi, R., Mancini, P. and Donato, L. (1969) Deconvolution techniques for the analysis of short-term metabolic studies with radio-iodinated albumin. In *Physiology and Pathophysiology of Plasma Protein Metabolism*, G. Birke, R. Norberg and L. O. Plantin, eds., Pergamon Press, p. 29
5. Bianchi, R., Federighi, G., Giagnoni, P., Giordani, R. Navalesi, R. and

Donato, L. (1969) Patterns of serum albumin metabolism in renal diseases. In *Physiology and Pathophysiology of Plasma Protein Metabolism*, G. Birke, R. Norberg and L. O. Plantin, eds., Pergamon Press, p. 195

6. Zierler, K. (1964) Basic aspects of kinetic theory as applied to tracer-distribution studies. In *Dynamic Studies with Radioisotopes*, U.A.S.E.C., p. 55

7. Bianchi, R., Mariani, G., Pilo, A. and Toni, M. G. Measurement of albumin extravascular transfer rate and distribution from short term two tracer studies in man. In *Radioaktive Isotopes in Klinik und Forschung*, 11ed., R. Höfer, ed., Munich, p. 220

Discussion

DONATO

Did I understand right that only a relatively small part of the extravascular albumin pool is influenced by possible changes of the intra- to the extra-vascular transfer rate (FTR)? Is the greatest part of the extravascular mass little correlated to changes of FTR?

PILO

Correct. This is well shown, for instance, by a comparison between the data obtained in case 264 with those of case 259 (figure 7.4 and table 7.1 in the text). In the first subject the extravascular albumin mass exchangeable in 24 hours is about twice the corresponding value of the second one (FTR, respectively, 184% and 80% of the intravascular protein per day), while the total extra-vascular albumin mass is only 20% greater in the first case than in the second one.

HOFFENBERG

Could I ask how valid it is to use the early part of the plasma disappearance curve of radioactively labelled albumin for this analysis? So far as I understand, you are looking at the first 12 hours or so of the activity disappearance, and this would mean that this part of the curve is the most acceptable to the esti-mate of the labelled molecule distribution by your type of analysis.

PILO

Indeed, absolute purity and no loss of the tracer are necessary conditions to obtain reliable data by this method.

Moreover, while the exchangeable and renewed albumin masses are derived by experimental data up to 48 or 72 hours after the injection, the fractional transfer rate is computed by using the first few hours of the plasma disappear-ance curve, and this last value is reliable only if a very good tracer is used.

HOFFENBERG

I just wonder whether it is possible to make a tracer of that degree of purity? It is possible to use biologically screened labelled proteins in animal studies, but I do not think there is any way of doing this in human studies. Therefore, I do not think that it is valid to make that assumption.

BIANCHI

A tracer of good quality is obviously mandatory for short-term turnover studies in man, and I think we have succeeded in obtaining a tracer of this kind, not only by using careful and gentle purification techniques of the protein, but also by labelling it by the electrolytical procedure, which appears to be the less damaging one for albumin. This is proved by the close agreement we found between the values for fractional catabolic rate derived in the same subjects from experimental data of the first 24 hours and those obtained from data of 8 to 10 days.

SCHEINBERG

As a comment to the second question of Dr Hoffenberg, I would like to recall a special system to perform tracer turnover studies with biological screened material in humans. One can administer radioactive copper to a normal individual, let him release labelled ceruloplasmin into the plasma, then take his plasma and transfuse it into a patient with Wilson's disease, for instance, to investigate the turnover of the carrier protein in that situation.

ROSSING

I should like to agree completely with you, Dr Pilo, that the extravascular albumin mass is completely independent in steady-state conditions from the fractional transfer rate, which means that when transfer from intra- to extravascular compartments is high there is a high lymph return flow. However, I note that your range of FTR values is rather wide as compared with ours, and I think that the critical point in the reliable determination of transfer rate is the experimental definition of the very initial part of the plasma disappearance curve. So, my question is: How many experimental points do you have during the first 1 or 2 hours after injection of the tracer?

PILO

The curves presented here have been derived from about 6 experimental points in the first 2 hours. I agree with you that probably this is not a sufficient number of points for a reliable measurement of the initial slope, according to your technique to compute transcapillary escape rate. Actually, our experiments were performed mainly in order to measure albumin catabolism, and the data were subsequently analyzed to obtain the fractional transfer rate. However, since the derivative of the plasma activity curve at zero time gave rather scattered values, I used the following approach to compute transfer rate. Bearing in mind that the zero intercept of the function G (obtained by deconvolution) is analytically equivalent to the zero time derivative of the plasma disappearance curve, one can find that this intercept fluctuates considerably. In fact, if you have for instance two G functions that differ from each other in the first 1 or 2 hours only and you convolve them with the respective plasma curves, $p(t)$, you obtain extravascular activity functions, $e(t)$, that present only undetectable differences in the first part. So, you cannot determine the very initial part of G starting from extravascular activity. For this reason, I computed FTR as the area of the *G* function divided by the time interval considered. Now, the renewed mass during the time interval [0, *T*] divided by *T* tends to the zero intercept value, for *T* sufficiently small. Moreover, this value represents an average of the intra- to extravascular flow during this time interval, on the assumption that during this period there is practically no return of labelled material from the extravascular space.

DONATO

Perhaps we could clarify this point for the audience by saying that Dr Pilo is not discussing the zero-time intercept of the tracer disappearance curve in plasma, but the zero-time intercept of the function *G*, obtained by deconvolution of the extravascular into the intravascular tracer curves.

WALDMANN

Are you implying that you could not get abnormalities of the extravascular mass due to changes in the egress and return rate constants? Can you reliably measure a huge extravascular compartment if there is a disturbance in the rate of egression out from plasma into a series of extravascular compartments?

We have studied a series of patients with idiopathic oedema, and they developed enormous extravascular compartments with increasing values of the outflow of proteins into the extravascular compartment in the period of 12 hours or so (in the absence of proteinuria). I wonder whether in studying distribution kinetics there is a reason for excluding some model system, where you can hold certain elements as constants and define as errors of egress or return what it has been necessary and sufficient to define so?

PILO

I think that our approach is more general and extremely more stable than the compartmental one, since it is based on integral properties of the curve, not on intercepts and slope values of the exponentials you use to fit the plasma curve. With our procedure you can use the same mathematical approach to analyse a given activity curve sampled for 7 days or the same curve sampled for 24 hours only, and you will obtain the same results. If instead you use multicompartmental analysis you have to change the model, since in the first case you have three compartments, while in the second one you can detect two or perhaps only one compartment, and you do not have any correspondence between the model used and the experimental data. So, whenever you shorten your experimental interval you have to change the mathematical model and, moreover, you cannot give a physiological meaning to your compartments. You must rather speak in terms of invariants of the compartmental models which, I think, is just the same as speaking of integral properties of the curves.

DONATO

Could I try to reply on behalf of Dr Pilo to the first part of the question of Dr Waldmann? I do not think that the distribution data here presented and discussed imply that in pathological conditions there could not be any relationship between transfer rate and the size of extravascular mass. In fact, results have been shown only from normal subjects; such results indicate, as Dr. Rossing was confirming, that in normals there is no direct relation between transfer rate and size of the extravascular pool, but I could well imagine that in pathological circumstances this could occur.

8

The arginine-6-^{14}C (^{14}C-bicarbonate) method of measuring plasma protein synthetic rate and its simplification*

E. B. REEVE, W. LEW and T. CARLSON

Figure 8.1 introduces the subject of this paper which is how to simplify and make more generally usable the ^{14}C-bicarbonate method of measuring synthetic rate of plasma proteins synthesised by the liver. It shows measurements of the specific radioactivity of the arginine 6-carbon of plasma albumin and plasma fibrinogen obtained about 3 hours after one hour's intravenous infusion of 0.5 mC ^{14}C-bicarbonate. Rabbits were injected with 5 mg of cycloheximide per kg of body weight from 1 to 48 hours before the start of the infusion and the specific radioactivities are expressed as fractions of the mean levels we found in untreated healthy control animals. It is seen that the specific radioactivities (SA) of both albumin and fibrinogen arginine-6-C, are reduced almost to zero shortly after the injection, but both show some recovery 6 hours later; albumin SA remains unchanged 12 hours later but by this time fibrinogen arginine-6-C SA has rebounded to more than twice its initial levels; by 24 hours albumin arginine-6-C SA has returned to initial levels while fibrinogen arginine-6-C has fallen to 1.5 times initial levels; and by 48 hours both albumin arginine-6-C and fibrinogen arginine-6-C SA are near initial levels. This is a rather remarkable contrast in the responses of two different liver-synthesised plasma proteins to an agent that blocks ribosomal translation of mRNA[1,2]. We should like to be able to interpret such results in terms of the underlying albumin and fibrinogen synthesis. To obtain measurements of synthesis using the current fullscale ^{14}C-bicarbonate methods[3-7] in groups of animals would be almost prohibitive in the time and effort required.

*This research was supported by Grant HL-02262 from the National Heart Institute, Grant RR-51 from the General Clinical Research Centers Program of the Division of Research Resources, National Institutes of Health; and grants from the Colorado Heart Association, Denver, Colorado.

E. B. Reeve, W. Lew and T. Carlson

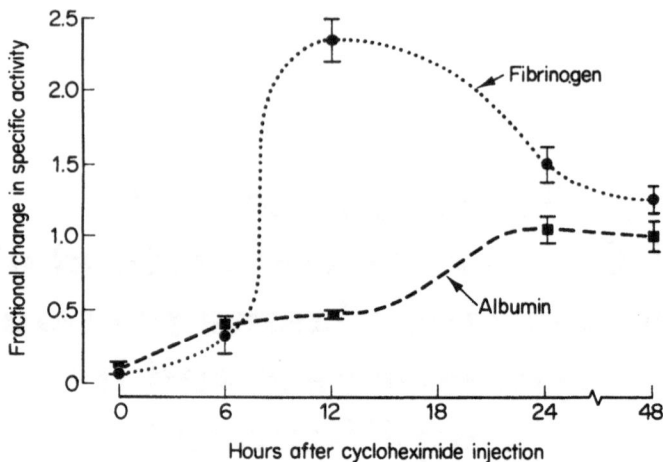

Figure 8.1. The fractional change in the SA of plasma albumin and plasma fibrinogen arginine-6-carbon in rabbits at various times after the injection of 5 mg cycloheximide per kg body weight. Uninjected rabbits average a normalised SA of 1.0. The SAs were determined on proteins obtained from plasma samples withdrawn approximately 4 hours after the start of a 1-hour ^{14}C-bicarbonate infusion. The bars about the points show a range of $\pm$ one standard error.

To determine the synthetic rate of a plasma protein requires measuring the total radioactivity of a labelled amino acid that enters that protein over a given time and the average specific radioactivity of that amino acid at the synthetic site over the same time[8]. Most current methods use arginine labelled in the 6-carbon as the labelled amino acid[9]. This is formed in the liver following ^{14}C-bicarbonate infusion[10] and its specific activity is determined from that of newly synthesised urea carbon[8]. However, it is known that the specific radioactivity of newly formed urea-^{14}C is not easy to measure[11–13]. Measurement is made difficult by a large reservoir of tissue fluid urea[13]; by continual loss of urea from the body primarily through kidney excretion and intestinal bacterial metabolism; by variable renal urea content dependent on water balance and ADH secretion[6] and by quite rapidly variable levels of urea cycle enzymes which are inducible for instance by the levels of dietary protein[14]. To circumvent these difficulties, several kinetic approaches have been tried[3,4,5,7,11] but these depend on assumptions which are not always justified[6] while the necessary measurements greatly complicate the ^{14}C-bicarbonate method. Thus, in practice, urea does not give convenient estimation of the specific radioactivity of arginine-6-C at the protein synthetic site. We adopted another approach (*see* table 8.1) and used the specific radioactivity of blood total CO_2 following ^{14}C-bicarbonate infusion to indicate specific radioactivity of the arginine-6-C at the synthetic site[6]. To illustrate the method, figure 8.2 shows a standardised ^{14}C-total CO_2 specific radioactivity curve obtained by

Table 8.1. ^{14}C-Bicarbonate input method of measuring protein synthetic rate[6]

1. Infuse ^{14}C-bicarbonate intravenously at constant rate for 1 hour.
2. Draw 4 blood samples at 15, 30, 45 and 60 min during the infusion, and 4 blood samples in the next 3 hours.
3. Analyse each blood sample for blood ^{14}C-total CO_2 and total CO_2 content. Determine mean SA of blood total CO_2 (mean SA blood CO_2) over experimental time, T h.
4. Analyse a zero time and final blood sample for urea content and ^{14}C-urea activity.
5. Isolate the protein under study from the final blood sample, digest with 6 N HCl and determine the SA of the arginine 6-carbon (SA arg-6-C). Measure the protein concentration in the plasma.
6. Determine protein synthetic rate from:

$$\text{Mg protein day}^{-1} = \frac{\text{SA arg-6-C} \times \text{mg protein in plasma} \times F \times 24}{\text{mean SA blood } ^{14}\text{C-bicarbonate carbon} \times T}$$

where F corrects for the arginine content of the protein, the rates of catabolism and distribution to the interstitial fluids of the protein, the delay in passage of newly synthesised protein through the liver cell and the small differences between the SA of ^{14}C-bicarbonate and arginine-6-[14]C.
7. Fractional synthetic rate (fraction of a protein in the plasma synthesised per day) is given by

$$\frac{\text{SA arg-6-C} \times F \times 24}{\text{mean SA blood } CO_2 \times T}$$

averaging results of 4-hour experiments on 4 animals. The curve is based on measurements from 4 blood samples withdrawn over the hour's infusion of ^{14}C-bicarbonate and 4 samples withdrawn during the following 3 hours. The mean value over the 4 hours is obtained by integrating the curve by square counting and dividing by the time. This mean has been shown by simultan-

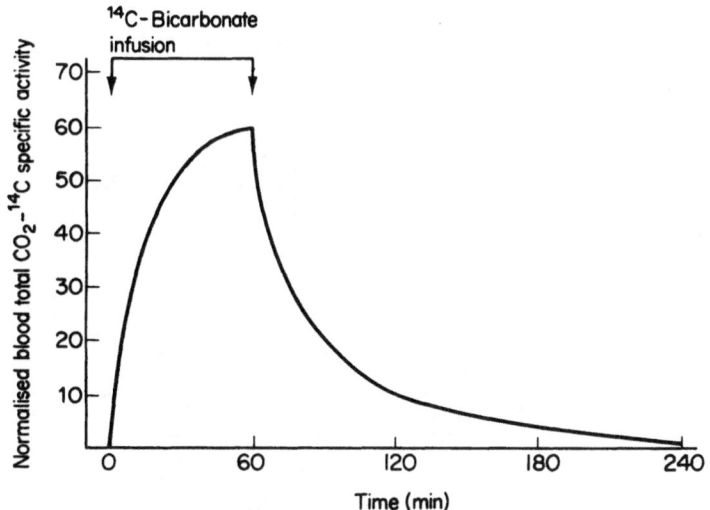

Figure 8.2. Average normalised blood ^{14}C-bicarbonate SAs in 4 healthy rabbits. These were calculated from the expression

$$\frac{\text{blood (dpm)}}{\mu\text{Eq. HCO}_3^- \text{ (ml)}} \times \frac{3.85 \times 10^3}{^{14}\text{C-HCO}_3^- \text{ injected (dpm)}}$$

eous comparison with the urea form of the method, to estimate closely mean SA of arginine-6-C at the synthetic site[6].

This new approach uses the specific radioactivity of a metabolite entering, rather than leaving, the synthetic site to measure synthetic site specific radioactivity and this poses kinetic problems. We now examine these.

Consider as a simple but sufficient analogue of the urea cycle a chain of three linked compounds, C_B the blood total CO_2 carbon, C_C the carbon of a single intermediate compound originating from C_B, and C_A the liver arginine-6-carbon pool derived from C_C. Figure 8.3a shows this simple chain, $C_B \rightarrow C_C \rightarrow C_A$ connected by the rate constants l_1, l_2, and l_3 with leaks governed by the rate constants m_1, m_2 and m_3. Note that *C indicates radioactive carbon, C non-radioactive carbon. We now impose on this analogue our experimental conditions. Thus an input of ^{14}C-bicarbonate is infused into compartment C_B (figure 8.3a) at a constant rate, I, for the first hour of the experiment but for the remaining three hours it is zero. At the same time non-radioactive carbon in the form of CO_2 released during metabolism continually enters compartment C_B at a constant rate M throughout the whole experiment. We now need a representative mathematical function which will describe the behaviour with time of the first components of the chains *C_B and C_B. This is obtained by picturing the blood total CO_2 pool, C_B, as a single pool with entry rate M and loss rate $(m_1 + l_1)C_B$ where m_1 is loss primarily through the lungs and $m_1 \gg l_1$. Though not completely accurate this description is adequate and allows formulation of the blood total CO_2 carbon SA by simple functions. Given this information we now derive the time-dependent specific radioactivities of C_B, C_C and C_A which allows us to test under what conditions the blood specific radioactivity, *C_B/C_B gives a good measure of the specific radioactivity at the synthetic site, *C_A/C_A.

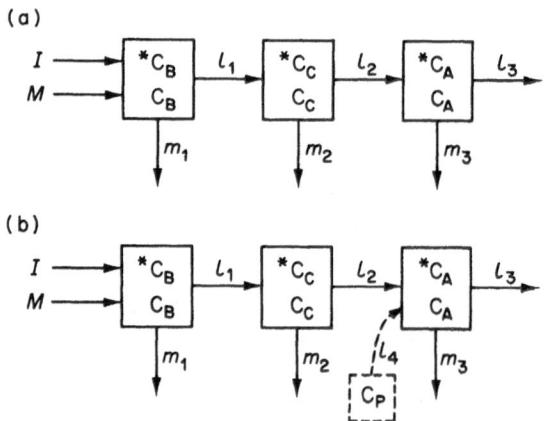

Figure 8.3. Elementary systems which present the main features of the urea cycle intermediates. Species labelled with an asterisk are radioactive, those unlabelled are not radioactive. (a) A system without later entry of unlabelled intermediates, and (b) with such entry.

During ^{14}C-bicarbonate infusion at the constant rate I ^{14}C dpm $\min^{-1}$ the total radioactive species in the compartments of figure 8.3a are given by the following differential equations with their solutions:

$$*C_B' = I - (l_1 + m_1)*C_B, \text{ with solution} \tag{1}$$

$$*C_B = I(1 - \exp(-\gamma_1 t))/\gamma_1 \tag{1a}$$

$$*C_C' = l_1 *C_B - (l_2 + m_2)*C_C, \text{ with solution} \tag{2}$$

$$*C_C = l_1 I\left[\frac{1}{\gamma_1\gamma_2} - \frac{\exp(-\gamma_1 t)}{\gamma_1(\gamma_2 - \gamma_1)} + \frac{\exp(-\gamma_2 t)}{\gamma_2(\gamma_2 - \gamma_1)}\right] \tag{2a}$$

$$*C_A' = l_2 *C_C - (l_3 + m_3)*C_A, \text{ with solution} \tag{3}$$

$$*C_A = l_1 l_2 I\left[\frac{1}{\gamma_1\gamma_2\gamma_3} - \frac{\exp(-\gamma_1 t)}{\gamma_1(\gamma_2 - \gamma_1)(\gamma_3 - \gamma_1)} + \frac{\exp(-\gamma_2 t)}{\gamma_2(\gamma_2 - \gamma_1)(\gamma_3 - \gamma_2)}\right.$$
$$\left. - \frac{\exp(-\gamma_3 t)}{\gamma_3(\gamma_3 - \gamma_1)(\gamma_3 - \gamma_2)}\right] \tag{3a}$$

In the above and following equations $\gamma_1 = l_1 + m_1$, $\gamma_2 = l_2 + m_2$, $\gamma_3 = l_3 + m_3$, while the primes represent d/dt.

Considering the non-radioactive species during the entry of metabolic CO_2 at the rate M mg CO_2 $\min^{-1}$ from the tissues into the blood stream

$$C_B' = M - (l_1 + m_1)C_B \tag{4}$$

and in the steady state (ss)

$$C_B(\text{ss}) = M/\gamma_1 \tag{4a}$$

$$C_C' = l_1 C_B - (l_2 + m_2)C_C \tag{5}$$

$$C_C(\text{ss}) = l_1 M/\gamma_1\gamma_2 \tag{5a}$$

$$C_A' = l_2 C_C - (l_3 + m_3)C_A \tag{6}$$

$$C_A(\text{ss}) = l_1 l_2 M/\gamma_1\gamma_2\gamma_3 \tag{6a}$$

Inspection of the ^{14}C blood total CO_2 specific activity curve of figure 8.2 shows that the compartment containing $*C_B$ does not reach a steady state during the one-hour constant rate infusion so that the SAs of the non-steady-state solutions during this hour must be examined. Since C_B, C_C and C_A are in steady states these are

$$*C_B/C_B = \frac{I}{M}(1 - \exp(-\gamma_1 t)) \tag{7}$$

$$*C_C/C_C = \frac{I}{M}\left[1 - \frac{\gamma_2 \exp(-\gamma_1 t)}{\gamma_2 - \gamma_1} + \frac{\gamma_1 \exp(-\gamma_2 t)}{\gamma_2 - \gamma_1}\right] \tag{8}$$

$$*C_A/C_A = \frac{I}{M}\left[1 - \frac{\gamma_2\gamma_3 \exp(-\gamma_1 t)}{(\gamma_2-\gamma_1)(\gamma_3-\gamma_1)} + \frac{\gamma_1\gamma_3 \exp(-\gamma_2 t)}{(\gamma_2-\gamma_1)(\gamma_3-\gamma_2)} - \frac{\gamma_2\gamma_3 \exp(-\gamma_3 t)}{(\gamma_3-\gamma_1)(\gamma_3-\gamma_2)}\right] \quad (9)$$

From previous studies[6] $\gamma_1 \simeq .025$ min^{-1} when acting on the ~35 mM total CO_2 pool of a healthy 2.5 kg rabbit while the lower limits of γ_2 and γ_3 are defined by the flow of CO_2 carbon in the urea cycle and the size of the pools of urea cycle intermediates. The former is given by the value of approximately 40 mM urea formed in the 2.5 kg rabbit liver per day[13] while studies on the rat liver[10] show very low concentrations of urea cycle intermediates. If, as is likely, these concentrations are true for rabbit liver, minimal values for γ_2 and γ_3 of 8 to 25 min^{-1} can be calculated. Thus $\gamma_1 \ll \gamma_2$ and $\gamma_1 \ll \gamma_3$ and equation 8 becomes

$$*C_C/C_C \simeq \frac{I}{M}\left[1 - \exp(-\gamma_1 t) + \frac{\gamma_1}{\gamma_2}\exp(-\gamma_2 t)\right] \quad (10)$$

and equation 9 becomes

$$*C_A/C_A \simeq \frac{I}{M}\left[1 - \exp(-\gamma_1 t) + \frac{\gamma_1}{\gamma_2}\frac{\gamma_3}{(\gamma_3-\gamma_2)}\exp(-\gamma_2 t) - \frac{\gamma_2}{\gamma_3-\gamma_2}\exp(-\gamma_3 t)\right]$$

$$(11)$$

The minimal values calculated for γ_2 and γ_3 superimpose further conditions on equations 10 and 11. Thus any terms in these equations containing $\exp(-\gamma_2 t)$ or $\exp(-\gamma_3 t)$ will approach zero in less than a minute. As a result the specific radioactivities $*C_B/C_B$, $*C_C/C_C$, $*C_A/C_A$ are very close to I/M $(1-\exp(-\gamma_1 t))$ and are thus essentially equal. Because of the small size of the pools of all urea cycle intermediates[10] and the rapid flow of carbonate carbon through the urea cycle, were more compartments introduced in figure 8.3a between C_B and C_A to represent all the urea cycle intermediates, similar arguments would apply and the specific activity of each intermediate would be essentially equal to $I/M(1-\exp(-\gamma_1 t))$.

Figure 8.3b shows a further complicating factor: entry of unlabelled C_p, e.g. carbon released by the breakdown of liver protein, into compartment C_A at a rate $l_4 C_p$. The steady state solution for C_A (*see* equations 6 and 6a) is now $C_A = l_1 l_2 M/\gamma_1\gamma_2\gamma_3 + l_4 C_p/\gamma_3$. The last term may be written $\gamma_1\gamma_2 l_4 C_p/\gamma_1\gamma_2\gamma_3$ and if we let $\gamma_1\gamma_2 l_4 C_p = l_1 l_2\Delta$ then $C_A = l_1 l_2(M+\Delta)/\gamma_1\gamma_2\gamma_3$. Ordinarily Δ is very small compared with M while the ratio $(M+\Delta)/M$ measures the minor dilution of urea cycle intermediate CO_2-carbon caused by protein catabolic release of arginine into the liver arginine pool.

The specific activity of C_A now becomes

$$*C_A/C_A = \frac{I}{M+\Delta}(1-\exp(-\gamma_1 t)) \quad (12)$$

Thus the net effect of steady-state entry of unlabelled carbon by other than urea cycle paths into the urea cycle carbonate carbon will be proportional to the sum of the metabolic carbon entry rate plus the entry rate of the diluting carbon, $(M + \Delta)$, divided by the metabolic carbon entry rate, M.

We have so far considered only the SAs of C_B and C_A during the infusion of ^{14}C-bicarbonate. Analysis shows that after the ^{14}C-bicarbonate infusion has stopped the specific radioactivity $*C_B/C_B$ is very close to

$$\frac{I}{M}(1 - \exp(-\gamma_1 t)) + \frac{\gamma_1 C_B(0) \exp(-\gamma_1 t)}{M}$$

which is also true of $*C_A/C_A$. Thus the two can be taken as identical. The effects of dilution of SA from entry of unlabelled carbon are analysed by the methods applied above. Examination of the urea cycle of figure 8.4 shows that

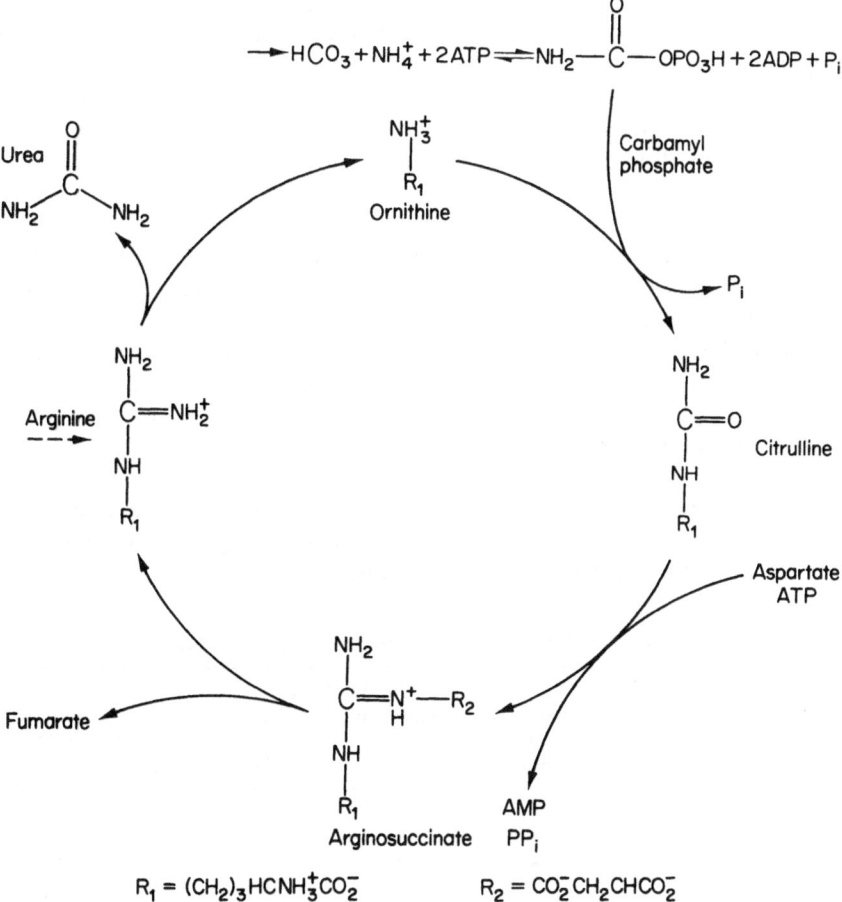

Figure 8.4. The urea cycle intermediates showing sources of diluting arginine-6-C indicated by the dashed arrows. The carbon originating primarily from metabolic CO_2 is indicated in the formulae with a large C.

only dilution effects from entry of unlabelled carbon are likely to cause errors in the assumption that $*C_A/C_A = *C_B/C_B$. Thus infused radioactive bicarbonate carbon is the sole source of arginine-6-^{14}C, while blood total CO_2 carbon is the chief source of the arginine-6-carbon; but the latter may be diluted by some liver metabolic carbon and perhaps some recycled liver arginine-6-carbon. Carbon dilution from the above sources is not great, for in healthy rabbits the specific activity of arterial blood total CO_2 carbon averaged about 3.5% more than that of urea carbon[6]. Thus, the SA of the input of blood total CO_2 carbon to the liver protein synthetic site gives a fair estimate of the SA of the arginine-6-carbon at the synthetic site.

Applicability and simplification of input form of ^{14}C-bicarbonate method

In the remainder of this paper we draw on studies made in rabbits given 5 mg cycloheximide per kg body weight. This toxic antibiotic arrests protein synthesis by combining with the ribosome–mRNA complex and has profound metabolic effects which allow a test of the performance of the blood ^{14}C-bicarbonate method in unfavourable circumstances.

Applicability

The ^{14}C-bicarbonate method requires an adequate flow of carbon between blood bicarbonate and liver-synthesised arginine-6-C. As a test of this we measure the specific radioactivity of the plasma urea carbon in the final blood sample withdrawn 4 hours after the start of the ^{14}C-bicarbonate infusion. Table 8.2 shows studies of this SA in experiments in which ^{14}C-bicarbonate was infused at various times after cycloheximide injection. Though the cycloheximide has profound effects on blood urea levels and the standard deviations are quite large, there are no significant differences between the mean blood urea SAs in the different groups of animals. These and similar studies demonstrate that the urea cycle enzyme pathways function effectively from 1 to 48 hours after injecting cycloheximide.

Table 8.2. Effect of cycloheximide on plasma urea SAs.

Cycloheximide Dose (mg kg BW^{-1})	Time Interval* (h)	Number of Animals	Plasma Urea Concentration (mg ml^{-1})		Specific Activity (dpm μg^{-1})	
			Mean	SD	Mean	SD
0	Control	5	0.214	0.023	52.4	6.1
5	0	4	0.370	0.052	76.1	8.5
5	12	4	0.300	0.127	52.9	13.8
5	24	4	0.150	0.059	49.7	7.3

*The interval shown is that between injecting the cycloheximide and starting the ^{14}C-bicarbonate infusion. The plasma samples for urea measurement were obtained 4–4.5 hours later at the end of each experiment.

Simplification

It would save much time if we could dispense with the measurements required to determine mean arterial blood ^{14}C-total CO_2 SA. Table 8.3 shows mean standardised total CO_2-C SAs obtained in normal rabbits, and in two groups of animals receiving cycloheximide. The results are expressed in terms of a standardised dose of ^{14}C-bicarbonate and show that there is no significant difference between the means of the three groups of animals. This is surprising and reassuring. The chief causes of differences are likely to be found in variations in respiration. Our animals were given small doses of Vistaril^R and minimal doses of Nembutal^R and with careful handling seemed to show remarkably constant mean blood total CO_2 carbon SAs. Thus with careful experimentation, if looking for reasonably large changes in synthetic rate of a liver-synthesised plasma protein, one could dispense with measurements of blood total CO_2-C SAs and substitute instead previously determined mean values.

Table 8.3. Mean blood ^{14}C-bicarbonate SAs over the 4 hours following a 1-hour infusion of ^{14}C-bicarbonate. (Based on 8 blood samples).

Dose of Cycloheximide mg kg BW^{-1}	Number of Animals	Interval* (h)	Mean Blood ^{14}C-Bicarbonate SA† Mean	SD
0	4	—	19.15	2.58
5	4	12	18.87	1.44
5	5	24	19.17	2.68

*Interval between injecting cycloheximide and starting ^{14}C-bicarbonate infusion.

$$\dagger\frac{1}{240}\int_0^t \frac{\text{blood dpm. }1000}{\text{ml. mEq blood HCO}_3}\, dt \bigg/ \frac{\text{dpm }^{14}\text{C-HCO}_3 \text{ injected}}{3.85 \times 10^3}$$

A better way should be to take a few blood samples during the ^{14}C-bicarbonate infusion and calculate the mean SA from these. To test this approach table 8.4 applies the methods of multiple regression to the data on which table 8.3 is based. Table 8.4 shows that if one analyses the 30-minute blood sample only and uses the regression equation relating the values of this to the mean SA determined from all 8 blood samples, the precision of prediction of mean total CO_2-C SA is about doubled, i.e. the variance shown by a group of matched controls is halved.

Table 8.4 also shows that using the regression equation based on the 30- and 120-minute blood total CO_2-C SA measurements the variance of prediction is reduced by 85%, while using the regression equation based on 30-, 120- and 45-minute blood total CO_2-C SA measurements the variance is reduced by 95%. Thus one can get a good estimate of mean blood total CO_2-C SA from

Table 8.4. Use of 1, 2 or 3 blood sample ^{14}C-bicarbonate SAs to estimate mean SA determined from 8 blood samples. Studies based on multiple correlation.

Best Sample(s) (min)	Intercept	Regression Coefficients	Correlation Coefficient	% Reduction Σ Squares*
30	10.7	0.164	0.723	52.2
30	4.69	0.200	0.916	83.9
120	—	0.451	—	—
30	0.71	0.131	0.978	95.6
120	—	0.465	—	—
45	—	0.125	—	—

*See text.

measurements on three blood samples. Clearly substituting one to three analyses of blood total CO_2-CSA for 8 analyses would save much labour.

Discussion

The above analysis shows that the ^{14}C-bicarbonate method of measuring the synthetic rate of plasma proteins can be considerably simplified. In the simple form preliminary experiments are made to demonstrate that the pathways of the urea cycle enzymes are intact. Then following a one-hour infusion of ^{14}C-bicarbonate the mean SA of the blood total CO_2 carbon is obtained either as an average from a group of previous experiments or from one to three measurements and regression equations such as those shown in table 8.4. The SA of the protein under study is determined after isolating the protein from a plasma sample obtained four hours after the start of ^{14}C-bicarbonate infusion. This SA is corrected by multiplying by the correction factor F (table 8.1) for the arginine content of the protein, for catabolism and extravascular distribution of the protein and for the delay occasioned by passage of the protein through the liver to enter the blood stream.

Taking this approach we can now interpret the results of figure 8.1 by rewriting the ordinate label as 'fractional change in synthetic rate'. As shown in table 8.1, fractional synthetic rate is given by the quotient of SA of protein arginine 6-carbon and SA of blood total CO_2. This quotient is turned into a rate by multiplying by $24/T$ and corrected by multiplying by F. Thus, if we use mean values from suitable control experiments to obtain the SA of blood total CO_2, then all that is necessary to obtain approximate values for fractional synthetic rate is a measurement of the arginine-6 carbon SA in the protein under study. Interpreted in this way, figure 8.1 shows the very interesting results that injecting cycloheximide initially inhibits both albumin and fibrinogen synthesis but whereas the former gradually recovers to normal levels the latter rapidly shows a marked overshoot which then gradually falls to normal. Clearly the mechanisms governing albumin and fibrinogen synthesis in the liver cell show marked differences, knowledge of which should throw new light on the mechanisms in mammalian cells governing protein synthesis.

Summary

Current methods of measuring synthetic rate of liver-synthesised plasma proteins depend on measuring the specific activity (SA) of arginine-6-^{14}C at the synthetic site. Though in theory the SA of newly synthesised urea carbon gives a satisfactory measure of this, in practice urea carbon SA is difficult to measure. As an alternative we have proposed using the mean SA of blood 'total CO_2' during and three hours following a one-hour intravenous infusion of ^{14}C-bicarbonate. A kinetic analysis given here and earlier studies[6] show that this alternative is satisfactory. Methods are given for establishing that this 'input' form of the arginine-6-^{14}C method is applicable in specific experimental conditions. It can be simplified without serious loss of its precision.

References

1. Obrig, T. G., Culp, W. J., McKeehan, W. L. and Hardesty, B. The mechanism by which cycloheximide and related glutarimide antibiotics inhibit peptide synthesis on reticulocyte ribosomes. *J. Biol. Chem.*, **246** (1971), 174–181

2. Sisler, H. D. and Siegel, M. R. (1967) Cycloheximide and other glutarimide antibiotics. In: *Antibiotics*, Vol. I, *Mechanism of Action*. D. Gottlieb and P. D. Shaw, eds., Springer Verlag, New York, p. 283–307

3. Koj, A. and McFarlane, A. S. Effect of endotoxin on plasma albumin and fibrinogen synthesis rates in rabbits as measured by the ^{14}C-carbonate method. *Biochem. J.*, **108** (1968), 137–146

4. Tavill, A. S., Craigie, A. and Rosenoer, V. M. The measurement of the synthetic rate of albumin in man. *Clin. Sci.*, **34** (1968), 1–28

5. Wochner, R. D., Weissman, S. M., Waldmann, T. A., Houston, D. and Berlin, N. I. Direct measurement of the rates of synthesis of plasma proteins in control subjects and patients with gastrointestinal protein loss. *J. Clin. Invest.*, **47** (1968), 971–982

6. Reeve, E. B. and McKinley, J. E. Measurement of albumin synthetic rate with bicarbonate-^{14}C. *Am. J. Physiol*, **218** (1970), 498–509

7. Rothschild, M. S., Oratz, M., Zimmon, D., Schreiber, S. S., Weiner, I. and Van Caneghem, A. Albumin synthesis in cirrhotic subjects with ascites: studies with carbonate-^{14}C. *J. Clin. Invest.*, **48** (1969), 344–350

8. Reeve, E. B., Pearson, J. R. and Martz, D. C. Plasma protein synthesis

in the liver: method for measurement of albumin formation *in vivo*. *Science*, **139** (1963), 914–916

9. McFarlane, A. S. Measurement of synthesis rates of liver-produced plasma proteins. *Biochem. J.*, **89** (1963), 277–290
10. Ratner, S. Enzymes of arginine and urea synthesis. *Adv. Enzymol.*, **39** (1973), 1–90
11. McFarlane, A. S., Irons, L., Koj, A. and Regoeczi, E. The measurement of synthesis rates of albumin and fibrinogen in rabbits. *Biochem. J.*, **95** (1965), 536–540
12. Regoeczi, E., Irons, L., Koj, A. and McFarlane, A. S. Isotopic studies of urea metabolism in rabbits. *Biochem. J.*, **95** (1965), 521–532
13. McKinley, J. E., Gilbert, D. B., Chao, P. Y. and Reeve, E. B. Urea metabolism and distribution in rabbits treated with neomycin. *Am. J. Physiol.*, **218** (1970), 491–497
14. Schimke, R. T. Control of enzyme levels in mammalian tissues. *Adv. in Enzymol.* **37** (1973), 135–187

Discussion

TAVILL

Could I ask a question related to the cycloheximide experiment? The drug may have an almost instantaneous effect on protein synthesis yet its effect on urea cycle enzymes may be delayed as they turn over with a finite half-life following inhibition of *de novo* synthesis. Will this upset the normal relationship between the specific radioactivities of newly synthesised protein and urea?

REEVE

What we notice about our plasma urea studies is that though urea concentrations vary considerably, depending on the interval of sampling after cycloheximide injection, specific activity of the urea carbon varies much less. This suggests that in spite of loss of a proportion of urea cycle enzymes enough of these enzymes persist for the method to work.

JAMES

We would entirely agree with this approach of using a constant rate infusion. Do not your results of equivalent SAs in the urea and bicarbonate pool imply that you must indeed have very rapid exchange of bicarbonate across the liver cell, such that the turnover or production rate of carbon within the liver is only a very minor component? Have you tested this out under different circumstances where exchange rates might be somewhat different?

REEVE

I think our results show very rapid exchange of CO_2 across the liver cell. We calculate that liver CO_2 production is less than 10% of total body CO_2

production. We have not made studies of liver CO_2 exchange in abnormal liver metabolic states.

JAMES

But you are actually interested, are you not, in the SA in the liver itself. This is related to two components, namely the unlabelled carbon released from metabolism of, say glucose, fat and protein, and the exchange rate of labelled carbon entering from the plasma.

REEVE

We should picture plasma $^{14}CO_2$ and CO_2 exchanging very rapidly with the liver carbonate, because of the plasma and liver carbonic anhydrase, and being diluted by the addition of liver metabolic carbonate to the extent of at most 10 per cent. Because of blood and liver carbonic anhydrase this exchange should be almost instantaneous. Naturally the liver metabolic carbonate will be unlabelled.

MUNRO

A minor technical point: you mentioned cycloheximide. In all animals cycloheximide, like actinomycin, is liable to stimulate adrenocortical activity and secretion. Since corticosteroids have a considerable effect on liver metabolism, experiments should be repeated with adrenalectomised animals, in case there are effects which lead you to believe there may be some interference from that source.

DONATO

The constant infusion approach is very nice and is traditional in circulation studies. But in circulation studies one has to assume that infusion is performed long enough until equilibrium is attained in the compartment in which you want to measure flow. Therefore, I suppose that you imply that you have attained ^{14}C-bicarbonate equilibrium when measuring albumin or fibrinogen synthesis, do you not?

REEVE

No, I don't think so. All we use a constant rate infusion for is to get a defined plasma input. We can then measure the area under the plasma SA curve and divide by time to get a mean value of plasma carbonate SA. As our analysis shows, on the assumption that the system is linear this mean will be the same as, or close to, the mean SA of the 6-carbon entering arginine. We are not saying there is a ^{14}C-carbonate steady state.

DONATO

Then you are not trying to create some sort of plateau of SA. All you are doing is to obtain for yourself a mean SA that is easily measurable.

REEVE

That is correct.

JAMES

Yes, indeed, you should theoretically get exactly the same result. But have you in fact shown that you do get the same result using your shorter infusion as compared with a constant rate infusion of ^{14}C-bicarbonate for a more prolonged time to ensure a genuine isotopic steady state?

REEVE

No, all we have done is shown as noted earlier that answers given by the urea form of the method are within a few per cent of the answers given by the bicarbonate input form.

JAMES

You cannot infuse over too short a time because you are then in trouble with defining the time course of your SA.

REEVE

Yes. It is essential to know accurately the shape of the plasma SA curve. That is why one cannot use rapid ^{14}C-bicarbonate injection.

ANDERSEN

Why do not you use a logarithmic rate infusion pump that will get you to the plateau rapidly?

REEVE

We have tried, not too successfully, to build one but have not enough money to buy one.

9

The applicability of the deconvolution technique after repeated administration of labelled serum albumin

F. VITEK

Introduction

Compartmental analysis, deconvolution as well as other techniques, can be applied for quantification of serum albumin kinetics and catabolism, starting from plasma and whole-body or excretion curves under steady-state conditions[1-3]. As demonstrated previously, the main advantage of the deconvolution technique lies in the fact that the analysis starts from the beginning of the curves measured, when measurement errors are relatively low. It has been shown that the weighting function, $G(t)$, of albumin transfer from intra- to extravascular spaces after labelled albumin input into the intravascular pool at time $t = 0$ has some important properties: its initial value $G(0)$ equals the overall fractional albumin outflow from intra- to extravascular spaces, and its integral from zero to infinity equals the ratio of extravascular to intravascular albumin mass.

However, when the plasma and excretion or whole-body activity curves seem to be 'reasonable' we cannot distinguish whether steady-state conditions hold during the whole period of study. On the other hand, it is well known that such a relatively simple cause as for example vomiting of a patient results in a deformation of the plasma activity curve, which makes difficult its approximation by a continuous sum of exponentials. In such a case the disturbance of steady-state conditions is quite clear and the assumption of a steady state during the whole period of study becomes unacceptable.

For these reasons, I have tried to apply the deconvolution technique for the study of albumin kinetics under conditions in which some amount of labelled albumin is already distributed in the system under investigation. In other words, this approach would enable analysis of the curves to be started eventually at an arbitrary time of the experiment to check that the parameters of the system did not change significantly during the experiment. Experiments with a discontinuity in the plasma activity curve could be evaluated in two or more steps, assuming steady-state conditions during these time intervals, and re-peated injections of the same label would be possible as well.

Proposed computational approach

Assuming the initial activity of labelled albumin in the intravascular pool, $X_P(0) \neq 0$, and in the extravascular spaces, $X_{ex}(0) \neq 0$ are known at time selected for $t = 0$, it can be demonstrated that the mean fractional catabolic rate, k, can be calculated from the value of the integral of plasma activity curve over the time interval from the chosen zero up to infinity as follows

$$k = \frac{X_p(0) + X_{ex}(0)}{\int_0^\infty X_p(t)\,dt} = \frac{X_{WB}(0)}{\int_0^\infty X_p(t)\,dt}$$

where $X_{WB}(0)$ is the whole-body activity at time $t = 0$.

Thus, under steady-state conditions, repeated administration of labelled albumin does not disturb the determination of the mean fractional catabolic rate.

However, a more complicated situation occurs in determining the weighting function, $G(t)$, of albumin transfer from intra- to extravascular spaces under conditions of non-zero extravascular activity at time $t = 0$. In cases when the extravascular pool can be expected to be a single compartment, the expression for the weighting function contains a delta function at time $t = 0$ with the amplitude of $X_{ex}(0)/X_p(0) = a$, and the single exponential weighting function falls with the slope of $k_{in}(1 + a)$, where k_{in} is the fractional inflow into the plasma pool from extravascular spaces. The coefficient of this exponential is a function of all exchange rates of the system. The weighting function is then as follows

$$G(t) = a\delta(t) + \{k_{out}(1+a) + a(k - k_{in}(1+a))\} \exp(-k_{in}(1+a)t)$$

where k_{out} is the fractional outflow from intra- to extravascular spaces.

Assuming the initial distribution of activities between intra- and extra-vascular spaces is known, and the catabolic rate is determined independently, the overall fractional outflow from intra- to extravascular spaces can be calculated from the coefficient, k_{in}, obtained from the slope.

The assumption of a single extravascular pool is scarcely justified for man but it is justified for some experimental animals, for example for rats. Thus, this technique could be used for quantification of the changes of the param-eters of albumin kinetics in the same animal after the disturbance of the system, instead of using a different label.

Naturally, the ratio of extravascular to intravascular albumin mass can be calculated in this simple case from the integral of the weighting function, using a correction for the distribution at time $t = 0$ as follows

$$\frac{N_{ex}}{N_P} = \int_0^\infty G(t) \, dt - \frac{X(0)}{X_{WB}(0)} \cdot \frac{k}{k_{in}}$$

This approach cannot be recommended for the study of systems with a more complicated composition of extravascular spaces because the initial distribution of activity is not known at $t = 0$. Other techniques, especially simulation of the system with time-dependent exchange rates, can be recommended in these cases.

References

1. Donato, L., Matthews, C. M. E., Nosslin, B., Segre, G. and Vitek, F. Application of tracer theory to protein turnover studies. *J. Nucl. Biol. Med.*, **9** (1966), 3
2. Vitek, F., Bianchi, R. and Donato, L. The study of distribution and catabolism of labelled serum albumin by means of an analogue computer technique. *J. Nucl. Biol. Med.*, **10** (1967), 121
3. Vitek, F., Bianchi, R., Mancini, P. and Donato, L. (1969) Deconvolution techniques for the analysis of short-term metabolic studies with radio-iodinated albumin. In *Physiology and Pathophysiology of Plasma Proteins*, G. Birke, R. Norberg and L. O. Plantin, eds., Pergamon Press, p. 29

10

Protein turnover studies in the assessment of liver function

V. M. ROSENOER

The characteristic electrophoretic pattern of the plasma proteins in cirrhosis of the liver, with the low concentration of serum albumin and raised concentration of globulins, has long suggested that the serum albumin concentration and the ratio of the concentrations of albumin to globulin could be considered as quantitative indices of liver function. Indeed, Post and Patek[1] were able to relate the serum albumin level to prognosis in patients without ascites.

Let us re-examine these propositions. First, the consideration of globulins as a class, although convenient, is inappropriate in so far as the numerous protein species and the multiple extrahepatic sites of synthesis of most of the immunoglobulins will give a composite measure which has only a tenuous relationship to the functional ability of the liver itself[2]. Certainly, there is a rise in immunoglobulins in most forms of chronic active liver disease[3]. However, irrespective of their aetiological significance, whether they are of primary importance or merely epiphenomena in the unfolding of the disease defence interaction, the raised immunoglobulin levels must surely reflect the activity of the disease process rather than the functional activity of the liver itself. Turnover studies of these proteins might well be informative, but it should be noted that after secretion at their tissue sites of synthesis, immunoglobulins are not primarily distributed by way of the plasma; nor are they of necessity primarily metabolised in a pool in rapid exchange equilibrium with the plasma[4]. It is probable then that the standard labelled globulin turnover techniques, in which the labelled protein is given intravenously, do not accurately reflect the ongoing total body turnover of immunoglobulins.

Liver synthesised proteins include albumin, fibrinogen, ceruloplasmin and transferrin. The relations between liver disease and serum concentrations of albumin, transferrin and ceruloplasmin are considered in figures 10.1–10.3. Clearly, the most interesting changes are seen with albumin. In view of the early and ongoing interest in this protein, which in its daily synthesis normally utilises one-sixth of the daily dietary nitrogen, emphasis will be placed on this liver produced protein.

The sensitivity of serum transferrin concentration as an index of liver

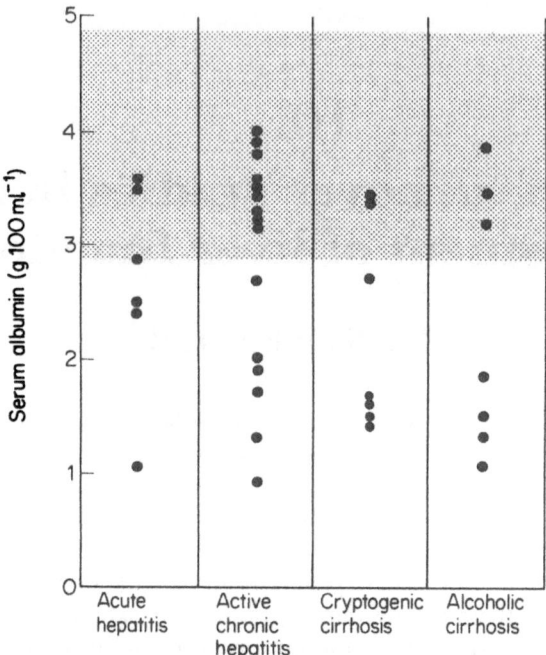

Figure 10.1. Serum albumin in liver disease. Shaded area represents normal range.

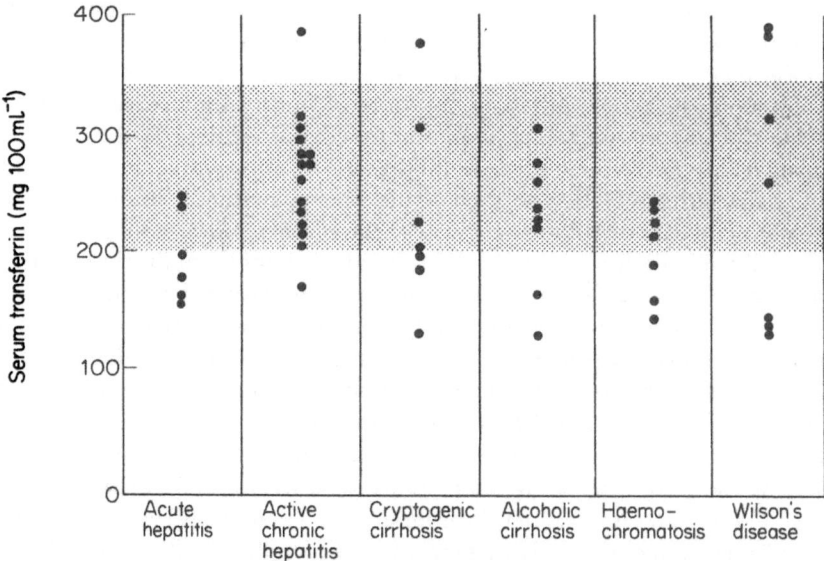

Figure 10.2. Serum transferrin in liver disease. Shaded area represents normal range.

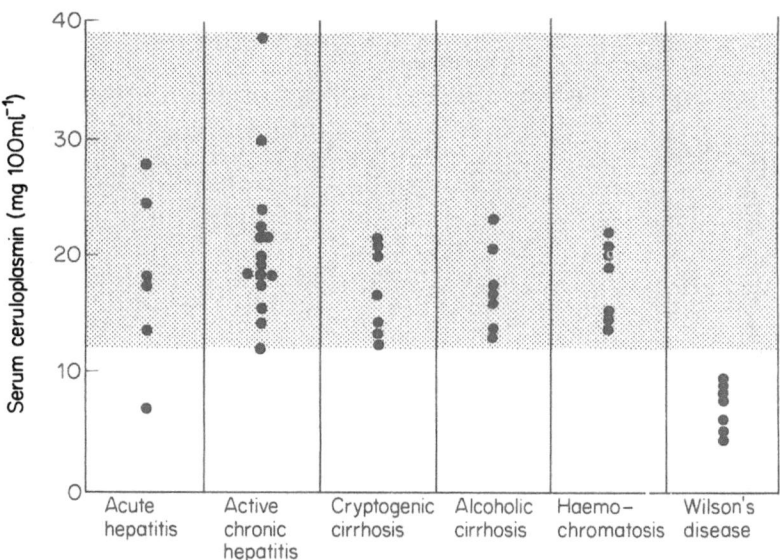

Figure 10.3. Serum ceruloplasmin in liver disease. Shaded area represents normal range.

dysfunction is poor. The correlation with serum albumin concentration is in fact significant, but barely useful (figure 10.4). Ceruloplasmin concentration in non-Wilsonian liver disease is of no value as an index of liver protein synthetic capacity, as monitored by the serum albumin concentration, despite the known liver synthetic site of this protein (figure 10.5). In this context, it is noteworthy that although more than 95% of urea is synthesised in the liver[5], there is no correlation in cirrhotics between the synthetic rate of urea and that of albumin (figure 10.6).

Recently, Hällén and Laurell[6] studied the plasma level of 21 proteins in 40 cases of liver cirrhosis in remission. Of the acute phase reactants, haptoglobins and orosomucoid were normal or decreased, while α_1-antitrypsin, fibrinogen, and C reactive protein were normal or slightly increased. Prealbumin, prothrombin, and B_2-glycoprotein I were most frequently subnormal. Of the plasma proteins studied, prealbumin seemed to be the most sensitive indicator of impaired liver function (figure 10.7).

The catabolic rate of albumin in cirrhotics is clearly related to the circulating serum albumin concentration (figure 10.8). In the steady state we may infer that the synthetic rate is markedly impaired in those cirrhotics with low circulating albumin concentrations so that the observation of Post and Patek of the prognostic value of serum albumin concentration might be expressed in terms of the albumin synthetic capacity. However, it is necessary to emphasise the obvious: the serum protein concentration reflects both the total plasma pool of protein and the volume in which it is distributed. In cirrhotics, the

V. M. Rosenoer

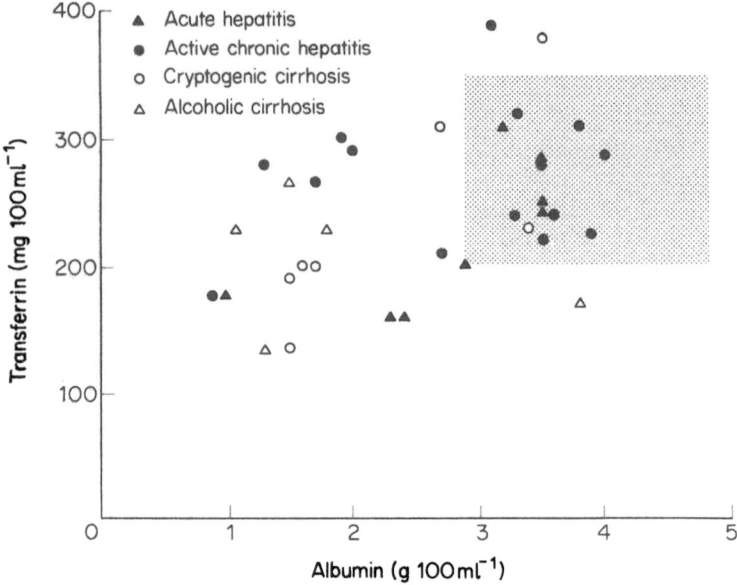

Figure 10.4. Relation between serum transferrin concentration and serum albumin concentration in a series of patients with liver disease. Shaded area represents normal range.

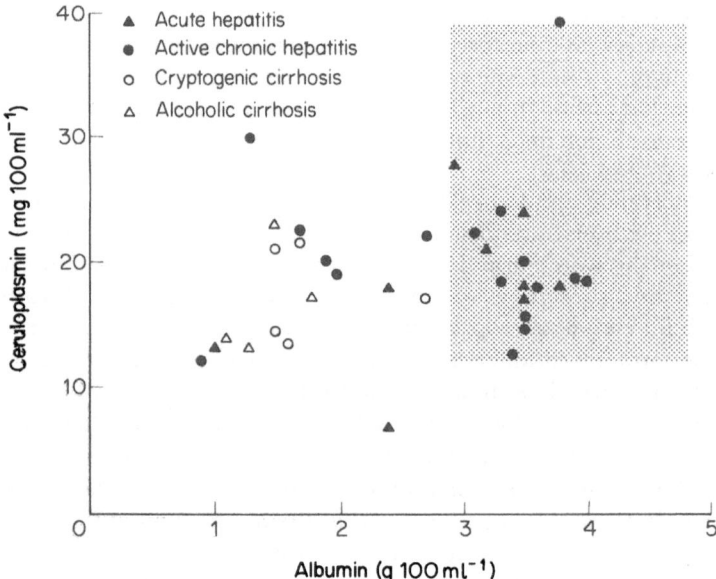

Figure 10.5. Relation between serum ceruloplasmin concentration and serum albumin concentration in a series of patients with liver disease. Shaded area represents normal range.

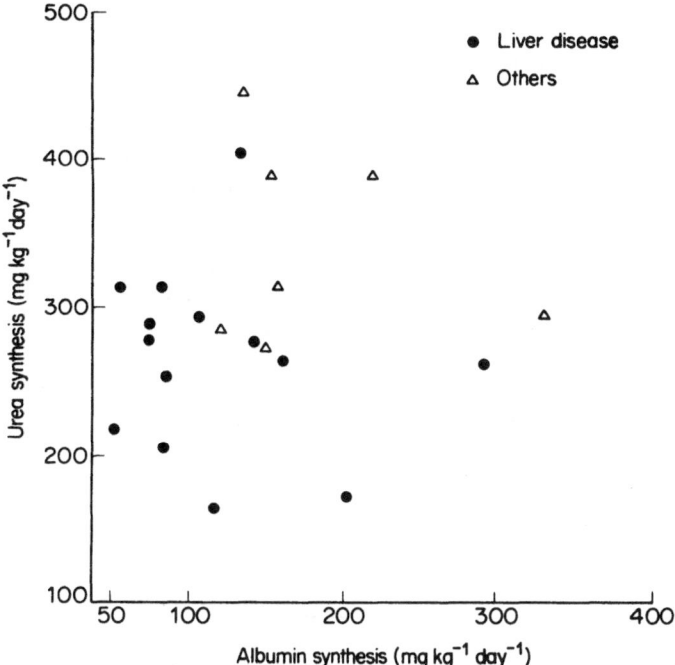

Figure 10.6. Relation between urea synthesis rate (determined using ^{14}C urea) and albumin synthesis rate (using ^{14}C sodium carbonate) in a series of patients with liver and other diseases (data from reference 5).

plasma volume is expanded by about 25% (figure 10.9) so that the degree of depletion of intravascular albumin is far less than would be suggested by the significant reduction in serum albumin concentration. The total body albumin pool is only slightly, though significantly, lower in cirrhotics than in control subjects (table 10.1).

The prognostic value of the serum albumin concentration, in fact, is not as good as Post and Patek earlier suggested. Certainly there is a trend, but in a series of patients dying from liver failure (data from references 5, 7, 8), the serum albumin concentration predicted very poorly who would die in less than or more than 6 months (figure 10.10). The albumin synthetic rates, either measured directly by the ^{14}C method or indirectly by the ^{131}I albumin catabolic measurement were little better in aiding this assessment (figure 10.11). The synthesis rates in cirrhotics were impaired and in general cirrhotics synthesising albumin at lower rates had more serious disease measured in terms of survival time. However, it is equally clear that the scatter is wide and the predictive value of the test is low.

Bianchi and his colleagues[16] recently reported albumin catabolic and synthetic studies in a series of 21 cirrhotics. They confirmed the near normal

V. M. Rosenoer

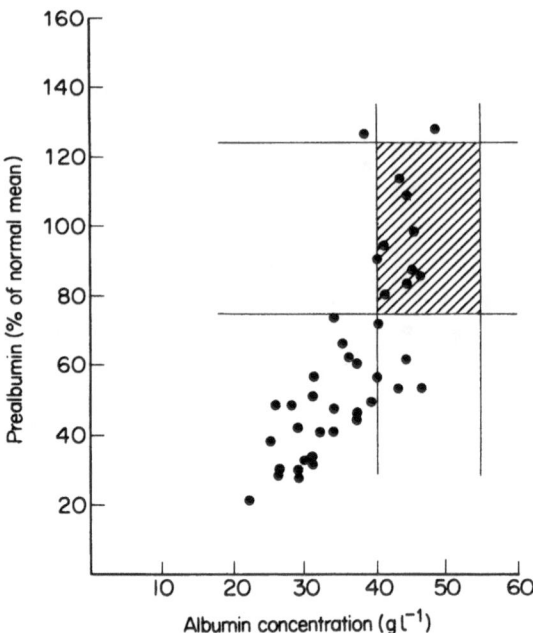

Figure 10.7. Relation between plasma prealbumin concentration and serum albumin concentration in a series of patients with liver disease. Shaded area represents normal range (from reference 6).

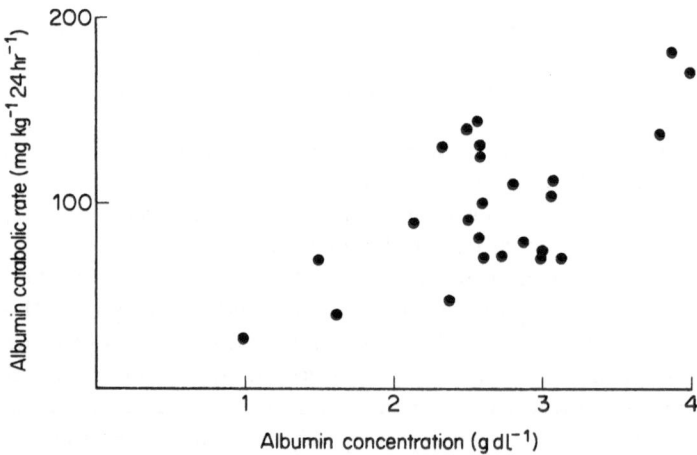

Figure 10.8. Relation between the albumin catabolic rate, determined using radioiodinated human albumin, and the plasma albumin concentration in a series of cirrhotic patients.

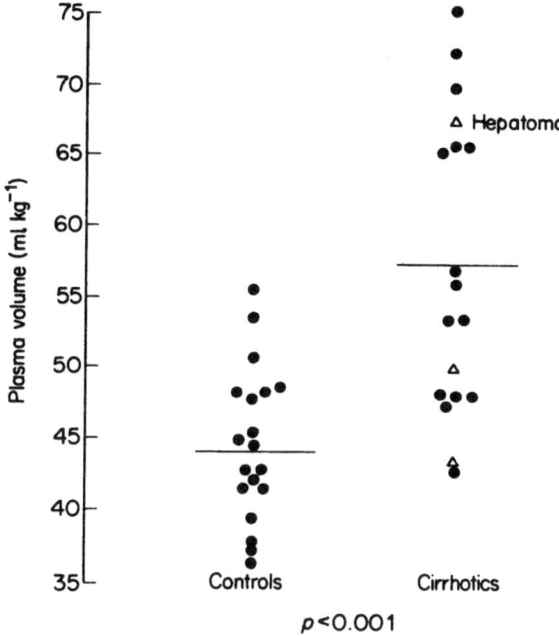

Figure 10.9. Plasma volume, determined using radioiodinated albumin, in cirrhotics and normal controls;△ represents 3 cirrhotic patients with hepatoma. Horizontal line in each group represents the mean.

Table 10.1. Albumin distribution in cirrhotic and control subjects*

	Cirrhotics (n = 21)	Control (n = 28)
Plasma volume (ml kg^{-1})	56±1.7†	46±1.1
Serum albumin (g dl^{-1})	2.70±0.15	4.31±0.11
I.v. albumin (g kg^{-1})	1.48±0.08	1.96±0.07
Total albumin (g kg^{-1})	3.97±0.35	4.66±0.16

*Data from reference 8.
†Mean±s.e.

intravascular albumin mass, despite the lowered serum albumin concentration and noted the markedly reduced extravascular albumin mass in patients with ascites. The decreased albumin catabolism in these patients was better correlated with the extravascular albumin mass than with the total body albumin pool. They concluded that the serum albumin concentration was a poor index of protein depletion.

In their study, although there were significant correlations between the plasma prothrombin content and the absolute catabolic rate of albumin and

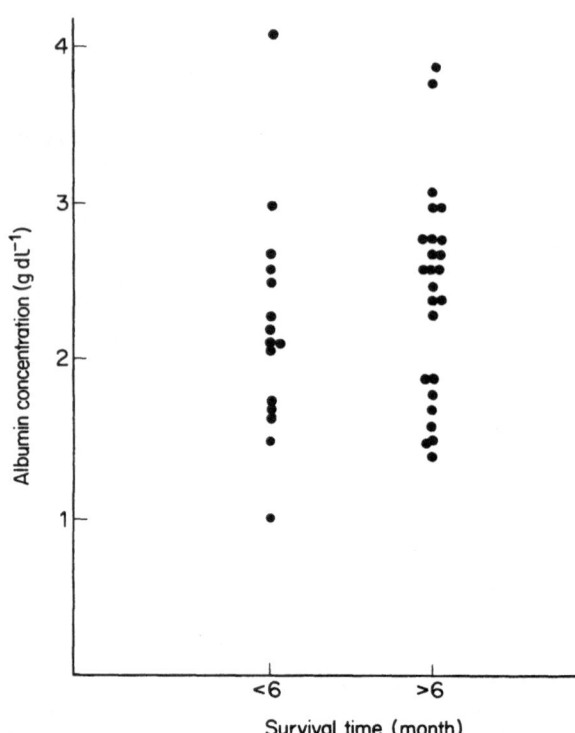

Figure 10.10. Plasma albumin concentration in cirrhotics. No significant difference was found between the plasma albumin concentration in patients surviving less than or more than 6 months (data from references 5, 7, 8).

between the serum aspartate amino transferase (SGOT) and the ratio of extra-to intravascular albumin, there was no significant correlation between SGOT and the synthesis rate of albumin. The authors concluded that the most satisfactory information on the disease status was given by the plasma pro-thrombin content and SGOT.

Albeit disappointing, this result is hardly surprising. Even the sick liver cell is responsive to extraneous influences and only when it has ceased to function at all could the synthesis rate of albumin reflect its functional capacity un-influenced by these extraneous factors.

The supply of amino acids must surely affect the rate of synthesis just as Rothschild and his colleagues[9] have demonstrated that feeding, fasting and perfusing with enhanced concentrations of tryptophane and ornithine would significantly influence the albumin synthesis rate of the isolated perfused rabbit liver. Certainly, this must be true *in vivo* for without adequate amino acid building blocks even the best liver will fail to synthesise albumin at a rate consistent with its functional capacity. In figure 10.12, which comes from a study of a patient with enteric bacterial overgrowth before and after treatment

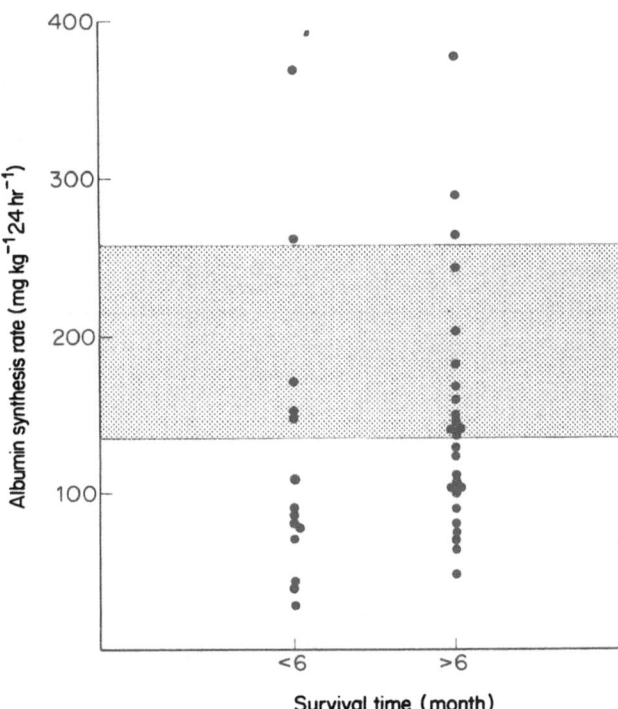

Figure 10.11. Albumin synthesis rate in cirrhotics determined either from the radioiodinated albumin catabolic rate or from the ^{14}C carbonate method. No significant difference was found between patients surviving less than or more than 6 months (data from references 5, 7, 8).

with appropriate antibiotics, it can be seen that restoration of the plasma amino acid concentrations to near normal levels restored the albumin and fibrinogen synthetic rates to supranormal levels[10]. In many patients with liver disease, with special reference to the malnourished alcoholic cirrhotic, this factor cannot be ignored. To measure the albumin synthetic rate in the malnourished can give but a limited picture of liver's functional ability. Alcohol itself will adversely affect liver function in terms of albumin synthesis, yet this is certainly a reversible effect, reversed in part by modifying the perfusing amino acid concentrations[9].

The hormone environment is clearly important. When excess thyroid is given in doses of 6–15 grains per day, there is a rapid increase in the amount of albumin degraded, but no change in albumin pool size, so that it may be inferred that albumin synthesis must have increased to keep pace with change in degradation[11]. Similar changes were noted in patients with spontaneous thyrotoxicosis[12]. In patients with myxoedema, not only is the absolute synthetic rate low, but the fractional rate also falls and turnover values as low as 1.9% of the total pool per day have been observed[13].

V. M. Rosenoer

Plasma protein synthesis rates

	Before	After	Normal range of catabolic rate
	(mg kg^{-1} day^{-1})	(mg kg^{-1} day^{-1})	(mg kg^{-1} day^{-1})
Albumin	40	323	136–257
Fibrinogen	16	103	29–63

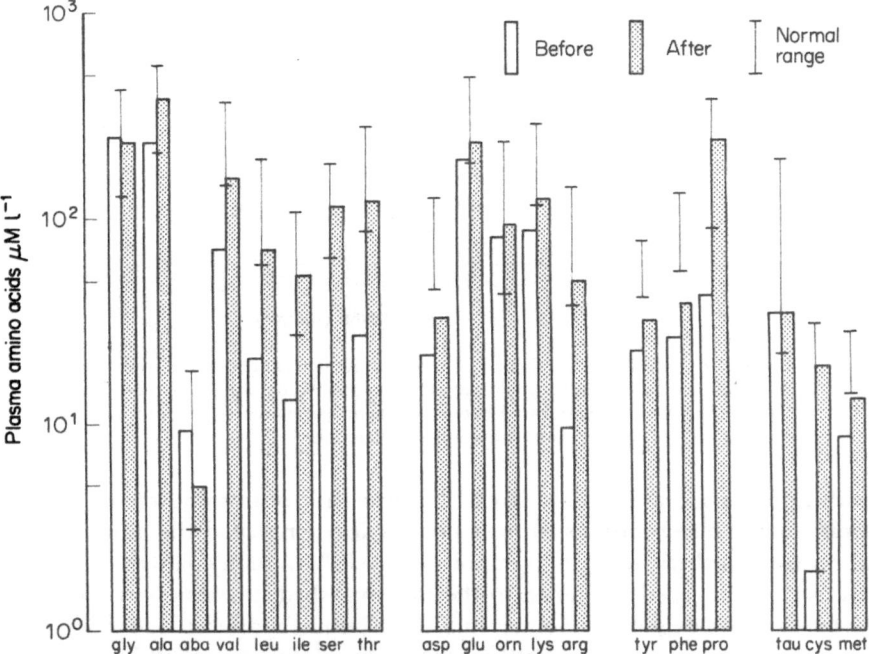

Figure 10.12. Plasma protein synthesis in a patient with the 'Blind Loop' syndrome before and after treatment with intestinal antibiotics (tetracycline and neomycin). After elimination of the intestinal bacterial overgrowth, the plasma amino acids returned to near normal concentrations: the synthesis rates of both albumin and fibrinogen, determined using ^{14}C carbonate, reached supranormal values (reference 10).

The effects of excess cortisone and prednisolone on albumin synthesis have also been studied. There is a marked increase in absolute and fractional degradative and synthetic rates[14]. Cain *et al.*[15] confirmed that the administration of prednisolone to cirrhotics enhanced the rate of albumin synthesis measured by the ^{14}C technique (figure 10.13). Although this could have been a non-

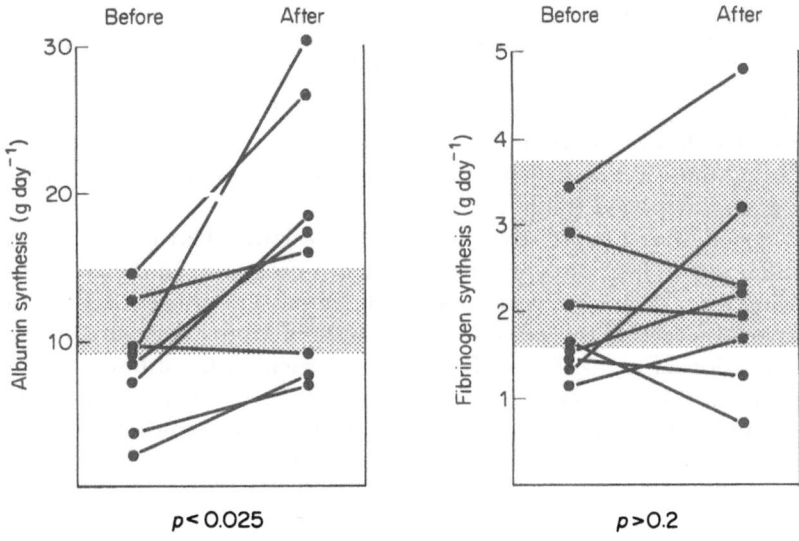

Figure 10.13. The effect of prednisolone therapy on the rate of synthesis of albumin and fibrinogen in cirrhotics using the ¹⁴C carbonate method. The patients were treated with 1 mg kg⁻¹ prednisolone daily for 13 days after the initial study. Whereas, the increase in albumin synthetic rate was significant, no significant increase in fibrinogen synthesis was found (data from reference 15).

specific effect of steroids in enhancing the rate of protein catabolism and supplying thereby more amino acid precursors to the liver to increase the rate of albumin synthesis, it might be expected that this nonspecific action would equally affect the rate of synthesis of other liver produced proteins. However, fibrinogen synthesis rates measured in the same patients were not enhanced.

Finally, we must consider whether there is a feedback mechanism whereby

Table 10.2. Albumin synthesis rates before and after 300 g Albumin infusion*

Patient	Diagnosis	Plasma albumin concentration (g 100 ml⁻¹)		i.v. albumin pool (g)		Synthesis rate (mg kg⁻¹ 24 h⁻¹)	
		Before	After	Before	After	Before	After
T.L.	Primary biliary cirrhosis	2.9	4.0	69.6	108.0	144	104
E.W.	Cirrhosis	3.3	4.6	117.0	136.7	114	127
L.W.	Cirrhosis	1.8	3.5	68.6	123.9	63	77
N.P.	Steatorrhoea	3.4	5.3	71.7	122.7	124	99
J.R.	Cirrhosis	3.0	4.3	119.8	195.8	108	120

*Intravenous infusion of 300 g albumin (salt poor, 50 g day⁻¹ for 6 days). Albumin synthesis rate determined by the ¹⁴C carbonate method. No consistent change was detected in the rate of albumin synthesis. The changes recorded were within the experimental error of the method. (Data from reference 5).

the rate of albumin synthesis is influenced by the concentration of circulating albumin. The observations of Bauman *et al.*[13] in adult patients with hypo-albuminemia secondary to excessive renal losses of protein, suggested that albumin synthesis is not sensitive to low albumin levels *per se*. In acute studies in man Tavill *et al.*[5] were unable to demonstrate a significant change in the rate of albumin synthesis before and after the infusion of 300 g of albumin over a ten-day period (table 10.2).

In conclusion, I must stress the poor correlation between the albumin synthesis rate in liver disease and the patient's prognosis. In the clinic we are studying a multivariable system in which we have insufficient knowledge of the state of several of the variables at any instant. Under these circumstances, identification of a single transfer function—such as the albumin synthesis rate—is unlikely to characterise the complete system. It would be more appropriate to use the increasingly sophisticated tools that we now have available to identify the factors concerned in the regulation of albumin metabolism.

References

1. Post, J. and Patek, A. S. Serum proteins in cirrhos_s of the liver. *Arch. Intern. Med.*, **69** (1942), 67–82 and 83–89
2. Schultze, H. E. and Heremans, J. F. (1966) *Molecular Biology of Human Proteins*. Vol. I. *Nature and Metabolism of Extracellular Proteins*. Elsevier, New York
3. Feizi, T. Immunoglobulins in chronic liver disease. *Gut*, **9** (1968), 193–198
4. Alper, C. A., Peters, J. H., Birtch, A. G. and Gardner, F. H. Haptoglobin synthesis. I. *In vivo* studies of the production of haptoglobin, fibrinogen and γ-globulin by the canine liver. *J. Clin. Invest.*, **44** (1965), 574–581
5. Tavill, A. S., Craigie, A. and Rosenoer, V. M. The measurement of the synthetic rate of albumin in man. *Clin. Sci.*, **34** (1968), 1–28
6. Hällén, J. and Laurell, C. B. Plasma protein pattern in cirrhosis of the liver. *Scand. J. Clin. Lab. Invest.*, **29** (Supplement 142) (1972), 97–103
7. Rothschild, M. A., Oratz, M., Zimmon, D., Schreiber, S. S., Weiner, I. and Van Caneghem, A. Albumin synthesis in cirrhotic subjects with ascites studied with carbonate ^{14}C. *J. Clin. Invest.*, **48** (1969), 344–350
8. Hasch, E., Jarnum, S. and Tygstrup, N. Albumin synthesis rate as a measure of liver function in patients with cirrhosis. *Acta Med. Scand.*, **182** (1967), 83–92
9. Rothschild, M. A., Oratz, M. and Schreiber, S. S. Alcohol, amino acids and albumin synthesis. *Gastroenterology*, **67** (1974), 1200–1213
10. Jones, E. A., Craigie, A., Tavill, A. S., Franglen, G. and Rosenoer, V. M. Albumin and urea synthesis in the 'blind loop' syndrome. *Gut*, **9** (1968), 466–469
11. Rothschild, M. A., Bauman, A., Yalow, R. S. and Berson, S. A. The

effect of large doses of dessicated thyroid on the distribution and metabolism of albumin ^{131}I in euthyroid subjects. *J. Clin. Invest.*, **36** (1957), 422

12. Lewallen C. G., Rall, J. E. and Berman, M. Studies on iodoalbumin metabolism. II. The effects of thyroid hormone. *J. Clin. Invest.*, **38** (1959), 88

13. Bauman, A., Rothschild, M. A., Yalow, R. S. and Berson, S. A. Distribution and metabolism of ^{131}I labelled human serum albumin in congestive heart failure with and without proteinuria. *J. Clin. Invest.*, **34** (1957), 1359

14. Rothschild, M. A., Schreiber, S. S., Oratz, M. and McGee, H. L. The effect of adrenocortical hormones on albumin metabolism studied with ^{131}I albumin. *J. Clin. Invest.*, **37** (1958), 1229

15. Cain, G. D., Mayer, G. and Jones, E. A. Augmentation of albumin but not fibrinogen synthesis by corticosteroids in patients with hepatocellular disease. *J. Clin. Invest.*, **49** (1970), 2198–2204

16. Bianchi, R., Mariani, G., Pilo, A. and Toni, M. G. Serum albumin turnover in liver cirrhosis. *J. Nuclear Biol. Med.*, **18** (1974), 20–29

Discussion

MILLER

The effects of glucocorticoids on albumin synthesis can also be demonstrated in the isolated perfused liver. In this system such effects can hardly be attributed to increased protein catabolism and increased supply of amino acids. To demonstrate an effect of steroids on the synthesis of the acute phase reactants, it is mandatory that the hormone be present continuously.

I am concerned that so much effort has been focussed on albumin—a protein which is not essential for the maintenance of normal human life—as demonstrated in the few but well documented cases of total analbuminaemia. I would suggest that a more appropriate protein, whose synthesis is much more delicately dependent on the integrity of hepatic function, would be prothrombin. A precise measurement of the rate of prothrombin synthesis would be of enormous interest.

ROSENOER

Certainly, we should consider prothrombin but albumin must be given the consideration it justly deserves. One-sixth of the daily dietary nitrogen goes into the synthesis of albumin and, despite the analbuminaemics, it would be extraordinary if this dietary intake was merely being recycled through a functionless delay pathway on its way to excretion.

MILLER

You are describing the storage function of albumin. Albumin has a longer half-life than free amino acids and so, from this point of view, it represents an attempt on the part of the body to use more economically available amino acids. It is an important but not essential function.

ENWONWU

In discussing the effects of prenisolone on the liver, did you say there was no increase in the free amino acid pool?

ROSENOER

No. I stated that if the increase in albumin synthesis, in response to prednisolone, was a non-specific response to an increased supply of amino acids we would have expected an enhanced synthesis rate of fibrinogen. This was not observed.

TAVILL

I would like to consider the different responses of the various liver produced plasma proteins to changes in liver functional capacity. Why is there a lack of correlation between the effects of disease on albumin concentration and on transferrin concentration? Is it related to their turnover times? Is a protein with a shorter turnover time such as transferrin more affected than one with a longer turnover time such as albumin?

ROSENOER

In that case, in chronic liver disease, the transferrin concentration should fall earlier and to a greater degree than the albumin concentration. But I must stress here that other factors will interfere. Certainly, iron metabolism is disturbed in liver disease and will affect transferrin synthesis. I await your paper on this.

TAVILL

Could I suggest that the stability of the messenger RNA for the individual proteins might be one other factor?

ROTHSCHILD

The response of albumin and fibrinogen synthesis to steroid therapy should certainly be considered in those terms. The synthesis rate and longevity of specific RNAs is clearly a field of increasing interest.

SCHEINBERG

In his talk, Dr Rosenoer threw in a parenthetical remark about fractional and absolute catabolic rates. In a situation in which the synthesis rate of albumin is depressed and a new steady state has been achieved, you cannot say that the catabolic rate has fallen. The catabolic rate should not be expressed in terms of mg kg^{-1} day^{-1} but as a fractional rate. It is a first-order reaction and the rate is constant.

ROSENOER

I am certainly unhappy about expressing catabolic rates as fractional rates in terms of pool size. It makes an implicit assumption of a first-order reaction, which is not always justified by the facts. In the case of albumin it may well be appropriate but with other plasma proteins it is definitely not. For this reason I prefer to express the catabolic and synthetic rates in absolute terms related to body size.

ROSSING

I think you should use the survival time. When the albumin concentration is low, those albumin molecules which have been synthesised persist longer in the body.

HOFFENBERG

The fractional catabolic rate of albumin can both increase and decrease. In severe protein depletion both the fractional and absolute catabolic rate is diminished. In hyperalbuminaemia states and in protein losing states the fractional catabolic rate is increased.

MILLER

Dr Hoffenberg, have you ever come across a clearly authenticated case of hyperalbuminaemia that was not associated with dehydration?

HOFFENBERG

No. However, in albumin infusion studies in animals and in man, the rates of catabolism of albumin increase enormously. In dogs and rabbits there is albuminuria. I think these mechanisms protect human beings from sustained hyperalbuminaemia.

ROSSING

We have seen a patient—a woman with a plasma cell malignancy not synthesising myeloma globulins—who had an expanded intravascular albumin pool; with serum albumin concentrations determined both immunochemically and electrophoretically of between 6.0 and 6.7 g dl^{-1}. Albumin turnover studies revealed an elevated fractional catabolic rate. The synthetic rate was between 3 and 4 times normal. There was minimal albuminuria.

HOFFENBERG

So this patient would fulfil the criteria of hyperalbuminaemia. In conclusion, I should not like this conference to go on record with Dr Miller's remarks about the unimportance of albumin. In the presence of low albumin concentrations, the analbuminaemics apart, one does find oedema. Certainly, albumin deserves further study.

PART 3

Regulation of Synthesis and Catabolism of Plasma Proteins

11
Life and death of plasma proteins*

I. H. SCHEINBERG

The work that I shall report on today has occupied much of the effort in our laboratory for the past six or seven years. It has represented a collaborative effort of Anatol Morell, Irmin Sternlieb, Richard Stockert, C. J. A. van den Hamer, Gregory Gregoriadis, all working in my laboratory, and Gilbert Ashwell and his collaborators in Bethesda[1-10].

We had been working for many years on ceruloplasmin, chiefly in relation to Wilson's disease, but also as a model metal-and-plasma protein. Although we also knew it to be a glycoprotein, with several terminal sialic acid residues, we had paid very little attention to the structure or possible functions of the molecule's carbohydrate moiety, until about six years ago: Morell had treated ceruloplasmin with neuraminidase and obtained an asialoprotein which possessed every essential characteristic of sialylated ceruloplasmin.

Figure 11.1 depicts the terminal sialic and penultimate galactose residues that characterise 8 or 9 of the approximately 10 carbohydrate chains on ceruloplasmin. The arrow represents the reaction that occurs when neuraminidase is allowed to interact with ceruloplasmin—or, apparently, with any other plasma glycoprotein. The figure also shows that the galactose residue that is exposed by this enzymatic reaction can be converted by galactose oxidase into an aldehyde function which can be reduced with tritiated borohydride. This sequence of reactions results in the introduction of tritium into the desialylated protein, a procedure that has proven quite useful, and to which we shall refer later in this paper.

For some time we were content to study asialoceruloplasmin only chemically. The protein proved not only to be virtually identical to the native sialylated protein (table 11.1), but was also indistinguishable immunochemically. The only differences that we could detect between the two forms of ceruloplasmin were in their sialic acid content, and the consequent lowered electrophoretic mobility. We were able, also, to crystallise asialoceruloplasmin—the first time that a glycoprotein whose carbohydrate moiety had been modified had been crystallised. The crystals were distinctly different in form from those

*The research reported in this chapter was supported, in part, by grant AM 1059, National Institute of Arthritis, Metabolism and Digestive Diseases of the U.S. Department of Health, Education and Welfare and by the Foundation for the Study of Wilson's Disease, Inc.

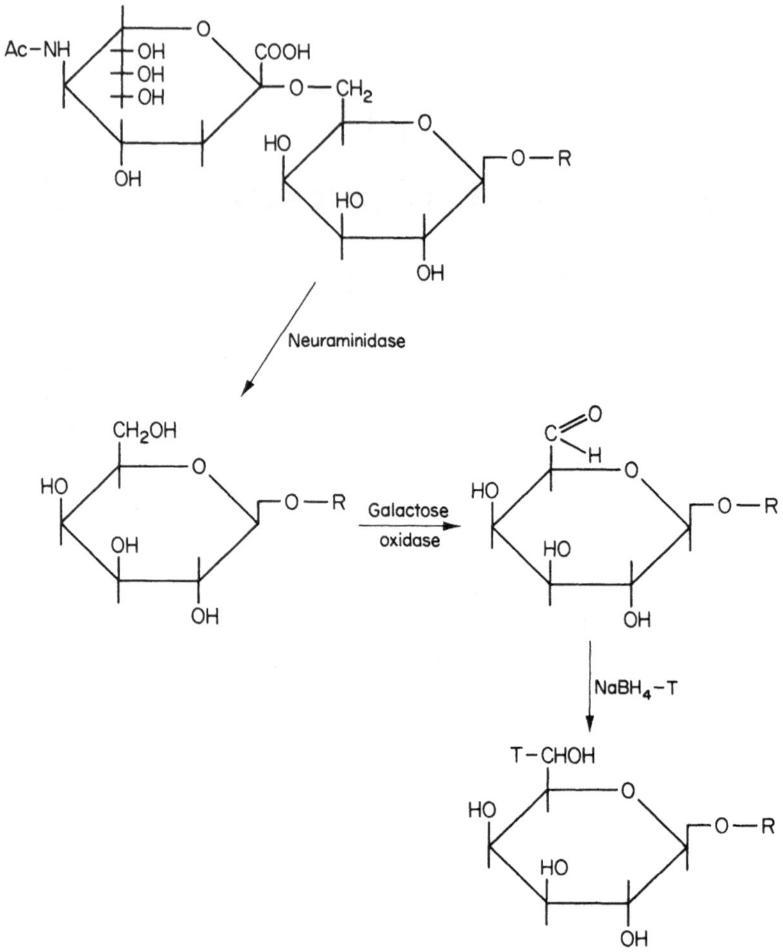

Figure 11.1. Three chemical reactions capable of being carried out on glycoproteins terminating in sialic acid and penultimate galactose.

of native ceruloplasmin. But intense interest on our part was only aroused when Morell injected labelled, homologous, desialylated ceruloplasmin into a rabbit, and determined its physiological half-life.

Figure 11.2 shows that the physiological half-life of asialoceruloplasmin is only several minutes (curve 2 in figure 11.2) whereas the half-life of native ceruloplasmin is 55 hours (curve 1 in figure 11.2). This rapid disappearance of the desialylated protein is not an artefact due to splitting of the label from the protein since the same result is observed whether the label is tritium (curve 4a, figure 11.2) or ^{64}Cu (curve 4b).

If the terminal galactose residues of asialoceruloplasmin are oxidised, as shown in figure 11.1, or if they are removed by the action of β-galactosidase, the physiological half-lives of these modified proteins are almost as long as those of the native protein (curves 3 and 5, respectively, figure 11.2). In the experiments in which the asialoceruloplasmin was doubly labelled—each

Table 11.1. Characteristics of crystallised ceruloplasmin and of asialo-ceruloplasmin[1].

Characteristic	Crystallised ceruloplasmin Solution B-8-101	Asialoceruloplasmin Solution B-8-201
$A_{1cm,610}^{1\%}$	0.684	0.640
$A_{1cm,280}^{1\%}$	15.03	(15.00)
$A_{610}:A_{280}$	0.046	0.043
Copper, total (%)	0.288	0.283
Copper, free (%)	0.012	0.005
Copper, protein-bound (%)	0.276	0.278
$A_{1cm,610}$ /μg protein-bound copper	0.025	0.023
$A_{1cm,280}$ /μg protein-bound copper	0.545	0.540
Enzymatic activity	0.0272	0.0244
Sialic acid (%)	1.81	0.11
Hexoses (%)	3.93	3.75

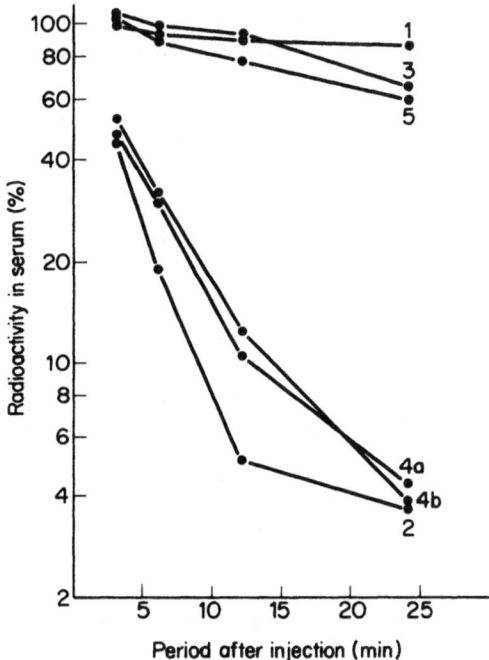

Figure 11.2. Disappearance from the serum of ^{64}Cu-labelled native and modified rabbit ceruloplasmin. Each point is the average value obtained from two animals. 1, ceruloplasmin; 2, asialoceruloplasmin; 3, oxidised ceruloplasmin; 4, asialoceruloplasmin doubly labelled with tritium (4a) and [64]Cu (4b); 5, asialo-agalactoceruloplasmin[2].

molecule carrying both tritium and ^{64}Cu—the close parallelism of the two curves of disappearance indicates that the entire protein is disappearing from the plasma (curves 4a and 4b, figure 11.2).

Table 11.2 shows the distribution of these various modifications of ceruloplasmin, as well as of the native protein, in several organs and blood, 24 minutes after injection of the homologous protein into a rabbit. When asialoceruloplasmin is injected about 80% is recovered in the liver, with 5% remaining in blood and the remainder being distributed in much smaller concentrations in the tissues sampled. Note, too, that oxidised ceruloplasmin and asialoagalactoceruloplasmin are not removed as rapidly, or to so large an extent, as asialoceruloplasmin.

Our first conclusion from these physiological findings was that asialoceruloplasmin simply represented a denatured form of the native protein and was therefore removed from the blood as a foreign protein by the reticuloendothelial system. Benacerraf had previously shown unequivocally that denatured albumin, for example, was removed from plasma in just this manner[11].

But when we injected tritiated asialoceruloplasmin into a rabbit, radioautographic pictures of the liver revealed the tritium to be present in hepatocytes and not in the reticuloendothelial Kupffer cells of the liver (figure 11.3).

We also demonstrated that asialoceruloplasmin was at least partly removed from the circulation without further modification since 24 minutes after

Table 11.2 Distribution of ^{64}Cu and ^{3}H 24 min after injection of labelled native or modified rabbit ceruloplasmin[2].

Preparation	Protein injected mg	Label	Percentage of total injected label recovered in					
			Serum %	Liver %	Kidneys %	Spleen %	Lungs %	Heart %
1. Ceruloplasmin								
Rabbit 1, 2.20 kg	2.14	^{64}Cu	88.5	7.0	1.60	0.05	0.90	0.30
Rabbit 2, 2.00 kg	2.41	^{64}Cu	86.7	11.0	1.10	0.13	1.70	0.40
2. Asialoceruloplasmin								
Rabbit 3, 2.25 kg	2.18	^{64}Cu	4.3	79.5	0.39	0.025	0.26	0.07
Rabbit 4, 2.25 kg	1.98	^{64}Cu	3.1	81.2	0.15	0.010	0.07	0.04
3. Oxidised asialoceruloplasmin								
Rabbit 5, 2.25 kg	1·76	^{64}Cu	72.0	24.2	1.50	0.11	0.87	0.44
Rabbit 6, 2.40 kg	1.55	^{64}Cu	62.5	31.6	0.87	0.14	1.30	0.35
4. Asialoceruloplasmin labelled with ^{64}Cu and with tritium								
Rabbit 7, 2.15 kg	1.60	^{64}Cu	4.7	83.3	0.22	0.004	0.10	0.03
Rabbit 7, 2.15 kg		^{3}H	4.7	84.7				
Rabbit 8, 2.30 kg	1.63	^{64}Cu	3.0	82.5	0.24	0.004	0.08	0.04
Rabbit 8, 2.30 kg		^{3}H	4.1	87.0				
5. Asialoagalactoceruloplasmin								
Rabbit 9, 2.20 kg	2.17	^{64}Cu	61.5	34.2	0.78	0.07	0.13	0.14
Rabbit 10, 2.25 kg	1.43	^{64}Cu	59.5	36.0	1.60	0.13	0.91	0.40

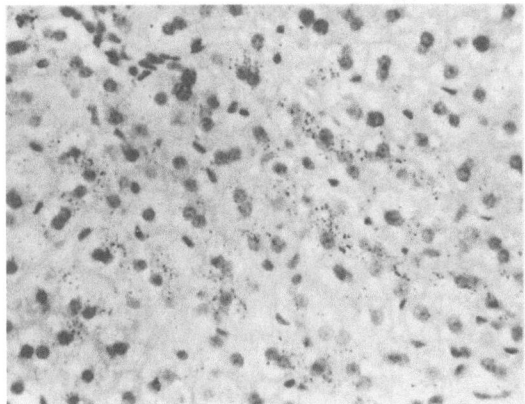

Figure 11.3. Radioautograph of the liver of a rabbit into which tritiated asialo-ceruloplasmin had been injected. The dots, representing radioactivity, appear in hepatocytes and not in Kupffer cells that are distinguished by their linearly elongated nuclei[2].

injection into a rabbit, one-third of it could be recovered immunochemically, and exhibited the same ratio of ^{3}H to ^{64}Cu as the injected protein (table 11.3). A species difference in the catabolic sequence of asialoceruloplasmin could be shown when the same experiment was repeated in a rat. In this animal, as figure 11.4 shows, the ratio of ^{3}H to ^{64}Cu falls sharply with time in immuno-chemically recovered doubly labelled protein. It is clear here that terminal galactose is split off from the protein before the copper is removed, and that such removal of galactose does not interfere with immunochemical precipitation of the protein.

The change with time in the subcellular localisation of the radioactive copper of singly labelled asialoceruloplasmin was followed in the liver for 30 minutes after injection into a rat. The proportion of radioactive copper in the mitochondrial–lysosomal fraction rose for 15 minutes after injection, a clue that catabolism of asialoceruloplasmin was taking place in the lysosomes (figure 11.5). In subsequent experiments we were able to discriminate between mitochondria and lysosomes, and it was clear that it is in the lysosomes that catabolism indeed occurs.

Table 11.3 Immunochemical recovery of doubly labelled asialoceruloplasmin from rabbit liver[2].

Sample	^{3}H content dpm	^{64}Cu content cpm	Ratio, ^{3}H content to ^{64}Cu content	Recovery ^{3}H %	^{64}Cu %
Injected asialoceruloplasmin	2.3×10^6	5.3×10^4	43.4		
Recovered asialoceruloplasmin	7.6×10^5	1.75×10^4	43.8	33	33

I. H. Scheinberg

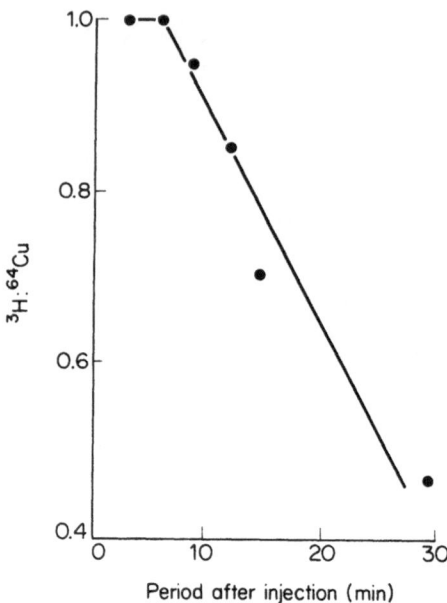

Figure 11.4. Ratio of tritium to ^{64}Cu in immunochemically recovered asialo-ceruloplasmin from livers of rats into whom doubly labelled asialoceruloplasmin had been injected[6].

Are the phenomena we observed with respect to ceruloplasmin and asialo-ceruloplasmin unique to this pair of proteins or are they true of other plasma proteins—virtually all of which, with the single exception of albumin, are glycoproteins? Figure 11.6 shows that the desialylated derivatives of orosomucoid, fetuin, haptoglobin, thyroglobulin, and α_2-macroglobulin behave in essentially the same manner, with respect to their rapid disappearance from plasma following injection, as does asialoceruloplasmin. We have presumed that they are also taken up by the liver but have only indirectly demonstrated this. The half-lives of the sialylated native congeners of all these proteins are measured in hours and days, of course, rather than in the minutes shown in this figure. Asialotransferrin seems to be an exception, but a more detailed study by Regoeczi indicates that it too has a shortened half-life when sialic acid is removed from the native molecule[12].

We have recently turned our attention to the liver to determine the molecular basis of the transfer of asialoceruloplasmin from plasma into hepatocytes and lysosomes. Table 11.4 summarises our work to date. We have isolated a protein, from the particulate fraction of rabbit liver homogenate, which specifically binds asialoprotein. In partially purified material, 24 ng of asialoorosomucoid were bound per 200 mcg of particulate lyophilised material. Note that fully sialylated, native orosomucoid is not bound and that binding requires a pH greater than 5.6, calcium—EDTA abolishes binding—and,

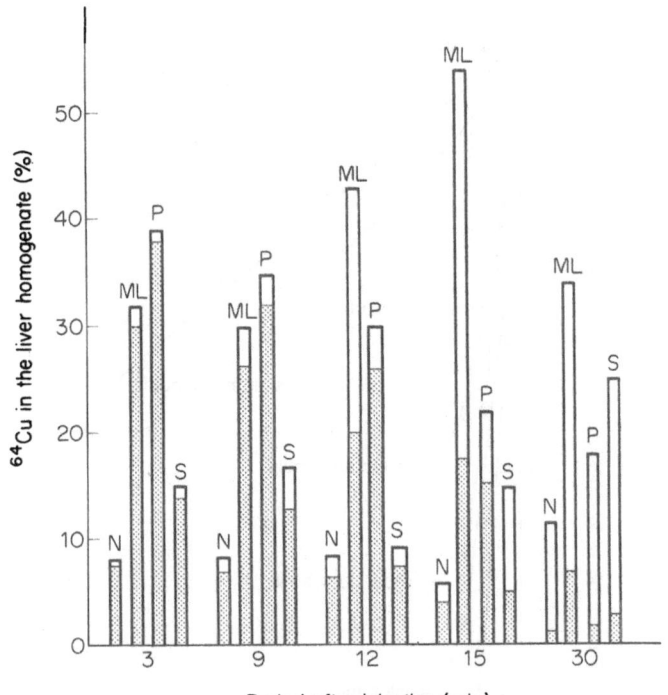

Figure 11.5. The proportion of radioactive copper in various subcellular fractions of the livers of rats killed at varying intervals following injection of singly labelled asialoceruloplasmin. N = nuclear fraction; ML = mitochondrial–lysosomal fraction; P = microsomal fraction; S = cytosol[2].

Table 11.4. Binding of desialylated human orosomucoid by the particulate fraction of rabbit liver homogenate[8].

Labelled human glycoprotein*	Treatment of particulate fraction after suspension	Glycoprotein bound (ng)
[125]I-asialoorosomucoid	None	24.2
	Acidified (pH 5.6)	0.2
	EDTA (20 mM)	0.1
	Digested with neuraminidase†	1.7
[125]I-orosomucoid	None	0.4

*Asialoorosomucoid, 12100 cpm ng^{-1}; orosomucoid, 5100 cpm ng^{-1}.
†Tubes incubated for 2 h at 37°C with 0.005 units of *Clostridium perfringens* neuraminidase prior to the addition of ^{125}I-asialoorosomucoid.

most surprisingly, the *presence* of sialic acid on the binding protein. There is thus a reciprocal relationship between the two components of this binding reaction; sialic acid must be absent from the plasma glycoprotein that is bound, and must be present on the receptor protein of liver.

I. H. *Scheinberg*

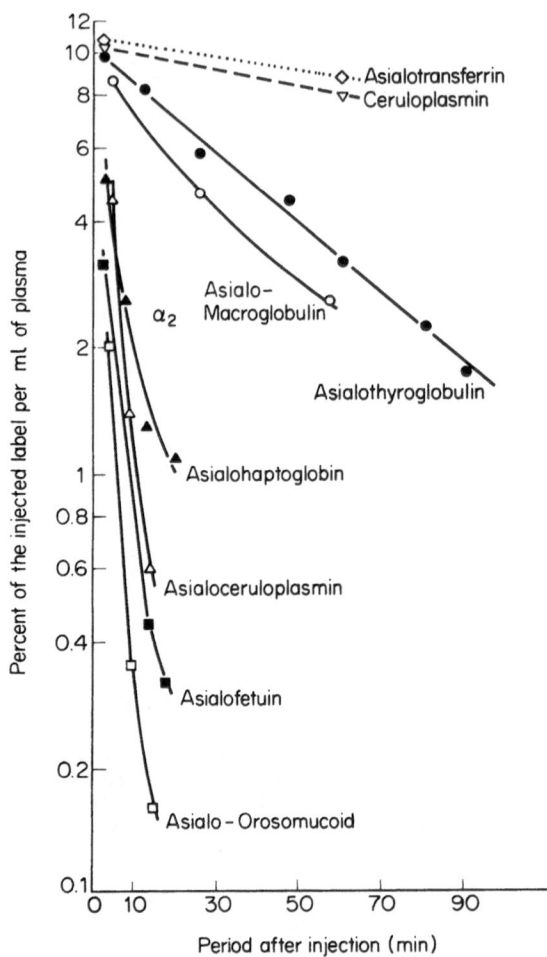

Figure 11.6. Survival of intravenously injected, labelled ceruloplasmin and 7 desialylated proteins in plasma in the rat[7].

These facts are consistent with an hypothesis of a mechanism underlying the dynamic equilibrium of plasma proteins: only after sialic acid is split from a plasma glycoprotein is the protein removed from the circulation by the liver where it is catabolised.

The need for a mechanism of this type is obvious. If plasma proteins circulated indefinitely they would be denatured, sooner or later, by the variety of physical and chemical traumatic influences that occur intravascularly, and thereby suffer loss of their functional specificities. The mechanism postulated above would assure continual removal of proteins, by the action of *in vivo* neuraminidases, to forestall such denaturation.

References

1. Morell, A. G., Van Den Hamer, C. J. A., Scheinberg, I. H. and Ashwell, G. Physical and chemical studies on ceruloplasmin. IV. Preparation of radioactive, sialic acid-free ceruloplasmin, labelled with tritium on terminal D-galactose residues. *J. Biol. Chem.*, **241** (1966), 3745–3749

2. Morell, A. G., Irvine, R. A., Sternlieb, I., Scheinberg, I. H. and Ashwell, G. Physical and chemical studies on ceruloplasmin. V. Metabolic studies on sialic acid-free ceruloplasmin *in vivo*. *J. Biol. Chem.*, **243** (1968), 155–159

3. Morell, A. G., Sternlieb, I. and Scheinberg, I. H. Physical and chemical studies on ceruloplasmin: Crystallisation of desialised human ceruloplasmin asialoceruloplasmin. *Science*, **166** (1969), 1293–1294

4. Hickman, J., Ashwell, G., Morell, A. G., Van Den Hamer, C. J. A. and Scheinberg, I. H. Physical and chemical studies on ceruloplasmin. VIII. Preparation of N-acetylneuraminic acid-1-^{14}C-labelled ceruloplasmin. *J. Biol. Chem.*, **245** (1970), 759–766

5. Van Den Hamer, C. J. A., Morell, A. G., Scheinberg, I. H., Hickman, J. and Ashwell, G. Physical and chemical studies on ceruloplasmin. IX. The role of galactosyl residues in the clearance of ceruloplasmin from the circulation. *J. Biol. Chem.*, **245** (1970), 4397–4402

6. Gregoriadis, G., Morell, A. G., Sternlieb, I. and Scheinberg, I. H. Catabolism of desialylated ceruloplasmin in the liver. *J. Biol. Chem.*, **245** (1970), 5833–5837

7. Morell, A. G., Gregoriadis, G., Scheinberg, I. H., Hickman, J. and Ashwell, G. The role of sialic acid in determining the survival of glycoproteins in the circulation. *J. Biol. Chem.*, **246** (1971), 1461–1467

8. Morell, A. G. and Scheinberg, I. H. Solubilisation of hepatic binding sites for asialo-glycoproteins. *Biochem. Biophys. Res. Comm.*, **48** (1972), 808–815

9. Sternlieb, I., Morell, A. G. and Scheinberg, I. H. The remarkable selectivity of hepatocytes in the uptake of glycoproteins. *Gastroenterology*, **64** (1973), 1049–1052

10. Stockert, R. J., Morell, A. G. and Scheinberg, I. H. Mammalian hepatic lectin. *Science*, **186** (1974), 365–366

11. Benacerraf, B. (1958) Quantitative aspects of phagocytosis. *Proceedings of the Symposium on Liver Function*, Brauer, R. W., ed., American Institute of Biological Sciences, Washington, pp. 205–227

12. Regoeczi, E. and Hatton, M. W. C. Studies of the metabolism of asialo-transferrins: The mechanism for the hypercatabolism of human asialo-transferrin in the rabbit. *Canad. J. Biochem.*, **52** (1974), 645–651

Discussion

WALDMANN

When I first saw these observations, I concluded that they explained some of the observations of some of us interested in erythropoietin that were made some 15 years ago. In these studies it was shown that if you desialylate erythropoietin, it had no biological activity whatsoever *in vivo*, but still had full activity *in vitro*. Probably desialylation revealed a galactose residue leading to rapid hepatic clearance *in vivo*. I do have a couple of questions. One is: Cohen and Mannik some years ago desialylated immunoglobulin and found there was no alteration in its survival[1]. Do you feel that IgG does not participate in the phenomenon you are discussing? As a second question: Some proteins that have required very vigorous extraction for their purification, such as carcino embryonic antigen, have had an exceedingly rapid catabolism and clearance into the liver. I wonder what else other than neuraminidase might remove enough sialic acid to reveal a galactose, and thus cause this short *in vivo* survival?

SCHEINBERG

The last question first. We do not know any other process except prolonged hydrolysis under controlled conditions that will just take off sialic acid and expose galactose. We prefer to use neuraminidase to remove the sialic acid.

The comment you first made about the hormones has been followed up almost *pari passu* with these studies. Ashwell has done a good deal of this. Many of the preparations that are used therapeutically have been prepared without any regard for whether or not sialic acid has been split off during processing. The differences in activity, from batch to batch, of hormones from the same pharmaceutical house are probably attributable to different amounts of sialylation. This results in remarkable divergence in *in vitro* and *in vivo* hormonal activity. There is a difference in survival of native and desialylated immunoglobulin when one looks more closely, just as Dr Regoeczi found with transferrin.

ROTHSCHILD

How much sialic acid does have to be taken off before the recognition function comes into action? Is it the same proportionally for all the plasma glycoproteins? Second, is there any specificity in the membrane receptor for the individual asialoprotein, or is it non-specific in recognising any asialoprotein?

SCHEINBERG

Van den Hamer made a very careful study of two different neuraminidases and ceruloplasmin. He showed that, when he took off 20% of ceruloplasmin's sialic acid, the resulting ceruloplasmin had essentially the same short half-life that fully desialylated ceruloplasmin had. By working with neuraminidases that attacked different sialic acid residues, of the 9 or so in each molecule, he was able, using a Poisson distribution function, to show that the removal of any single sialic acid did not shorten the half-life but the removal of *any* two sialic acids resulted in virtually instantaneous removal of the ceruloplasmin from the circulation.

We have not made a similar quantitative study of any other protein but it looks as though removal of a small number of sialic residues is sufficient, and it does not look as though there is specificity of the particular sialic residues that need to be removed.

As far as the specificity of the hepatic plasma membrane binding protein goes, the problem is to account for the different rates of removal. Thus, for example, orosomucoid is preferentially removed from the circulation. One can do an experiment in which one injects asialoorosomucoid into the rabbit or rat, and then follows this with labelled asialoceruloplasmin. In such a situation the asialoceruloplasmin will survive just like a fully sialylated ceruloplasmin until practically all of the asialoorosomucoid is removed. You can interpret this, if you like, as a preference of the binding protein for asialoorosomucoid—a kind of quantitative specificity. One can construct a table of the order in which various desialylated proteins prevent the hepatic uptake of others.

TAVILL

Dr Scheinberg, I understand you to say that there is a strong chemical bond between the receptor and the asialoprotein. Do you mean that there is a covalent bond formed?

SCHEINBERG

What would you call the bond between acetate ion and hydrogen ion? Is that electrostatic or covalent? It is obviously stronger than the bond between chloride and hydrogen—which have exactly the same electrostatic attraction —probably because there are other kinds of forces acting. The bond between asialoprotein and membrane receptor is highly susceptible to changes in ionic constitution. Calcium ions must be present and if you decrease the pH to 5.6, or below, the bond is broken. So it probably involves an electrostatic component and the kind of poorly understood forces that make a weak base or acid different from a strong one.

TAVILL

The implication of my question is whether the binding shows a susceptibility to dissociation by physiological factors.

SCHEINBERG

We have not been able to do dialysis experiments to make a Scatchard plot, for example, and calculate K or n for the reaction. Within the past year we have solubilised the protein; it is found in the particulate matter of the homogenates and localised to plasma membrane. We are studying its binding, not only *of* other proteins but also *to* cells. It appears to be capable of binding certain cells which have carbohydrate chains on their surfaces.

Reference

1. Cohen, G. L., Mannik, M. Catabolism of gamma-6-globulin with reduced interchain disulphide bonds in rabbits. *J. Immunol.*, **96** (1966), 683

12

In vivo behaviour of asialotransferrins

E. REGOECZI and M. W. C. HATTON

In recent years, Ashwell, Morell, Scheinberg and their colleagues established a hepatic pathway for the elimination of a number of desialylated glycoproteins (e.g. the asialo derivatives of ceruloplasmin, orosomucoid, haptoglobin, fetuin, α_2-macroglobulin and thyroglobulin) from the circulation. In essence desialylation is effective through the exposure of penterminal galactosyl groups which mediate the binding of the·modified protein to a glycoprotein on the plasma membrane of hepatocytes. In addition to the paper given by Scheinberg (p. 121), several excellent review articles[1,2] are available as a source of detailed information regarding the hepatic clearance mechanism for asialoglycoproteins.

Our interest in asialotransferrin originally arose from the apparent discrepancy between the weak interaction of asialotransferrin with the hepatocyte on one hand and the structure of the carbohydrate chains of transferrin on the other. Thus experiments with plasma membrane preparations obtained from rat liver showed that the affinity of human transferrin for the receptors was several orders of magnitude smaller than that of asialoorosomucoid and other asialoglycoproteins tested[3,4]. In accordance with this finding *in vitro*, the disappearance from the plasma of intravenously injected transferrin appeared to be unaffected by desialylation in short-term (30–60 min) survival studies[5,6]. This is surprising since the structures of the carbohydrate chains proposed by Wagh *et al.*[7] for α_1-acid glycoprotein and by Jamieson *et al.*[8] for transferrin bear a close resemblance to one another at least as far as the primary sequences of the distal chain segments are concerned.

It was for these reasons that a *de novo* appraisal of the biological behaviour of asialotransferrin with different methodology was desirable. In view of the relatively low affinity of asialotransferrin for the hepatic plasma membrane, the plasma protein turnover approach was adopted, comprising measurements with simultaneously administered control and asialo preparations over extended periods of time. Numerous questions required clarification. Is the liver over a longer period able to recognise and preferentially eliminate asialotransferrin from the circulation? Are there any kinetic differences among asialotransferrins from various species? Do heterologous asialotransferrins behave metabolically in the same way as homologous asialotransferrin? And, finally, does desialylation of circulating transferrin occur as a physiological process or is asialotransferrin only a product of human creativity?

At present not all the questions posed above can be answered unequivocally. Nevertheless, the information now available on asialotransferrins is broad enough to be reviewed within the general framework of plasma glycoprotein–hepatic plasma membrane interactions. Animal and laboratory techniques as well as details of results have been given in earlier publications[9–12].

Two types of metabolic response of transferrin to desialylation

All mammalian transferrins (i.e. bovine, canine, human, porcine and rabbit) studied so far in our laboratory responded to desialylation by an increase in their catabolic rates, the extent of which was variable—and two main types were distinguishable. Although the difference between the types was primarily of a kinetic nature, there is good reason to believe that they are not due to identical biological mechanisms. Since the key issue involved here is whether or not an asialotransferrin is preferentially attracted to the liver, we will refer to the two types of asialotransferrin catabolism below as hepatic and non-hepatic forms.

Elimination of asialotransferrin by the hepatic mechanism

This type of clearance is characterised by a marked difference between the plasma radioactivities of simultaneously injected transferrin and asialotransferrin. One way of studying this metabolic form is by injecting human asialotransferrin into the rabbit. A typical experiment is illustrated in figure 12.1, showing that rapid elimination of the asialotransferrin commences without any delay. This gives rise to the development of a rapidly widening gap between the curves representing the control and the asialoproteins, already apparent within the first few hours of the experiment. The initial half-life of human asialotransferrin in the plasma of rabbits is usually 1.5–3 h, but when the level of the dose in the plasma reaches 8–15% of the starting value, a gradual slowing down becomes evident even if desialylation is apparently complete. The reason for this phenomenon is under investigation.

Plasma levels of trichloroacetic-acid-soluble breakdown products arising from labelled human transferrin in the rabbit are also markedly different depending whether or not the protein has been desialylated (figure 12.2).

Bezkorovainy et al.[13,14] established some years ago that siderotransferrin is more stable towards denaturing agents than is apotransferrin. As the transferrin used in our studies (Behringwerke/Marburg, Germany) was >98% iron-free, it seemed important to establish whether the fast elimination of human asialotransferrin in the rabbit was in any way referable to the 'asialo–

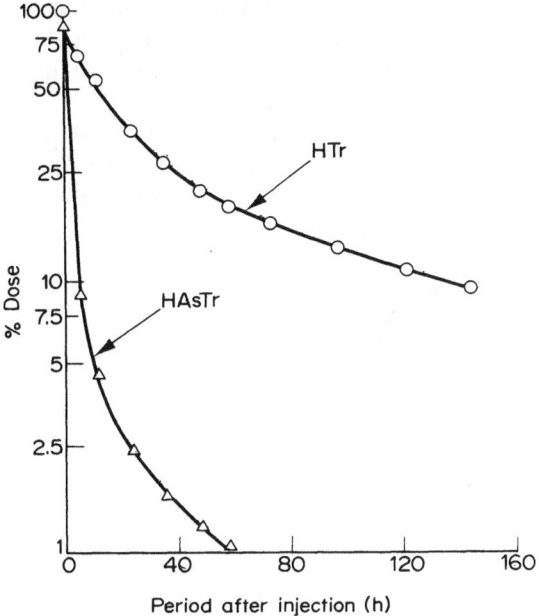

Figure 12.1. Protein-bound radioactivities in the plasma of a 4.5-kg rabbit at various intervals following the injection of a mixture of human ^{125}I-transferrin (HTr) and 100%-asialo human ^{131}I-transferrin (HAsTr).

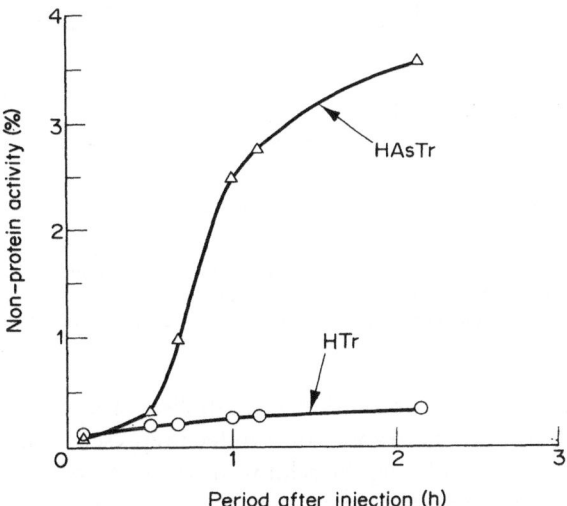

Figure 12.2. Levels of non-protein radioactivities in the plasma of the rabbit at various intervals following a mixed injection of human ^{125}I-transferrin (HTr) and 100%-asialo human ^{131}I-transferrin (HAsTr). Values are expressed as percentages of the total respective radioactivity in a plasma sample which was taken 5 minutes after the injection.

apo' combination. A sample of 90% desialylated human apotransferrin was therefore divided into two and each half labelled with ^{125}I and ^{131}I, respectively. The portion labelled with ^{125}I was then converted into siderotransferrin and, after dialysing away the excess chelated iron, both preparations were injected into two rabbits. One of these experiments is depicted in figure 12.3 from which it is quite clear that the disappearance of human asialotransferrin from the circulation of rabbits is equally fast irrespective of whether the injected protein is in the sidero or apo form.

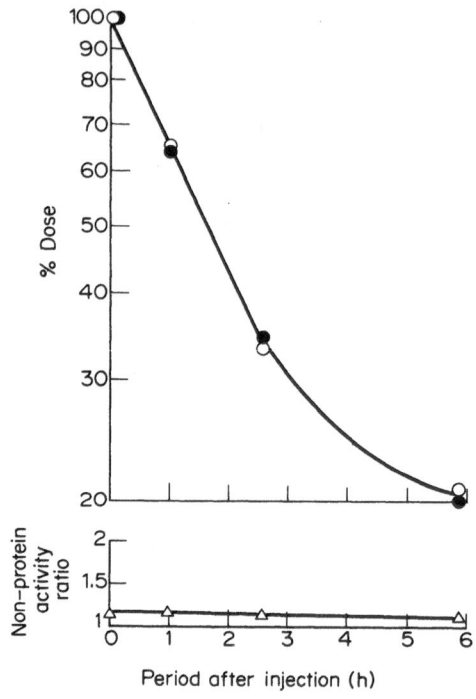

Figure 12.3. Behaviour of simultaneously injected human ^{125}I-asialosiderotransferrin (●) and ^{131}I-asialoapotransferrin (○) in a rabbit, both proteins lacking 90% of their sialyl residues. The upper diagram shows circulating protein-bound radioactivities, the lower diagram is a plot of ^{125}I/^{131}I non-protein radioactivity ratios in the plasma.

Proof that the type of asialotransferrin catabolism observable after injecting human asialotransferrin into rabbits does indeed represent elimination via the liver was obtained by analysing the relative distribution of human transferrin and asialotransferrin among various organs. Tissue-to-plasma radioactivity ratios of simultaneously injected asialo and control proteins indicated a preferential uptake of asialotransferrin by the liver but not by lung, kidney or spleen. The lack of altered affinity of asialotransferrin for lung and spleen is noteworthy because these organs are rich in macrophages and, as shown recently by O'Shea *et al.*[15], cultures of both pulmonary and splenic macrophages readily ingest and hydrolyse transferrin. The fraction of the dose recoverable from the livers of 6 rabbits which had been sacrificed and partially exsanguinated at various intervals after injection (0.5–2 h) averaged at 21% for asialotransferrin as contrasted by 8% for control transferrin. Of the former, 10.5% (=2.2% of the dose) was present in a trichloroacetic-acid-soluble state, the corresponding value for the latter being 3.7% (=0.3% of the dose).

The above post-mortem findings posed the question whether asialotransferrin may, under circumstances similar to the one just described, be subject to elimination via the Ashwell–Morell pathway[1]. To see this more clearly, we studied the behaviour of partially desialylated transferrin preparations and of galactose-oxidised human asialotransferrin in rabbits, and our results[10] were fully comparable to those obtained by other workers with ceruloplasmin[6,16]. By applying the hypothetico-deductive method of reasoning[17] it must therefore be concluded that our experiments failed to provide sufficient evidence upon which the identity of both mechanisms could be refuted.

Elimination of asialotransferrin by the non-hepatic mechanism

This form of elimination is characterised by a small difference between the plasma curves of simultaneously injected asialotransferrin and control protein as may be seen from the example shown in figure 12.4. The curves in this figure depict the catabolism of canine transferrin and canine asialotransferrin in a rabbit. It can be observed that the two protein-bound radioactivity curves are

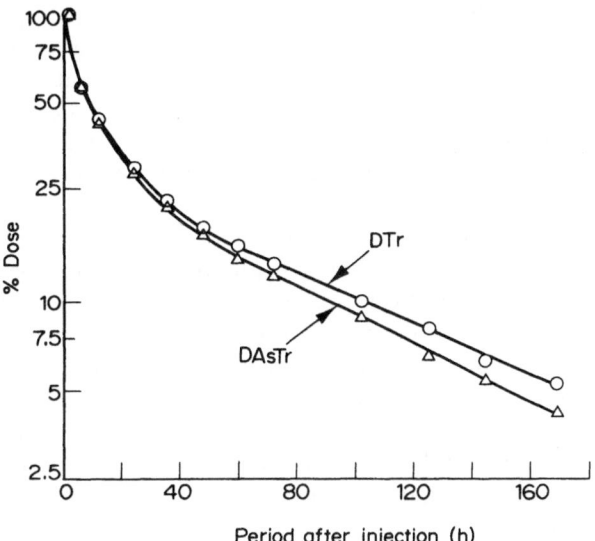

Figure 12.4. Catabolism of simultaneously injected dog ^{125}I-transferrin (DTr) and 95%-asialo dog ^{131}I-transferrin (DAsTr) in a 3.1-kg rabbit. Curves denote plasma protein-bound radioactivities.

almost coincident initially but a slight divergence develops later affecting mainly the terminal slopes. Comparison of the data in figure 12.4 with that in figure 12.1 leaves no doubt that the metabolic fate of canine asialotransferrin in the rabbit is quite different from that of human asialotransferrin.

We studied asialotransferrins from several other species (cattle, pig and rabbit) in groups of rabbits and found that they all behaved similarly to canine asialotransferrin[9,12]. Evaluation of the data by compartmental analysis[18] showed in fact that, within experimental error, all four asialoproteins had the same capillary transfer rate and the same extravascular distribution volume as their parent proteins but their fractional catabolic rate was somewhat higher. The difference in fractional catabolic rates between asialo and control preparations ranged between 13% and 20%, and no statistically significant differences were established in this respect among the four species examined.

From a technical point of view, a difference of this order of magnitude between the fractional catabolic rates of two proteins is not easily established, considering that the standard deviation of the fractional catabolic rate of transferrin in rabbits is about $\pm 10\%$[12]. Had we studied asialo and control preparations singly in groups of recipients, it would probably have been necessary to extend the observations over very large numbers of animals in order to verify the pattern. The alternative approach, in which the two proteins are labelled with different isotopes and injected simultaneously into the same recipient, dispenses with the uncertainty arising from individual variations in transferrin catabolism and therefore is the preferred method. The small, but reproducible, difference between the terminal slopes of certain transferrins and their asialo derivatives is not considered a technical artefact for the following reasons: first, when two samples from a batch of rabbit control transferrin were labelled with different iodine isotopes the curves obtained in rabbits were indistinguishable, and secondly, reverse labelling of asialo and control transferrins (using ^{125}I instead of ^{131}I) did not alter the relative position of the plasma curves.

From a biological point of view, the significance of a difference of the magnitude under consideration is probably negligible. This conclusion is borne out from the role which has been proposed for carbohydrate attachments in plasma glycoproteins: according to Winterburn and Phelps[19], the oligosaccharide chains represent a 'hidden' code which, if activated by the loss of the terminal sialyl residue(s), determines the extracellular fate of the glycoprotein (= removal for catabolism). However, as indicated by our data, the changes which take place in the catabolic rate of certain transferrins after desialylation are clearly too small to be regarded as an expression of a biologically important regulatory mechanism.

Our belief is that this type of asialotransferrin catabolism is not mediated by a single receptor organ such as the liver but it is due to non-specific effects imparted onto the protein by the loss of its sialic acids (e.g. change in charge).

Hence our reference to it as a 'non-hepatic' form. It must be understood, however, that at present this is only a plausible assumption supported mainly by certain observations on the carbohydrate composition of transferrins[12] together with additional indirect evidence, but not strictly proven. The formidable obstacle in obtaining direct evidence is the marginal difference between the slopes. Even if the hypercatabolism of asialotransferrin were to take place exclusively in the liver, the quantity of excess radioactivity going through this organ at any one time over and above the activity associated with the control protein would be too small to be conclusively demonstrated. Furthermore, since the demonstration of slight divergencies between slopes requires measurements over several days (cf. figure 12.4), hepatectomy could not be considered a promising alternative.

Occurrence of the two clearance types

Two distinct mechanisms have been described above for the elimination of asialotransferrins from plasma: in one the Ashwell–Morell hepatic pathway plays a central role and in the other it apparently does not. The rationale behind this remarkable metabolic dichotomy has not yet been fully elucidated. Nevertheless it is clear that the type of elimination mechanism which is characteristic for any one instance depends on the species both as far as the origin of the transferrin and the recipient host are concerned.

On the basis of the experimental material collected to date it appears that mammalian asialotransferrins fall into one of the following two categories:

1. Those which are always eliminated by the 'non-hepatic' mechanism.

2. Those which are facultatively eliminated via the 'hepatic' pathway.

Bovine, canine, porcine and rabbit transferrins are examples of the first category, human transferrin is of the second one.

The biochemical reason for the existence of these two categories is connected with certain differences among transferrins with regard to their carbohydrate composition. Recent analyses of bovine, canine, human, porcine and rabbit transferrins in our laboratory showed both quantitative and qualitative differences in relation to species[12]. Thus the total carbohydrate content of human transferrin was estimated at 6.25% (wt/wt) as contrasted with values ranging from 2.75% to 3.05% for the other species, and the pattern was similar for the different carbohydrate classes examined, such as neutral sugars, hexosamines and sialic acid. This pattern, for reasons which are not obvious, is at variance with some data obtained by Hudson *et al.*[20], particularly as far as porcine transferrin is concerned.

However, probably more intriguing is a qualitative difference found to exist for the penterminal sugar residue. When human asialotransferrin was the substrate, galactose was readily released from the chain by galactosidases both from *Diplococcus pneumoniae* and Jackbean Meal. In sharp contrast, neither of these galactosidase preparations was able to liberate galactose from any of the other transferrins mentioned above. Furthermore, incubation of these asialo-transferrins with N-acetyl hexosaminidase from Jackbean Meal failed to yield N-acetyl galactosamine.

Species-related differences are apparent not only at the level of the carbo-hydrate composition of transferrins but also at the degree of recognition of an asialotransferrin by various host species. This conclusion is supported by the preliminary results of a comparative study with human asialotransferrin presently underway in a number of species.

Conclusions and outlook

Perhaps the most significant piece of information which studies with asialo-transferrins have so far contributed to the generalised concept of asialoglyco-protein clearance is the recognition that, from a kinetic point of view, this type of clearance exhibits a continuum of half-lives ranging from the immeasur-ably short (asialoorosomucoid) to more than one hour (human asialotrans-ferrin). This poses interesting theoretical questions on the biological signifi-cance of the pathway itself, but a treatise on this subject at this point of time would undoubtedly be premature.

Most asialoglycoproteins disappear from the plasma too fast to be used for comprehensive kinetic studies. Thus the relatively slow elimination rate of human asialotransferrin in the rabbit constitutes a distinct technical advantage without which the significance of species in relation to asialoglycoproteins elimination could not have been established.

The striking difference between the behaviour *in vivo* of human asialo-transferrin on one hand and of asialotransferrins from several other species (cattle, dog, pig, rabbit) on the other probably reflects evolutionary changes in transferrin as a glycoprotein. As we have shown[12], human transferrin is not only richer in carbohydrate attachments but there is also evidence for qualita-tive changes in the terminal region of the oligosaccharide chains. Which of these two factors is responsible for the rapid clearance of human asialotrans-ferrin through the hepatic clearance mechanism is not yet known. The need to remove 2 sialic acid residues from ceruloplasmin has been postulated for inducing its rapid elimination[16]. It is possible that the number of sialyl resi-dues in the non-primate transferrins examined is too small to achieve an effec-tive desialylated state, even by complete desialylation. However, this seems unlikely since it has been shown that the sialic acid content of bovine trans-ferrin isomorphs ranges from 0 to 5 residues per molecule[21]. Thus if the response to desialylation were only a matter of the number of residues affected, some fractions of bovine asialotransferrin should be eliminated rapidly, which clearly is not the case.

It seems more likely therefore that non-recognition by the hepatic asialo-glycoprotein receptor of the non-primate asialotransferrins will be explained by future identification of the sugar residue which is exposed by the removal of the sialic acid from these transferrins. Since sialic acid in glycoproteins is invariably linked to either galactose or N-acetyl galactosamine[22], the failure of our attempt with specific enzymes from various sources to liberate one of these sugars from several non-primate asialotransferrins probably signifies a variation in the galactose-N-acetyl hexosamine linkage rather than an absence of the galactose. As the Jackbean Meal preparation contained both α- and β-galactosidases an α-β-type variation can be ruled out, thus leaving a deviation from the common 1→4 linkage as a likely possibility. Further clarification of the carbohydrate structure of these transferrins can therefore be expected to contribute to the better understanding of the role of galactose as a 'cryptic determinant'[1] in the interaction of asialoglycoproteins with the hepatic membrane receptor.

References

1. Ashwell, G. and Morell, A. G. (1971) Galactose: a cryptic determinant of glycoprotein catabolism. In: *Glycoproteins of Blood Cells & Plasma*, G. A. Jamieson and T. J. Greenwalt, eds., Lippincott, Philadelphia, p. 173

2. Ashwell, G. and Morell, A. G. The role of surface carbohydrates in the hepatic recognition and transport of circulating glycoproteins. *Advances in Enzymology*, **41** (1974) 99

3. Pricer, W. E. and Ashwell, G. The binding of desialylated glycoproteins by plasma membranes of rat liver. *J. biol. Chem.*, **246** (1971), 4825

4. Van Lenten, L. and Ashwell, G. The binding of desialylated glycoproteins by plasma membranes of rat liver. Development of a quantitative inhibition assay. *J. Biol. Chem.*, **247** (1972), 4633

5. Morell, A. G., Irvine, R. A., Sternlieb, I., Scheinberg, I. H. and Ashwell, G. Physical and chemical studies on ceruloplasmin. V. Metabolic studies on sialic acid-free ceruloplasmin *in vivo. J. Biol. Chem.*, **243** (1968) 155

6. Morell, A. G., Gregoriadis, G., Scheinberg, I. H., Hickman, J. and Ashwell, G. The role of sialic acid in determining the survival of glycoproteins in the circulation. *J. Biol. Chem.*, **246** (1971), 1461

7. Wagh, P. V., Bornstein, I., Winzler, R. J. The structure of a glycopeptide from human orosomucoid (α₁–acid glycoprotein). *J. Biol. Chem.*, **244** (1969), 658

8. Jamieson, G. A., Jett, M. and Debernardo, S. L. The carbohydrate sequence of glycopeptide chains of human transferrin. *J. Biol. Chem.*, **246** (1971), 3686

9. Regoeczi, E., Hatton, M..W. C. and Wong, K.-L. Studies of the metabolism of asialotransferrins: potentiation of the catabolism of human asialotransferrin in the rabbit. *Canad. J. Biochem.*, **52** (1974), 155

10. Regoeczi, E. and Hatton, M. W. C. Studies of the metabolism of asialo-transferrins: the mechanism for the hypercatabolism of human asialotransferrin in the rabbit. *Canad. J. Biochem.*, **52** (1974), 645
11. Wong, K.-L., Charlwood, P. A., Hatton, M. W. C. and Regoeczi, E. Studies of the metabolism of asialotransferrins: evidence that transferrin does not undergo desialylation *in vivo*. *Clin. Sci. mol. Med.*, **46** (1974), 763
12. Hatton, M. W. C., Regoeczi, E. and Wong, K.-L. Studies of the metabolism of asialotransferrins: relationship between the carbohydrate composition of bovine, canine and porcine asialotransferrins and their metabolic behaviour in the rabbit. *Canad. J. Biochem.*, **52** (1974), 845
13. Bezkorovainy, A., Grohlich, D. The behaviour of native and reduced-alkylated human transferrin in urea and guanidine-HCl solutions. *Biochim. Biophys. Acta*, **147** (1967), 497
14. Bezkorovainy, A., Zchocke, R. and Grohlich, D. Some physical-chemical properties of succinylated transferrin, conalbumin, and orosomucoid. *Biochim. Biophys. Acta.*, **181** (1969), 295.
15. O'Shea, M. J., Kershenobich, D. and Tavill, A. S. Effects of inflammation on iron and transferrin metabolism. *Brit. J. Haemat.*, **25** (1973), 707
16. Van den Hamer, C. J. A., Morell, A. G., Scheinberg, I. H., Hickman, J. and Ashwell, G. Physical and chemical studies on ceruloplasmin. IX. The role of galactosyl residues in the clearance of ceruloplasmin from the circulation. *J. Biol. Chem.*, **245** (1970), 4397
17. Popper, K. (1968) *The Logic of Scientific Discovery*. Harper, New York, p. 30
18. Matthews, C. M. E. The theory of tracer experiments with ^{131}I-labelled plasma proteins. *Phys. Med. Biol.*, **2** (1957), 36
19. Winterburn, P. J. and Phelps, C. F. The significance of glycosylated proteins. *Nature*, **236** (1972), 147
20. Hudson, B. G., Ohno, M., Brockway, W. J. and Castellino, F. J. Chemical and physical properties of serum transferrins from several species. *Biochemistry*, **6** (1973), 1047
21. Stratil, A. and Spooner, R. L. Isolation and properties of individual components of cattle transferrin: the role of sialic acid. *Biochemical Genetics*, **5** (1971), 347
22. Neuberger, A. and Marshall, R. D. (1966) Structural analysis of the carbohydrate group of glycoproteins. In *Glycoproteins*, A. Gottschalk, ed., Elsevier, Amsterdam, p. 263

Discussion

TAVILL

I wonder if you could tell us something about the distribution of neuraminidase in mammals. Most of the present work has been done with bacterial neuraminidase, and my question is whether this enzyme exists in significant amounts in man and other animals, and if so in which tissues?

REGOECZI

As I was interested in this question myself, some months ago we conducted a computer-based search of this area. From the information thus obtained it appears that neuraminidase is present in many tissues. Two types of enzyme can be identified: one with a lower and another with a higher pH maximum (4–4.5 and 5.5–5.8, respectively).

The former occurs in lysosomes, the latter in cytosol. Obviously, neuraminidase in either of these locations could not be expected to play the role of a regulator of plasma glycoprotein catabolism. The only reference to neuraminidase activity in plasma I am aware of is by Warren and Spearing, who found that human and bovine Cohn fraction VI (but not unfractionated plasma) had weak sialidase activity with a pH optimum at 5.5[1]. Unfortunately, nobody seems to have taken interest in this observation and no attempts have been made to find the origin of the neuraminidase activity in Cohn fraction VI.

Reference

1. Warren, L. and Spearing, C. W. Mammalian sialidase (neuraminidase). *Biochem. Biophys. Res. Commun.*, **3** (1960), 489

13

Control of synthesis of retained and secreted liver proteins in relation to amino acid supply

H. N. MUNRO

Introduction

The objective of this contribution is to survey the current status of responses of liver protein synthesis to variations in the quantity and quality of amino acid supplied. This survey will distinguish between the synthesis of proteins retained in the liver and of proteins secreted into the blood. The picture will be integrated with other aspects of liver cell metabolism which occur concomitantly with these changes in protein synthesis. A more detailed account of the evidence has been published elsewhere[1].

Each day the liver monitors considerable amounts of absorbed amino acids. Furthermore, the liver undergoes wider fluctuations in amino acid supply than any other organ with the exception of the intestine. For example, the liver of man receives every day amino acids absorbed from 90 g or more of dietary protein and also at least 60 g of protein secreted into the gut endogenously in the form of digestive juices and shed mucosal cells[2]. The liver of man uses these amino acids to secrete only some 20 g of protein daily into the circulation as plasma proteins, and probably retains temporarily about twice this amount as liver enzymes after each meal (based on the studies of Elwyn[3] on dogs). Thus there is a gross excess of amino acids absorbed compared with the requirement of the liver for synthesis of its own secreted and retained proteins.

Influence of amino acid supply on liver protein synthesis

Changes in liver protein synthesis can be identified by administering a labelled amino acid and observing its incorporation into liver protein, provided that corrections are made for (a) dilution of the labelled amino acid by the free amino pool of the liver, and (b) for the amount of liver protein into which this labelled new protein is deposited[3]. When such comparisons of rates of liver protein synthesis between fasting and protein-fed rats are made, the fed animals show greater deposition of label in the liver protein[4]. An alternative index of the impact of amino acid supply on liver protein synthesis is provided by changes in the relative abundance of polyribosomes in the liver. Proteins retained within the cell are made on polyribosomes lying free in the cell sap, whereas secreted proteins are made on ribosomes attached to the endoplasmic reticulum through which the protein is transported for secretion[5]. In general, this rule holds well, but some proteins are made on both sites, as seems to be the case for liver ferritin[5]. Polysomes prepared from tissues by homogenisation are mostly derived from the free ribosome population, for technical reasons. When rats have been fasted overnight, the profile shows disaggregation into monomers with a reduced population of polyribosomes. It has been repeatedly demonstrated that, soon after feeding a meal containing protein to such an animal, the profile shows an increase in polyribosomes and a reduction in monomers. In consequence, polyribosome abundance in the rat consuming a normal diet undergoes a diurnal rhythm. It is only seen when the diet contains a significant level of protein[6,7]

These studies demonstrate that ribosome aggregation in the liver is sensitive to dietary protein intake. It has also been shown that aggregation is influenced by the supply of individual amino acids. In fasting rats, ribosome aggregation is dependent on the presence of tryptophan in the mixture fed[8-11]. The response to tryptophan does not depend on adrenocortical secretion and is not abolished by prior administration of actinomycin D. It was postulated that, in the fasting state, the rate of protein synthesis in the liver is limited by the level of one amino acid, normally tryptophan, and that it may be possible to lower the hepatic concentration of another free amino acid to the point at which *it* limits the rate of protein synthesis. Pronczuk *et al.*[12] therefore fed rats on diets that were grossly imbalanced in amino acid pattern in order to reduce the level of one free amino acid in the liver. In this way, rats made deficient in isoleucine or threonine responded to a meal containing the missing amino acid with aggregation of liver polyribosomes. Ip and Harper[13] have recently confirmed this, using rats fed on a threonine-deficient diet.

The effect of amino acid supply on synthesis of secreted liver proteins has occasionally been compared with the impact of amino acid availability on synthesis of retained proteins. This has been examined both with intact animals and in perfused livers and intact liver cells; the data have been reviewed elsewhere[1]. Here, we shall consider only evidence obtained *in vivo*. Sidransky *et al.*[14] found that administration of tryptophan to fasting rats caused little aggregation of membrane-attached liver ribosomes, and Peters and Peters[15] found no diurnal rhythm in albumin synthesis, whereas retained proteins showed this response to intermittent food intake. However, following a period of protein depletion, feeding of amino acids to rats results in stimulation of albumin synthesis to supernormal rates within one day[16,17]. One plausible explanation for the difference in responsiveness to amino acids is that albumin synthesis in the well-nourished rat is regulated in relation to the amount of albumin in the plasma and is thus prevented from responding to amino acid stimulation, whereas in the depleted animal, albumin production has fallen below this optimum and now responds to an influx of amino acids.

Nature of control by amino acid supply

The preceding evidence suggests that the supply of a single limiting amino acid can determine the rate of protein synthesis. The process of preparing the amino acid for insertion into the growing peptide chain on the ribosome is illustrated in figure 13.1, which displays the role of tRNA in this process: addition of three terminal nucleotides by nucleotidyl transferase, its charging with amino acids by ligases (activating enzymes), and finally its association with elongation factor I (EF1) in order to transfer the amino acid to the ribosome.

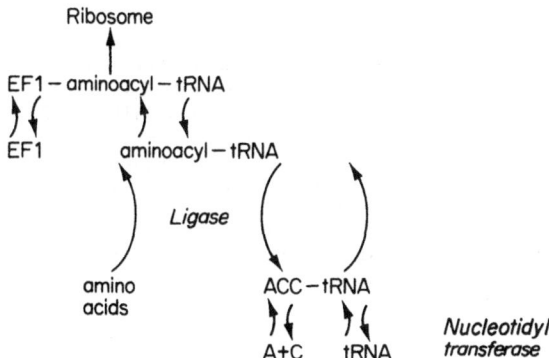

Figure 13.1. Steps in the ligation of amino acids to tRNA and transfer to the ribosome[1].

Elongation factor II (EF2) then moves the peptide chain to make room for entry of the next amino acid-charged tRNA. The problem of evaluating the site of a defect in cell sap factors leading to chain elongation is illustrated by some recent studies we have made on human placenta[18]. The precipitate formed by adjusting placental cell sap to pH 5 resulted in a preparation that was much less active in cell-free protein synthesis than that made from rat liver cell sap. This defect was traced to the presence of partly degraded tRNA in the placental cell sap so that charging of tRNA was much impaired. The pH 5 preparation from placenta was then examined as a source of chain elongation factors. Rat liver ribosomes were incubated with ^{14}C-leucyl-tRNA and either placental pH 5 or liver pH 5 preparation and incorporation of ^{14}C-leucine into the growing peptide chain was measured (table 13.1). The placental preparation was much less efficient in promoting peptide bond formation.

Table 13.1. Uptake of ^{14}C-leucyl-tRNA by ribosomes in presence of pH 5 fractions prepared from human placenta and from rat liver[18]*.

Tube No.	pH 5 Fraction	Elongation factors added	^{14}C-leucyl-tRNA incorporated (pmole)
1.	—	—	0.0
2.	—	EF1	6.6
3.	Placenta	—	2.9
4.	,,	EF1	6.4
5.	,,	EF2	3.4
6.	,,	EF1 + EF2	8.1
7.	Rat liver	—	7.9
8.	,,	EF1	8.1
9.	,,	EF2	6.3
10.	,,	EF1 + EF2	7.9

*The incubation mixture in a total volume of 0.1 ml contained 50 mM Tris HCl (pH 7.8), 5 mM $MgCl_2$, 2 mM dithiothreitol, 80 mM NH_4Cl, 0.2 mM GTP, 14.5 μg EF1 or 45 μg EF2, 60 μg washed liver polysomes, 140 μg of pH 5 enzyme protein, and 34 pmole of ^{14}C-leucyl-tRNA (1.9×10^4 cpm). The mixture was incubated for 20 min at 37°. The radioactivity in the fraction insoluble in hot TCA was measured in a Nuclear Chicago Isocap 300 scintillation counter. The results shown are the averages of two experiments.

This was not altered by addition of EF2, but the incubation mixture containing the placental preparation became as efficient as the liver preparation when EF1 was added. From subsequent studies, it was concluded that the defective tRNA of the placenta was responsible for failure of the EF1 to coprecipitate with it in the pH 5 enzyme precipitate, and evidence was obtained that EF1 and charged tRNA stabilise one another. We have concluded that charging of tRNA may be the common factor regulating and coordinating several aspects of translation. First, uncharged tRNA is known to inhibit initiation of peptide chains[19]. Secondly, the amount of charged tRNA in the cell may determine the stability and thus the quantity of EF1. Thirdly, the rate of chain elongation is affected by charging of tRNA[20]. The concept of the charging of tRNA as a multisite regulator in mammalian protein synthesis is comparable to its regulation of the bacterial metabolic regulator guanosine tetraphosphate, formed when the ribosomes of certain stringent bacteria are exposed to uncharged tRNA[21].

Amino acid supply and liver RNA metabolism

In parallel with changes in the protein content of the liver resulting from a change in the amount of protein in the diet, there are alterations in its RNA content, due mainly to changes in the number of ribosomes. We have evidence that this is at least partly due to variations in rate of ribosomal RNA breakdown with changes in amino acid supply[22]. These changes in RNA breakdown rate appear to affect the nucleotide pools and the rate of *de novo* biosynthesis of purine bases (figure 13.2). During a period of rapid reduction in the RNA content of the cell, it would be expected that free adenine and guanine nucleotides would accumulate and that this would turn off the biosynthesis of nucleotide bases through feedback inhibition of the first enzyme in the pathway.

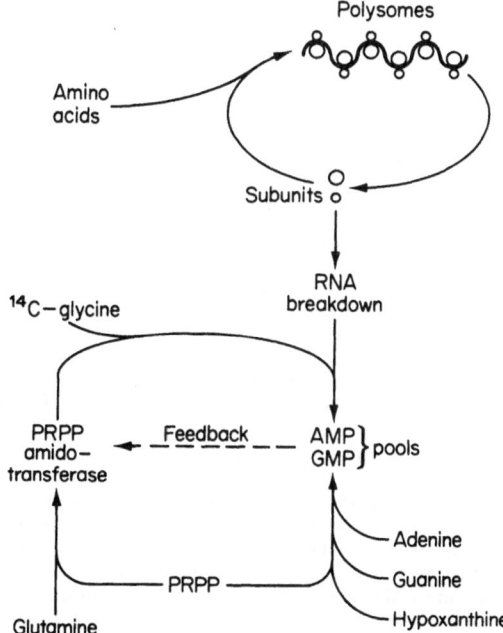

Figure 13.2. Diagrammatic scheme for interrelationships between amino acid supply and liver nucleotide metabolism. PRPP = phosphoribosyl pyrophosphate[23].

On the other hand, retardation of liver RNA breakdown resulting from amino acid intake releases the inhibition and stimulates *de novo* purine synthesis. This has in fact been confirmed[23] by observing changes in liver nucleotide metabolism resulting from diurnal rhythms in protein intake (figure 13.3). During the hours of darkness, when the rats were feeding, polyribosomes and RNA accumulated in the liver, and the free nucleotide pool diminished (here illustrated for guanine). There was also a diurnal

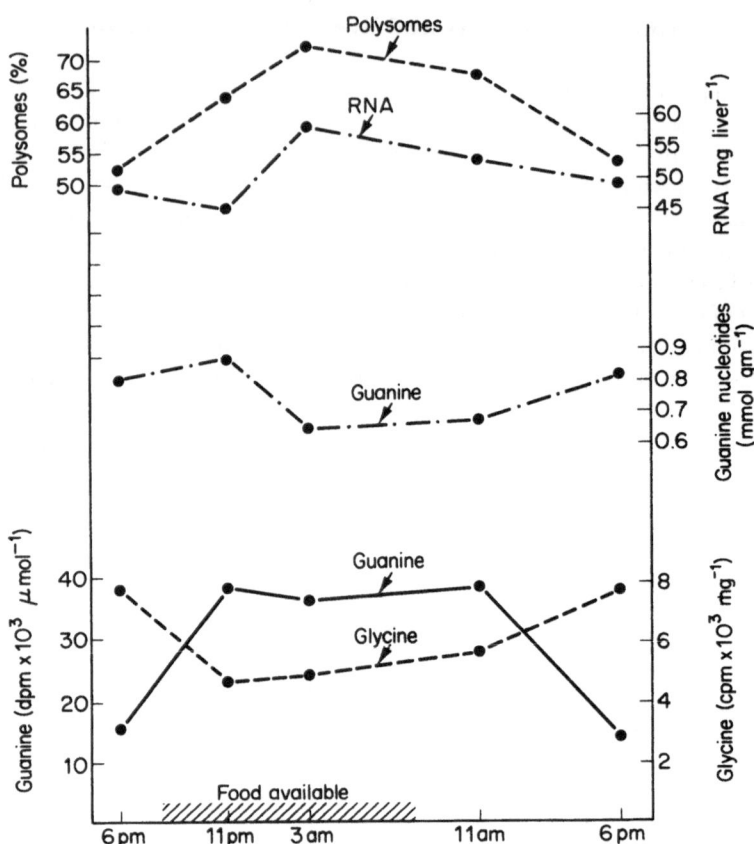

Figure 13.3. Diurnal rhythms in liver polysome abundance and nucleic acid metabolism in rats fed for 12 h per day. From above downwards, the lines represent successively the percentage of polysomes in the liver ribosome fraction, the total RNA content per liver, the total free guanine nucleotides. The lower two lines indicate the uptake by free glycine, and free guanine nucleotides of ^{14}C from a pulse dose of ^{14}C-glycine administered at different times of day. Each point is the mean value obtained from four animals[1].

rhythm in guanine biosynthesis from ^{14}C-labelled glycine that was maximal during the feeding period. This is compatible with reduction in feedback inhibition of the *de novo* pathway. These changes are all reversed when the rat ceases to eat (figure 13.3). In addition to changes in rate of ribosomal RNA breakdown, there is evidence from many tissues that nucleic acid synthesis responds to changes in amino acid intake. This evidence is summarised elsewhere[1], where it is concluded that these effects could be secondary to changes in the rate of synthesis by the polyribosomes of enzymes and other proteins (e.g. ribosomal proteins) which are involved in RNA synthesis, processing and incorporation into ribosomes.

The evidence reviewed above shows that amino acid supply exerts considerable control over various aspects of protein synthesis in the liver of the mammal. Figure 13.4 presents an integrated picture of these responses. When amino acids or protein are administered to the fasting animal, the liver responds with a rapid increase in protein synthesis, accompanied by aggregation of ribosomes. The impact of amino acid supply on the rate of protein synthesis may be determined through the degree of charging of tRNA that can affect both initiation and chain elongation. In addition, the completeness of charging of tRNA may coordinate the levels of soluble factors involved in protein synthesis, notably EF1, through stabilisation dependent on the presence of charged aminoacyl-tRNA. The amount of RNA in the liver and its turnover are affected by amino acid supply, and there are coordinated changes in *de novo* purine biosynthesis resulting from alterations in free nucleotide levels.

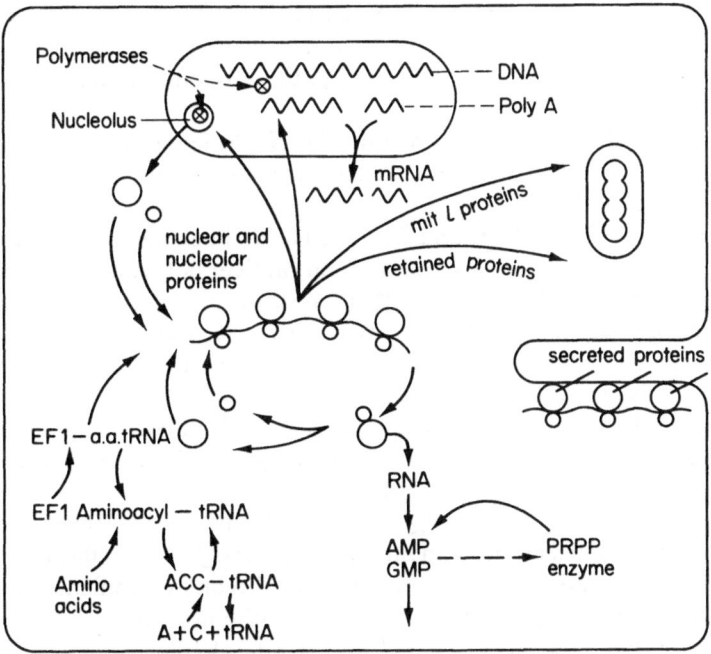

Figure 13.4. Mechanisms in the cell responding rapidly to changes in amino acid supply[1].

References

1. Munro, H. N., Hubert, C. and Baliga, B. S. (1975) Regulation of protein synthesis in relation to amino acid supply—a review in *Alcohol and Abnormal Protein Biosynthesis*, M. Rothschild, M. Oratz and S. S. Schreiber, eds., Pergamon, Oxford, p. 33

2. Munro, H. N. (1974) Protein hydrolysates and amino acids. In *Total Parenteral Nutrition*, P. L. White and M. E. Nagy, eds., Publishing Sciences Group, Acton, Mass, p. 59

3. Elwyn, D. (1970) The role of the liver in regulation of amino acid and protein metabolism. In: *Mammalian Protein Metabolism*, Vol. 4, H. N. Munro, ed., Academic Press, New York, p. 523

4. Clark, C. M., Naismith, D. J. and Munro, H. N. The influence of dietary protein on the incorporation of ^{14}C-glycine and ^{32}P into the ribonucleic acid of rat liver. *Biochim. Biophys. Acta*, **23** (1957), 587

5. Munro, H. N. and Steinert, P. (1975) The intracellular organisation of protein synthesis in *International Review of Science. Biochemistry Series*, Vol. 7, Chapter 10, H. R. V. Arnstein, ed., Butterworths, London

6. Fishman, B., Wurtman, R. J. and Munro, H. N. Daily rhythms in hepatic polysome profiles and tyrosine transaminase activity: role of dietary protein. *Proc. Natl. Acad. Sci. U.S.*, **64** (1969), 677

7. Symmons, R. A., Maguire, E. J. and Rogers, Q. R. Effect of dietary protein and feeding schedule on hepatic polysome patterns in the rat. *J. Nutr.* **102** (1972), 639

8. Fleck, A., Shepherd, J. and Munro, H. N. Protein synthesis in rat liver: influence of amino acids in diet on microsomes and polysomes. *Science*, **150** (1965), 628

9. Wunner, W. H., Bell, J. and Munro, H. N. The effect of feeding with a tryptophan-free amino acid mixture on rat liver polysomes and ribosomal ribonucleic acid. *Biochem. J.*, **101**, (1966), 417

10. Sidransky, H., Sarma, D. S. R., Bongiorno, M. and Verney, E. Effect of dietary tryptophan on hepatic polyribosomes and protein synthesis in fasted mice. *J. Biol. Chem.*, **243** (1968), 1123

11. Park, O. J., Henderson, L. M. and Swan, P. B. Effects of the administration of single amino acids on ribosome aggregation in rat liver. *Proc. Soc. Exp. Biol. & Med.*, **142** (1973), 1023

12. Pronczuk, A. W., Rogers, Q. R. and Munro, H. N. Liver polysome patterns of rats fed amino acid imbalanced diets. *J. Nutr.*, **100** (1970), 1249

13. Ip, C. C. Y. and Harper, A. E. Effect of threonine supplementation on hepatic polysome patterns and protein synthesis of rats fed a threonine-deficient diet. *Biochim. Biophys. Acta*, **331** (1974), 251

14. Sidransky, H., Verney, E. and Sarma, D. S. R. Effect of tryptophan on polyribosomes and protein synthesis in liver. *Am. J. Clin. Nutr.*, **24** (1971), 779

15. Peters, T., Jr. and Peters, J. C. The biosynthesis of rat serum albumin. *J. Biol. Chem.*, **247** (1972), 3858

16. Kirsch, R. E., Frith, L., Black, E. and Hoffenberg, R. Regulation of albumin synthesis and catabolism by alteration of dietary protein. *Nature*, **217** (1968), 578

17. Morgan, E. H. and Peters, T. Jr. The biosynthesis of rat serum albumin.

V. Effect of protein depletion and refeeding on albumin and transferrin synthesis. *J. Biol. Chem.*, **246** (1971), 3500

18. Hubert, C., Baliga, B. S., Munro, H. N. and Villee, C. A. Protein synthesis in a cell-free system prepared from human placenta. II. pH 5 enzyme inefficiency due to defects in tRNA charging with resulting loss of elongation factor 1. *Biochim. Biophys. Acta*, **374** (1975), 359

19. Vaughan, M. H. and Hansen, B. S. Control of initiation of protein synthesis in human cells. Evidence for a role of unchanged transfer ribonucleic acid. *J. Biol. Chem.*, **248** (1973), 7087

20. Munro, H. N. (1970) Free amino acid pools and their role in regulation. In: *Mammalian Protein Metabolism*, Vol. IV, chapter 34, Academic Press, New York, p. 299

21. Haseltine, W. A. and Block, R. Synthesis of guanosine tetra- and pentaphosphate requires the presence of a codon-specific, uncharged transfer ribonucleic acid in the acceptor site of ribosomes. *Proc. Natl. Acad. Sci.*, **70** (1973), 1564

22. Enwonwu, C. O. and Munro, H. N. Rate of RNA turnover in rat liver in relation to intake of protein. *Arch. Biochem. Biophys.*, **138** (1970), 532

23. Clifford, A. J., Riumallo, J. A., Baliga, B. S., Munro, H. N. and Brown, R. R. Liver nucleotide metabolism in relation to amino acid supply. *Biochim. Biophys. Acta*, **277** (1973), 443

Discussion

MILLER

I am a little tired of hearing myself bring up this matter of the functional significance of changes in polysome profile taken from organs like the liver, and the attempt to extrapolate those rather general observations to the synthesis of specific proteins.

I find it very difficult to accept the generalisation as being applicable to all proteins. The responses in terms of synthesis of specific plasma proteins are quantitatively not necessarily alike, and in fact can be quite strikingly different. Another point which I wish to make and that has bothered me ever since your original observations, Professor Munro, is the presumed key role of tryptophan in dominating protein synthesis that led us to do a series of studies with livers taken from fed rats and rats fasted for 18 h. We perfused livers for 12 h with continuous infusion of complete amino acid mixtures that were deficient solely in either tryptophan or threonins: we were unable to document any of the changes which one might be led to expect if the omission of a single amino acid such as tryptophan or threonin were indeed critically rate determining with respect to net synthesis of 5 specific plasma proteins. Although the change in polysome profiles produced in the laboratory must of course be real, what is indeed their functional significance, with respect to the synthesis of specific plasma proteins?

One final question which I would like you to elaborate on is in a table which you showed. You cited some work by Miller and Griffin in which tryptophan was used as a supplementary amino acid, and I recall no studies in which we we have used only a single tryptophan supplement.

MUNRO

In the experiment in which you deleted tryptophan—you did observe changes in albumin response?

MILLER

No, we did not. The only major change we could demonstrate in all of these studies was that tryptophan deficiency in the experiments with rats fasted for 18 h led to a significant negative nitrogen balance in the perfusion system. With respect to the biosynthesis of the 5 specific plasma proteins, the changes in albumin that we observed were quantitatively significant, but trivial. They were less than 15 or 10% of the total.

The point which I raised with the group was to what extent are we selecting out polysomes which are basically making short lived retained proteins by the techniques which are used?

MUNRO

The techniques for harvesting the total polysome profile emphasise the retained proteins, because the large envelopes of endoplasmic reticulum became discarded at the initial stages of emargination. We thus tend to look more at retained proteins than at the secreted ones in such studies. Secondly, if you do labelling studies on the liver, you are observing incorporation of labelled amino acids into retained proteins. Thus, both of these approaches tend to emphasise the retained rather than the secreted proteins.

MILLER

Now, you are just telling us that changes of polysome profiles are more concerned with the retained proteins, and yet of course the emphasis has been on albumin, which is not a retained protein.

MUNRO

We have not looked at albumin, we have been looking at proteins like tyrosine transferase, which also undergoes parallel changes.

14
The metabolism of tumour related proteins

T. A. WALDMANN, K. R. McINTIRE, D. W. DALGARD and R. H. ADAMSON

Introduction

Over the past few years there has been an intense interest and excitement in studies of groups of serum proteins associated with neoplasms. These include myeloma and Bence–Jones proteins as well as a group of proteins termed oncofetal proteins. The oncofetal proteins are present in the biological fluids of the fetus and of patients with certain forms of cancer but are absent or in exceedingly low concentration in the serum of normal adults. The demonstration of such proteins provides valuable leads for basic research concerned with the nature of malignant transformation. In addition, the quantitation of these proteins is of great value in the diagnosis and monitoring of the treatment of certain forms of cancer. This is expecially true when simultaneous *in vivo* and *in vitro* measurements of rates of biosynthesis of these protein markers are used in approaches that permit the determination of the total body burden of tumour cells. In these latter studies, the total number of tumour cells producing a protein marker can be quantitated by determining 3 sets of parameters: (a) the serum concentration of the protein marker; (b) the total body synthetic rate of the protein marker determined by *in vivo* turnover procedures, and (c) the synthetic rate of the protein marker per milligram of tumour or per tumour cell determined using *in vitro* culture techniques. This approach has been applied to the study of multiple myeloma in order to define the total body burden of myeloma cells in patients with this disorder[1,2]. Using repeated studies in the same patient it has been possible to define the kinetics of tumour growth prior to therapy and kinetics of tumour regression following chemotherapy.

We have directed our investigations to studies of a similar nature on another protein marker associated with malignancy, alpha-fetoprotein. Alpha-fetoprotein (AFP) is a 70 000 molecular weight protein with an α_1-electrophoretic mobility that is synthesised by parenchymal cells of the liver, yolk sac and gastrointestinal tract of the fetus[3,4]. It is the major serum protein of the early human fetus with the AFP concentration reaching its peak of about 3 mg ml^{-1} at the twelfth week of gestation. Subsequently the concentration of this protein decreases until it is approximately one per cent of the peak level at birth. By one year of age, the serum AFP concentration is reduced below 30 ng ml^{-1}, that is to less than one 10000th of the peak level[3-5]. AFP is not detectable in the serum of adults when conventional techniques of agar gel precipitation and countercurrent electrophoresis are utilised. However, with the most sensitive radioimmunoassay techniques it has been shown that normal adults have 1–16 ng of this protein per ml of serum. In 1963 Abelev *et al.*[6] demonstrated the finding of elevated alpha-fetoprotein levels with transplantable hepatomas of mice thus associating this fetal protein with a malignancy. Subsequently alpha-fetoprotein has been demonstrated to be elevated in a high percentage of patients with primary liver cancer and embryonal cell tumours of the testis[3,7] and in an occasional patient with gastrointestinal neoplasm[8-11]. In the majority of these studies of alpha-fetoprotein in disease the protein has been detected using the agar gel immunoprecipitation test. This test is quite simple but unfortunately the level of sensitivity of this procedure—about 3000 ng ml^{-1} is not sufficient for the detection of alpha-fetoprotein in the serum of many patients with cancer and is not sufficiently sensitive or quantitative to be used in actively monitoring cancer therapy where one must be able to detect the product of a few milligrams of tumour. We therefore have developed a quantitative double antibody radioimmunoassay test for alpha-fetoprotein. In the present study this sensitive double antibody radioimmunoassay technique was used to define the incidence of elevated alpha-fetoprotein levels in patients with various types of neoplastic disease and appropriate controls with benign disorders. In addition, studies of metabolism and *in vitro* biosynthesis of alpha-fetoprotein in non-human primates with carcinogen induced primary liver cancer were performed.

Materials and methods

Sera from 190 blood bank donors, 20 normal children aged 1 through 20, 340 patients with benign diseases and 709 patients with neoplastic diseases were obtained from patients at the National Institutes of Health or the Mayo Clinic and were stored at $-20°$ until assayed. The metabolic turnover studies were performed in animals of two species (*Macaca mulatta* and *Macaca fasicularis*). Alpha-fetoprotein synthesising hepatocellular carcinomas were produced in such animals using the carcinogenic compound N-nitrosodiethylamine. Administration of this compound was initiated shortly after birth and was continued until a predetermined exposure period had been fulfilled.

Serum alpha-fetoprotein assay
Double antibody radioimmunoassay procedure for quantitation of alpha-

fetoprotein has been described in detail previously[12]. Alpha-fetoprotein for use in the radioimmunoassay was purified from a pool of fetal plasma using the procedure of Nishi[13]. The standards used were dilutions of a human fetal serum pool which was originally quantitated using purified alpha-fetoprotein made available by Professor G. I. Abelev. The antiserum used in these studies was rabbit antihuman alpha-fetoprotein made available by Dr P. P. Sizaret[14]. The antiserum reacted with both human and monkey alpha-fetoprotein. The specificity of the antiserum and the purity of the antigen were tested by radioimmunoelectrophoresis using labelled alpha-fetoprotein and labelled albumin. The antiserum was monospecific and the antigen had no detectable contamination using this method. The double antibody radioimmunoassay technique used is sensitive to 0.5 ng of alpha-fetoprotein in the reaction mixture or 5 ng ml^{-1} of serum as the test is usually performed. We consider serum alpha-fetoprotein values over 40 ng ml^{-1} to be clearly abnormal since the serum concentration of all 210 control individuals over 1 year of age were below this level.

In vivo metabolism of alpha-fetoprotein

To define the metabolic parameters of native alpha-fetoprotein in monkeys, serum from a monkey with carcinogen induced hepatoma was injected intravenously into a normal adult monkey. The total dose of alpha-fetoprotein, 20 mg, was administered in a volume of 3.8 ml of whole serum. The alpha-fetoprotein level of the recipient monkey, 18 ng ml^{-1}, was insignificant compared with the level of 108 000 ng ml^{-1} present in the serum 7–10 minutes after the material was injected.

The metabolism of alpha-fetoprotein purified by the method of Nishi[13] was determined in similar fashion after intravenous injection of this dose to normal monkeys. Finally, purified monkey alpha-fetoprotein was labelled with ^{125}I and ^{131}I by the iodine monochloride method[15] for injection into the monkeys with hepatomas and elevated levels of serum alpha-fetoprotein.

In vitro biosynthesis of alpha-fetoprotein by hepatoma tissues

Fresh biopsy hepatoma specimens obtained from monkeys were minced under sterile conditions in Eagle's minimal essential medium with 10% fetal calf serum and washed in this medium. 20–50 mg of weighed tissue was placed in screw-capped culture tubes with 1 ml of the above medium plus 1 μC of ^{14}C labelled 1-leucine and 1-lysine (1:1). The tissue was incubated for varying periods of time and the incubation was then terminated by centrifugation of the cells at 3000 rpm. The alpha-fetoprotein synthesised and secreted by the hepatoma specimens were quantitated using the double antibody radioimmunoassay technique. In addition, the time course of incorporation of the radiolabelled precursor amino acids into alpha-fetoprotein was determined.

Analysis of the total body burden of the tumour

Estimates of the total body burden of tumour cells were determined from the serum alpha-fetoprotein concentrations and the estimates of *in vivo* and *in vitro* alpha-fetoprotein metabolism in monkeys bearing hepatoma tumours using the following formulation:

Total body tumour burden (mg)

$$= \frac{\text{Total body synthetic rate of AFP marker}}{\text{Synthetic rate of AFP per mg of hepatoma tissue}}$$

Where

Total body synthetic rate of AFP = serum AFP concentration (ng ml^{-1})

$$\times \text{ plasma volume} \times \text{AFP fractional catabolic rate}$$

and

Synthetic rate for AFP per mg of hepatoma tissue

$$= \frac{\text{AFP synthesised by hepatoma tissue } in\ vitro \text{ (ng day}^{-1})}{\text{Mg of hepatoma tissue in the } in\ vitro \text{ culture}}$$

Results

Serum alpha-fetoprotein levels in malignancy in man

Seventy-two per cent of the 130 patients with hepatocellular carcinoma in the present study had an elevated alpha-fetoprotein level (> 40 ng ml^{-1}) with a range in serum concentration of from 44 to over 5 million ng ml^{-1}. An elevated alpha-fetoprotein level was also very common in patients with testicular teratocarcinoma. It was present in 80 of the 101 patients with testicular teratocarcinoma, embryonal cell carcinoma or choriocarcinoma of the testis studied. Elevated alpha-fetoprotein levels were also noted in a lower frequency in association with other types of tumours. Thus, 23% of the 44 patients with pancreatic cancer, 18% of the 91 patients with gastric cancer, 5% of the 193 patients with colonic cancer and 7% of the 150 patients with lung cancer had serum alpha-fetoprotein levels of over 40 ng ml^{-1}. In contrast to these positive findings in patients with cancer of entodermal origin, virtually none of the patients with non-hepatic chronic diseases had alpha-fetoprotein levels over 40 ng ml^{-1}. There was only one patient with a minimally elevated level (52 ng ml^{-1}) in the group of 350 patients with an array of chronic diseases studied. Thus it appears that this radioimmunoassay for alpha-fetoprotein is of value in the diagnosis of malignancy in that the serum level is elevated in the majority of patients with hepatocellular carcinomas and testicular teratocarcinomas and in some patients with other malignancies but only very rarely elevated in patients with chronic non-malignant, non-hepatic diseases.

The radioimmunoassay for alpha-fetoprotein is of even greater value in monitoring the effectiveness of therapy in malignancy. In our studies in 5 patients with hepatocellular carcinoma and in 20 patients with testicular tumours, undergoing therapy, the product of a relatively small number of tumour cells was detectable with this radioimmunoassay when alpha-fetoprotein was undetectable as assessed by agar diffusion tests and when residual tumour could not be detected by other clinical parameters. In a number of patients there was an elevated serum alpha-fetoprotein titre following therapy but no clinical evidence of residual disease. In each of these patients there was a recurrence of the tumour within the subsequent 2 to 12 months of observation.

Metabolism of alpha-fetoprotein in normal adult monkeys

The parameters of alpha-fetoprotein metabolism were determined in adult monkeys by defining the time course of decline of alpha-fetoprotein from the plasma following intravenous administration of whole serum from hepatoma bearing monkeys. Serum containing 2×10^7 ng of alpha-fetoprotein was administered to a monkey with a serum alpha-fetoprotein level of 18 ng ml^{-1}. The mean half-life of survival was 3.5 days with a calculated fractional catabolic rate of 32.1% of the intravenous pool per day. The total circulating pool of alpha-fetoprotein in the recipient monkey with a serum concentration of 18 ng ml^{-1} was 864 ng kg^{-1}. The absolute catabolic rate was 276 ng kg^{-1} day^{-1} a value which equals the synthetic rate. Similar kinetics of alpha-fetoprotein survival were determined by other techniques. Thus, the serum alpha-fetoprotein levels of neonatal monkeys studied during the first two weeks of life declined with a half-life of 3.2 days, a value quite comparable to that observed with adult monkeys injected with alpha-fetoprotein. In addition, the serum alpha-fetoprotein concentration of a monkey following surgical removal of a well localised hepatoma tumour nodule declined with a half-life of 3.0 days. Finally, purified alpha-fetoprotein labelled with ^{131}I injected intravenously into a normal adult monkey had a half-life of survival of 3 days and a fractional catabolic rate of 30% of the intravenous pool per day. Despite the many differences in the four ways in which the die-away of alpha-fetoprotein was measured there was the common finding of a half-life of survival of alpha-fetoprotein of 3–3½ days and a fractional catabolic rate of approximately 30% of the intravenous pool of this protein per day.

In vitro biosynthesis of alpha-fetoprotein by monkey hepatoma tissue

Fresh tumour biopsy specimens from seven different monkeys with carcinogen induced hepatoma was shown by incorporation of ^{14}C-labelled amino acids to be synthesising alpha-fetoprotein for 2–6 days. Synthesis appeared to be linear for the first 24–48 h. The rate of synthesis of alpha-fetoprotein was quantitated by determining the amount of AFP in the culture using the radioimmunoassay procedure. In the animals producing measurable amounts of alpha-fetoprotein there was a synthetic rate of 3.3–7 ng of alpha-fetoprotein per mg of tumour per day.

Analysis of the total body burden of the tumour

The measures of serum alpha-fetoprotein concentration and *in vivo* kinetics of alpha-fetoprotein metabolism for a typical animal with a hepatocellular carcinoma are presented in table 14.1. This animal had a serum alpha-fetoprotein concentration of 1.2 mg ml^{-1} and a total circulating alpha-fetoprotein pool of 208 mg. The fractional catabolic rate for this protein was 32% of the intravenous pool per day and the total body synthetic rate 67 mg day^{-1}. The alpha-fetoprotein synthetic rate as determined from *in vitro* biosynthetic measures was 3.3 mg per g tumour tissue. Thus, the tumour load in this animal can be calculated to be 20 g as assessed using the formulae presented in the methods section.

Table 14.1. AFP synthesis and total body tumour load in a monkey with diethylnitrosimine induced hepatocellular carcinoma

Serum concentration (mg)	1.2
Plasma volume (ml)	173
Total circulating AFP (mg)	208
Fractional catabolic rate (fraction of intravenous pool per day)	0.32
Total body synthetic rate (mg)	67
AFP synthetic rate (mg/g tumour) (from *in vitro* culture study)	3.3
Tumour load (g)	20

Discussion

Several oncofetal proteins have been described that are elevated in the serum of certain patients with cancer. These proteins include carcinoembryonic antigen[16], alpha-fetoprotein[3,12], human chorionic gonadotropin β-chain[17], human chorionic somatomamotropin[18], β-fetoprotein[19] and gamma-fetoprotein[20]. The determination of such proteins in the serum has been of value in the diagnosis of cancer and has been of even greater importance in following the effectiveness of cancer therapy. When measures of *in vivo* and *in vitro* metabolism of serum tumour protein markers have been performed serially during the course of the patient's disease, it has been possible to define the kinetics of tumour growth. Such studies have been performed in the extensively investigated myeloma system in man[1,2,21]. It has become clear from these studies that the rate of tumour growth without therapy and the rate of tumour regression after chemotherapy follows a pattern of Gompertzian kinetics where this represents a progressive retardation of the rate of change of tumour size both in the circumstance of tumour growth and of therapy induced regression.

Evidence for the therapeutic ramifications of the kinetic analysis of tumour growth has been reported by others in a series of dramatic demonstrations with various animal tumours where the knowledge of the cell population and cell cytokinetics of these tumours has provided the necessary foundation for designing curative chemotherapy[22,23]. Similarly, insight into the nature of the growth pattern obtained in human multiple myeloma is already leading to alterations in the design of drug regimes that may lead to improvement in the treatment of this disorder.

In the present study we have demonstrated that serum alpha-fetoprotein concentrations are elevated in a high percentage of patients with hepato-cellular carcinomas and teratocarcinomas and embryonal cell carcinomas of the testis and in a smaller percentage of patients with entodermally derived tumours of other sites. In contrast, alpha-fetoprotein levels are normal in virtually all patients with non-hepatic benign disorders. In related studies the body burden of hepatocellular carcinoma was estimated in monkeys bearing carcinogen induced hepatomas. A half-life of survival of alpha-fetoprotein of approximately 3–3.5 days and a fractional catabolic rate of approximately 30% of the intravenous pool was obtained by studying the metabolism of alpha-fetoprotein using a variety of techniques. Synthesis of alpha-fetoprotein per mg of tumour tissue was studied using *in vitro* culture procedures. Using these *in vivo* and *in vitro* metabolic parameters the estimation of the tumour load in milligrams was possible. Further studies of the metabolic rate of alpha-fetoprotein where there are variations in the total body load and the establish-ment of rates of synthesis per cell will allow the estimation of total body tumour cell numbers and the changes that take place as a result of various modalities of tumour therapy. It is hoped that a comparable approach can be applied to the study of hepatocellular carcinoma and teratocarcinoma of the testis in man and that such studies will lead to more accurate monitoring of the presence of these tumours and greater insights into the kinetics of their growth and ulti-mately more effective therapy of these neoplasms.

Summary

Using a double antibody radioimmunoassay test alpha-fetoprotein was ele-vated (i.e., over 40 ng ml^{-1}) in the serum of 72% of patients with hepato-cellular cancer, 80% of patients with testicular teratocarcinoma and a smaller percentage of patients with entodermally derived tumours of other sites. In contrast to these positive findings in patients with cancer, serum alpha-fetoprotein levels were below 40 ng ml^{-1} in virtually all of the patients with non-hepatic benign diseases.

In related studies in monkeys, the total body burden of hepatocellular carcinoma was quantitated by determining three sets of parameters: the serum concentration of alpha-fetoprotein, the total body synthetic rate of alpha-fetoprotein determined by *in vivo* turnover procedures and the synthetic rate of alpha-fetoprotein per milligram of tumour determined using *in vitro* culture techniques.

References

1. Salmon, S. E. and Smith, B. A. *J. Clin. Invest.*, **49** (1970), 1114–1121
2. Sullivan, P. W. and Salmon, S. E. *J. Clin. Invest.*, **51** (1972), 1697–1708
3. Abelev, G. I. *Adv. Cancer Res.*, **14** (1971), 295–358
4. Gitlin, D., Perricelli, A. and Gitlin, G. M. *Cancer Res.*, **32** (1972), 979–982
5. Masopust, J., Kithier, K., Fuchs, V., Kotal, L. and Radl, J. (1967) in *Intrauterine dangers to the foetus*, Horsky, J. and Stembera, A. K., eds., Excerpta Medica Foundation, Amsterdam, pp. 30–35
6. Abelev, G. I., Perova, S. D., Khramkova, N. I., Postnikova, Z. A. and Irlin, I. S. *Transplantation*, **1** (1963), 174–180
7. Masopust, J., Kithier, K., Radl, J., Koutecky, J. and Kotal, L. *Int. J. Cancer*, **3** (1968), 364–373
8. Alpert, E., Pinn, V. W. and Isselbacher, K. J. *N. Engl. J. Med.*, **285** (1971), 1058–1059
9. Bourreille, J., Metayer, P., Sauger, F., Matray, F. and Fondimare, A. *Presse Med.*, **78** (1970), 1277–1278
10. Kozower, M., Fawaz, K. A., Miller, H. M. and Kaplan, M. M. *N. Engl. J. Med.*, **285** (1971), 1059–1060
11. Mehlman, D. J., Bulkley, B. H. and Wiernik, P. H. *N. Engl. J. Med.*, **285** (1971), 1060–1061
12. Waldmann, T. A. and McIntire, K. R. *Cancer*, **34** (1974) 1510–1515
13. Nishi, S. *Cancer Res.*, **30** (1970), 2507–2512
14. Sizaret, P. P., McIntire, K. R. and Princler, G. L. *Cancer Res.*, **31** (1971), 1899–1902
15. McFarlane, A. S. *Nature*, **182** (1958), 53
16. Gold, P. and Freedman, S. O. *J. Exp. Med.*, **122** (1965), 467–481
17. Braunstein, G. D., Vaitukaitis, J. L., Carbone, P. P. and Ross, G. T. *Ann. Intern. Med.*, **78** (1973), 39–45
18. Weintraub, B. D. and Rosen, S. W. *J. Clin. Endocr.*, **32** (1971), 94–101
19. Buffe, D., Rimbaut, C., Lemerle, J., Schweisguth, O. and Burtin, P. *Int. J. Cancer*, **5** (1970), 85–87
20. Edynak, E. M., Old, L. J., Vrana, M. and Lardis, M. P. *N. Engl. J. Med.*, **286** (1972), 1178–1183
21. Waldmann, T. A. and Strober, W. *Progr. Allergy*, **13** (1969), 1–110
22. Schabel, F. M., Jr. *Cancer Res.*, **29** (1969), 2384–2389
23. Skipper, H. E., Schabel, F. M., Jr., Mellet, L. B., Montgemery, J. A., Wilkoff, L. J., Lloyd, H. H. and Brockman, R. W. *Cancer Chemother. Rep.*, **54** (1970), 431–450

Discussion

REEVE

Tom, this is a very nice piece of work. What I would like to ask you is what evidence have you for constancy of synthetic rate by the tumour cells?

WALDMANN

By constancy I presume you are referring to the question of how constant the rate of marker protein synthesis per tumour cell is during the course of the patient's illness. We ourselves have not defined the rate of synthesis of a protein marker per tumour cell at different times in the same individual. However, Dr Sidney Salmon[1] has done such studies and has found that the rate of myeloma protein synthesis per tumour cell is exceedingly constant for a given patient throughout the course of the patient's natural disease as well as during periods following cancer chemotherapy. However, I do feel that a major concern that all of us have is whether measurements of *in vitro* biosynthetic rates truly reflect the rates of synthesis of these same cells under *in vivo* conditions.

ROTHSCHILD

Tom, I hope this is not too naïve a question but is there a relationship between the appearance of these markers and a kind of prematurity or basic nature of the cell of the tumour?

WALDMANN

Yes, there is a relationship between the state of maturation of the tumour cell and its production of an oncofoetal marker. For example, in patients with hepatocellular carcinomas alpha-fetoprotein is not produced by exceedingly anaplastic tumours or those that are exceedingly mature in their histological picture, but is produced by neoplasms with modestly immature cells. This observation is in accord with the view that in the normal maturational process liver cells go through an initial phase of development when they do not produce alpha-fetoprotein. Then, following a preliminary maturational phase they produce alpha-fetoprotein. As further maturation occurs, production of alpha-fetoprotein is repressed and is replaced by the production of more adult type serum proteins such as albumin.

HOFFENBERG

I think it is really excellent work and very fascinating. In one slide you showed that there could be a disassociation between protein production and growth and I wonder whether this could be related to therapy and whether you are suggesting that monitoring could be obscured by this. By this I mean whether therapy could inhibit specific protein production and at the same time not inhibit growth of tumour cells.

WALDMANN

I think that is a theoretical possibility. However, I feel that another explanation is more likely in these patients with the very complex teratocarcinomas of the testis where there are several different types of cells making up the tumour. In this case we feel that certain of the cells, the embryonal cells, are making alpha-fetoprotein. Whereas, other cells, the trophoblastic-like cells, are making human-chorionic gonadotropin beta-chain. In some patients following therapy one or the other of these populations is completely eliminated, whereas a few cells of the remaining type stay alive and it is this latter type that proliferates with recurrence of the tumour. For example, if a complex teratocarcinoma was treated and the trophoblastic elements completely destroyed but embryonal cells remained, then one would see persistent elevated levels of alpha-fetoprotein but normal levels of human-chorionic gonadotropin. Thus, there would still be a parallelism between the production of protein markers and the number of cells producing that type of marker.

TAVILL

With your very sensitive tests have you been able to screen people with liver disease for the development of hepatocellular carcinoma?

WALDMANN

We have performed very extensive studies of alpha-fetoprotein levels in patients with liver disease and found that patients with those diseases marked by active regeneration of liver cells may have serum levels of alpha-fetoprotein that are elevated to the range between 40 and 500 ng ml^{-1}. For example, a high percentage of patients with hepatitis develop elevated alpha-fetoprotein levels at the time that the enzymes SGOT and SGPT are dropping. That is at the time when there is active liver regeneration. We feel that progressively rising levels of alpha-fetoprotein in patients with chronic liver disease is a potentially ominous sign that suggests the possibility of the development of a hepatocellular tumour. We, however, do not have any serial studies in which a patient with liver disease developed a hepatocellular carcinoma to substantiate this view.

References

1. Salmon, S. E. and Smith, B. A. Tumour immunoglobulin synthesis and total body tumour cell number in IgG multiple myelomas. *J. Clin. Invest.*, **49** (1970), 1114

15

Studies on regulatory factors in transferrin metabolism in man and the experimental rat

ANNE MORTON, S. M. HAMILTON,
D. B. RAMSDEN and A. S. TAVILL

Introduction

The presence in extracellular fluid of the β-globulin, transferrin, with a stability constant for the binding of ferric iron of approximately 10^{23} M^{-1} guarantees that iron entering the plasma compartment remains completely protein-bound and immune from the formation of insoluble ferric hydroxide until delivered preferentially to the sites of haemoglobin synthesis[1]. In patients with a congenital failure of transferrin synthesis there is a disturbance both in the delivery of iron to sites of utilisation and in the mobilisation of iron from storage sites[2]. There is also evidence that transferrin may play a role in the later phase of iron absorption, namely its transport from the intestinal mucosa to the plasma[3]. This cumulative evidence indicates the importance of transferrin in the overall regulation of iron metabolism.

On the other hand, the turnover of iron and the degradation of transferrin normally appear to be mutually independent. The half-life of plasma iron in normal man is about 1.32 hours[4] in contrast to transferrin which has a half-life of 7–9 days[5-7]. It is clear that the degradation mechanism for transferrin is not intrinsically linked to its role in iron receipt and donation, so that it cannot be regarded as a 'suicidal' protein[6]. Whether regulation of degradation plays any part in the maintenance of normal levels of plasma transferrin remains uncertain, and studies relating iron utilisation to the rate of transferrin degradation have produced conflicting results[7,9].

In the case of transferrin synthesis, which occurs mainly in the liver[10,11], three regulatory mechanisms have been proposed.

1. The balance between tissue oxygen supply and demand is the theory proposed by Morgan[11] based on data obtained with rat liver slices. Haemorrhagic anaemia, hypoxia and thyroxine treatment all produced an increase in

the rate of incorporation of ^{14}C-leucine into transferrin, whereas propyl-thiouracil and polycythaemia produced the opposite effect.

2. The rate of erythropoiesis was suggested as a possible regulatory factor by Lane[13]. However, Morgan's early data[8] showing that suppression of the marrow's erythropoietic response with colchicine failed to prevent the rise in plasma iron binding capacity produced by experimental haemolysis in rats refutes this hypothesis.

3. The third suggestion is that transferrin synthesis is modulated by iron supply, mediated either through the iron stores or by means of the plasma iron concentration or both. Support for this hypothesis may be found in the frequently observed reciprocal relationship that exists between the plasma iron-concentration and the plasma total iron binding capacity and which occurs regardless of alteration in tissue oxygenation or changes in the rate of erythropoiesis (table 15.1). A similar relationship has been noted in non-anaemic blood donors[15]. Furthermore, the return of the raised plasma TIBC to normal on treatment of nutritional iron deficiency with iron before correction of the haemoglobin level is evidence against the primary role of anaemia in the control of transferrin metabolism[12,16]. Indications that it is the level of iron stores which is primarily responsible for such regulation in man has been provided by Weinfeld[17]. In the only group of conditions where this reciprocal relation between plasma Fe and TIBC does not prevail, namely in association with the chronic anaemia of inflammation, it is probable that a primary change in degradation rather than synthesis is responsible for the low plasma transferrin concentration[18,19]. Finally, our own observations on the role of thyroid hormone in the regulation of plasma transferrin concentration suggest that within the extremes of plasma thyroxine levels seen in human thyroid disease there appears to be no effect on plasma TIBC. In 166 patients with thyroid disease under the care of Dr Hoffenberg we were unable to find a correlation between the $T_4:T_3$ uptake ratio and the plasma TIBC (figure 15.1). The vast majority of patients with hyper- or hypothyroidism had a TIBC within the normal range. High and low TIBC levels occurred in both groups of thyroid disease, iron deficiency being the commonest association with a high TIBC and chronic inflammatory disease or malnutrition occurring commonly in those patients with a low TIBC. It seems likely that the changes in TIBC and in transferrin synthesis produced by high dosage thyroxine

Table 15.1. Examples in human disease states of the reciprocal relationship between serum iron and total iron binding capacity (TIBC)

Low iron and high TIBC	High iron and low TIBC
Iron deficiency anaemia	Pernicious anaemia (in relapse)
Late pregnancy	Refractory (sideroblastic) anaemia
Acute haemorrhage	Haemolytic anaemia
Pernicious anaemia (following treatment)	Aplastic anaemia
	Primary and secondary haemochromatosis

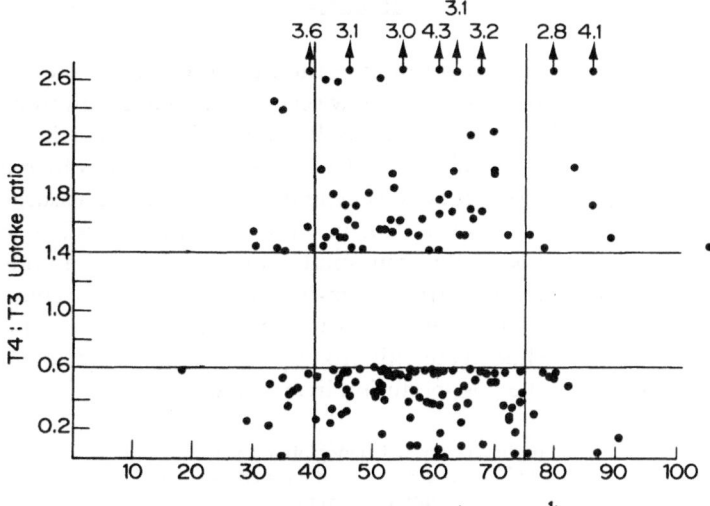

Figure 15.1. $T_4:T_3$ uptake ratio as a function of plasma total iron binding capacity in 166 patients with thyroid disease. The normal ranges are represented by the horizontal and vertical limits.

treatment in rats[11,14] represent a pharmacological effect on protein synthesis rather than indicating a physiological role in regulation. Tavill and Kershenobich[20], using rat liver slices to measure the incorporation of ^{3}H-lysine into transferrin, albumin and α_1-acid glycoprotein, have shown a greatly enhanced rate of incorporation into transferrin associated with nutritional iron deficiency. Restoration of the depleted iron stores with a single intravenous injection of iron reversed this enhanced rate of transferrin synthesis, while correction of anaemia by red cell transfusion produced an even greater enhancement in transferrin synthesis.

Experimental approach

Our recent research has been aimed at (1), defining the role of iron in the regulation of hepatic transferrin synthesis, (2) relating the rate of transferrin synthesis under varying conditions of iron supply to the uptake of iron by the liver and its incorporation into ferritin and (3) measuring the rate of transferrin degradation in relation to iron status.

Our approach has been experimental, using the isolated perfused rat liver and clinical, using conventional radioiodine labelled transferrin turnover techniques. For both approaches the effects of nutritional iron deficiency were studied. In the rat this was produced by feeding an iron-free diet from the time of weaning[21], while patients presenting with a microcytic, hypochromic anaemia without overt blood loss were studied.

Results

After about 10 weeks on a low iron diet the experimental rat shows a severe degree of anaemia associated with a low serum iron and a high iron binding capacity (table 15.2). By using the isolated rat liver and a heterologous perfusion system it was possible to make precise quantitative observations of the synthetic response to iron deficiency and to relate these measurements to the net uptake of iron by the liver. Protein synthesis was measured by the radial immunodiffusion method[22]. Net iron uptake was calculated from the specific radioactivity of the perfusate transferrin-bound ^{59}Fe and the hepatic radio-activity present after washing the liver with ice-cold 0.9% (w/v) saline at the end of the perfusion, while net ferritin iron uptake was determined in the heat stable supernatant of the post-perfusion liver homogenate.

Table 15.2 Relationship between haemoglobin level and serum concentrations of iron and iron binding capacity in the experimental rat, under varying conditions of iron supply

	(n)	Haemoglobin (g 100 ml^{-1})	Serum iron (μg 100 ml^{-1})	TIBC (μg 100 ml^{-1})
Controls	(10)	15.7 ± 0.4	228 ± 17	683 ± 37
Iron deficient (I.D.)	(20)	5.0 ± 0.3	47 ± 6	1021 ± 37
Refed I.D. (48 h)	(8)	9.0 ± 0.3	378 ± 62	918 ± 38

The absolute rates of albumin and transferrin synthesis show a ratio of about 5 : 1. In iron deficiency there is a fall in albumin synthesis, a rise in transferrin synthesis and a proportional fall in the albumin : transferrin synthesis ratio (figure 15.2, table 15.3). Since the liver : body weight ratio falls from 4.7% (SEM 0.15, $n = 16$) to 3.2% (SEM 0.07, $n = 22$) in iron deficiency, there is a

Table 15.3. Relationship between albumin and transferrin synthesis by the isolated perfused liver from rats undergoing varying dietary iron intake. Synthetic rates are expressed as mean ± SEM per standardised liver weight, standardised body weight or both

Experimental group		Albumin synthesis mg h^{-1}			Transferrin synthesis mg h^{-1}		
		per 10 g liver weight	per 300 g body weight	per 10 g liver weight per 300 g body weight	per 10 g liver weight	per 300 g body weight	per 10 g liver weight per 300 g body weight
	(n)						
Controls	(5)	4.04 ± 0.30	5.94 ± 0.59	3.75 ± 0.34	0.96 ± 0.1	1.45 ± 0.14	0.89 ± 0.08
Iron deficient	(5)	4.78 ± 0.16	4.60 ± 0.12	7.36 ± 0.38	2.22 ± 0.16	2.14 ± 0.16	3.41 ± 0.29
Iron refed 24 h	(4)	5.83 ± 0.47	5.69 ± 0.74	6.45 ± 0.98	1.46 ± 0.07	1.42 ± 0.11	1.72 ± 0.27
Iron refed 48 h	(3)	4.89 ± 0.59	4.91 ± 0.46	5.55 ± 0.97	1.17 ± 0.09	1.19 ± 0.18	1.29 ± 0.05

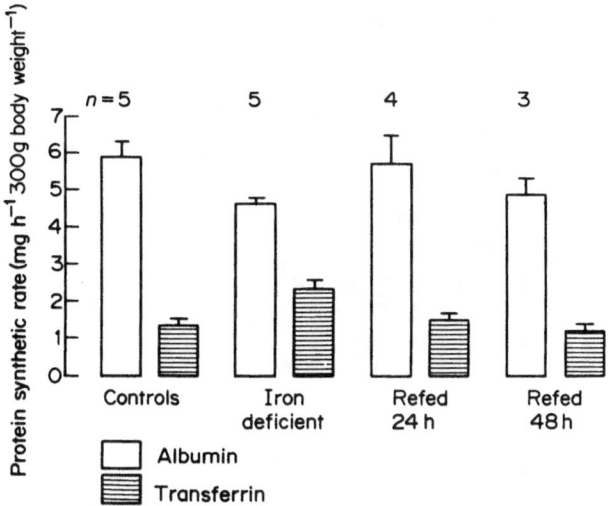

Figure 15.2. The protein synthetic rates of the isolated perfused liver as measured by the radial immunodiffusion method from rats subjected to varying dietary iron intake. n denotes the number of experiments and the bars represent one standard error of the mean.

rise in albumin synthesis when expressed per unit liver weight which is even greater when standardised to both unit body weight and liver weight. On refeeding with a diet containing added iron there was a rapid fall of the transferrin synthetic rate to normal, in the absence of any change in the liver:body weight ratio (mean 3.2%, SEM 0.1, $n = 9$). If one relates the transferrin synthetic rate to the haemoglobin level it is apparent that the rate has returned to normal before complete correction of the anaemia has occurred (figure 15.3).

The iron uptake experiments fall into three groups.

1. First, we examined the relationship of iron uptake to the percentage iron saturation of the circulating perfusate plasma in normal and iron deficient rats. In all situations we found that the uptake was proportional to the iron saturation both after 1 hour and 5 hours of perfusion. However, at all levels of saturation the rate of uptake was higher in iron deficiency, even after refeeding, than in controls (figure 15.4).

2. Taking only the livers which were allowed to perfuse for 5 hours it was apparent that the total net uptake of iron by the liver was greatest in association with the highest transferrin synthetic rates (figure 15.5).

3. On examining separately the total iron uptake and the uptake into ferritin we found that the ferritin radioactivity constituted about 15% of the total hepatic radioactivity at 1 hour rising to about 40% after 5 hours of perfusion. Although at both times the total uptake is enhanced in iron deficiency, that into ferritin is apparently enhanced only after 5 hours of per-

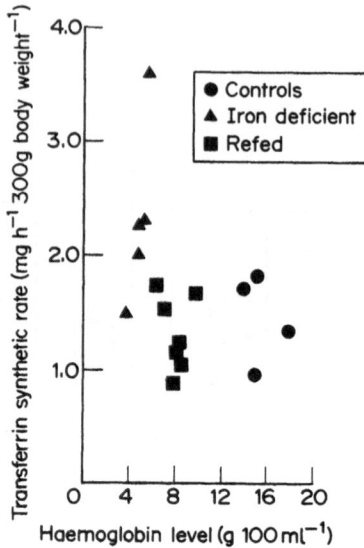

Figure 15.3. The transferrin synthetic rate of individual isolated perfused livers from rats subjected to varying dietary iron intake as a function of the haemoglobin level of the donor rat.

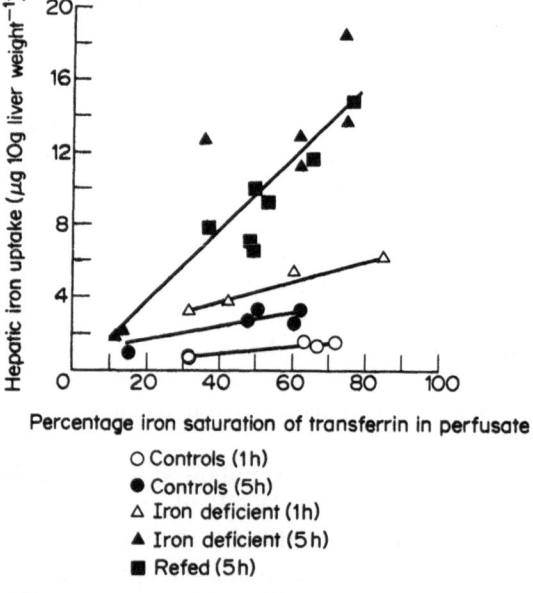

Figure 15.4. The relationship of the hepatic iron uptake by individual isolated perfused livers from rats subjected to varying dietary iron intake to the percentage iron saturation of the transferrin in the circulating perfusate plasma. Regression lines are drawn for each group. The iron deficient after 5 hours of perfusion and the refed after 5 hours of perfusion are grouped together.

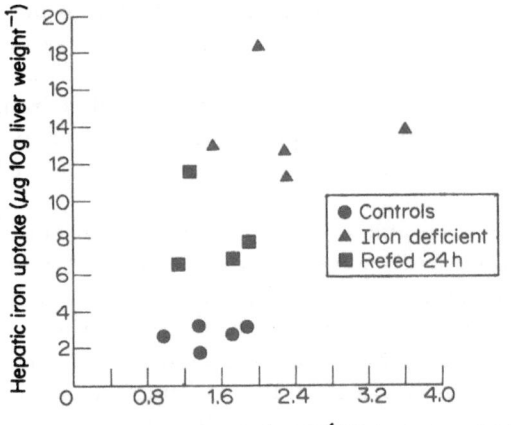

Figure 15.5. The relationship between the total hepatic iron uptake by individual isolated perfused livers from rats undergoing varying dietary iron intake to the transferrin synthetic rate.

fusion. At 1 hour there is no difference between iron deficient and control rats. After 24 hours of refeeding with iron both the total uptake and incorporation into ferritin have begun to fall (figure 15.6). Our preliminary conclusions are that an increased rate of transferrin synthesis in iron deficiency is associated with an increased avidity of the liver for iron. The increased total hepatic uptake of iron in the earliest phase of iron replacement (i.e. after one

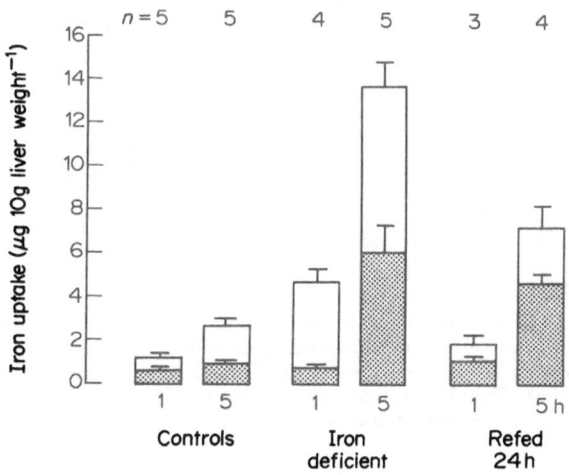

Figure 15.6. Total hepatic iron uptake (open columns) and the iron uptake into ferritin (stippled columns) after 1 and 5 hours of perfusion with perfusate of normal transferrin iron saturation. n denotes the number of experiments and bars indicate one standard error of the mean.

hour of perfusion with an iron containing perfusate) is not associated with increased incorporation of iron into apoferritin. It is possible that in iron deficiency there are increased numbers of vacant receptor sites for transferrin-bound iron and/or a depleted pool of intracellular labile iron. The enhanced uptake of iron into ferritin may represent binding to preformed apoferritin and induction of apoferritin synthesis. The latter seems more likely in view of the delayed response.

Finally, we have been able to study 6 patients with nutritional iron deficiency. Haemoglobin levels were between 6 and 8 g 100 ml^{-1}, serum iron levels were reduced and all showed the characteristic rise in plasma TIBC. The plasma half-life of radioiodine-labelled transferrin was in the normal range of 7–9 days, signifying a greatly increased absolute rate of degradation. In the steady state prior to treatment this indicates that synthesis is enhanced. Treatment with intramuscular iron (2 patients) total dose intravenous iron (2 patients) or oral iron (2 patients) resulted in a rapid fall in TIBC which occurred before complete restoration of the haemoglobin level. In all instances there was either no appreciable change or a prolongation in the survival of the labelled transferrin. Since the TIBC fell rapidly the specific radioactivity of the circulating transferrin fell at a reduced rate after treatment (figure 15.7).

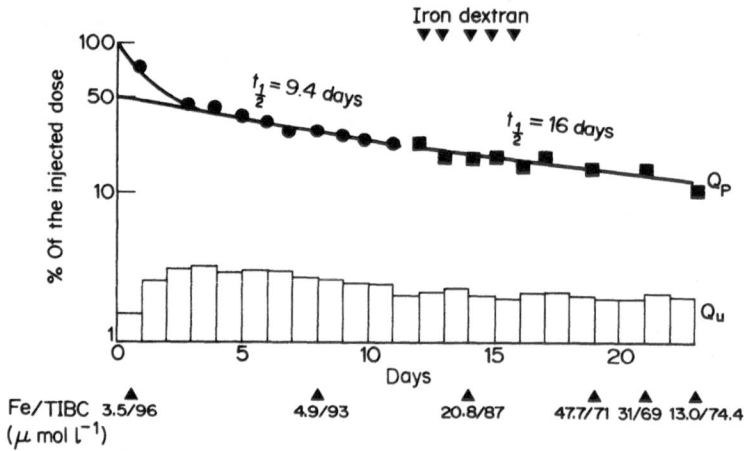

Figure 15.7. [^{131}I]transferrin turnover data in a patient with iron deficiency before and after treatment with intramuscular iron dextran. Q_p= plasma radioactivity (% of injected dose); Q_u= 24-hour urine radioactivity (% of injected dose). The plasma iron (Fe) and iron binding capacity (TIBC) are expressed in concentrations of μmol 1^{-1}.

While recognising the problem of making calculations in non-steady state conditions it is apparent that the fall in the plasma TIBC was not primarily the result of enhanced degradation but was more likely due to a relative inhibition of synthesis. It is possible that inhibition of synthesis is accompanied by a lower than normal rate of degradation which returns to baseline values only as the normal pool size is restored. The data is compatible with first-order control in most circumstances with modulation of the fractional rate of degradation only during major changes in the rate of synthesis.

Discussion

We would like to present our working hypothesis for an overall scheme of the relationship between iron, transferrin and ferritin metabolism in iron deficiency (figure 15.8). In the hepatocyte the negative feedback of iron supply on transferrin synthesis may be linked with the positive feedback on ferritin synthesis which has previously been demonstrated by Drysdale and Munro[23].

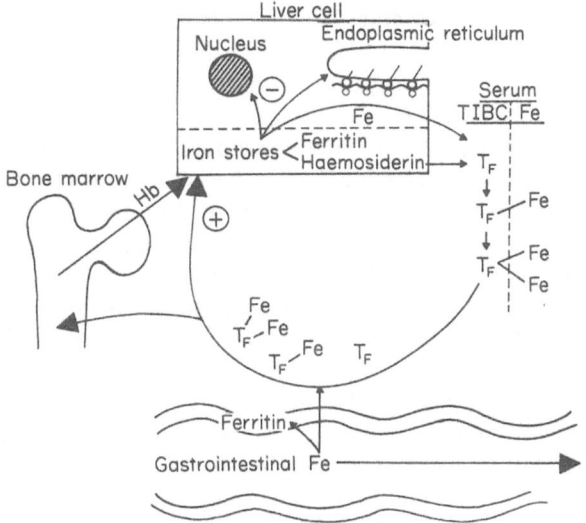

Figure 15.8. A working hypothesis for an overall scheme of the relationship between iron, transferrin and ferritin metabolism in iron deficiency linking the absorption of iron and its transport to its main site of utilisation, the bone marrow and to the storage depots in the liver.

In iron deficiency the low rate of ferritin synthesis in the liver and intestinal mucosa would result in a reduced utilisation of iron for storage purposes and less hold-up in the absorption of dietary iron. The increased rate of transferrin synthesis results in an increase in the plasma TIBC, and a relative preponderance in the plasma of transferrin molecules devoid of iron or containing only one iron atom. These may play a role in the increased mobilisation of iron stores and in the transport of iron across the intestinal mucosa. Finally, according to the Fletcher–Huehns hypothesis[24] this iron may bind to one of the two sites on the transferrin molecule which preferentially donates to primitive erythroid cells in the bone marrow, rather than being wastefully consumed in the formation of iron storage protein.

In conclusion, it is apparent from work on the structure and function of transferrin and ferritin that neither protein is a passive receptor or donor of either transport or storage iron, but rather that they are regulators of iron metabolism with closely integrated metabolic behaviour in which iron in turn plays an intrinsic role. The exact nature of this cooperative integration remains to be defined.

References

1. Aisen, P. Annotation: the role of transferrin in iron transport. *Brit. J. Haemat.*, **26** (1974), 159–163
2. Goya, N., Miyazaki, S., Kodate, S. and Ushio, B. A family of congenital atransferrinemia. *Blood*, **40** (1972), 239–245
3. Levine, P. H., Levine, A. J. and Weintraub, L. R. The role of transferrin in the control of iron absorption: studies on a cellular level. *J. Lab. Clin. Med.*, **80** (1972), 333–341
4. Katz, J. H. Iron and protein kinetics studied by means of doubly labelled human crystalline transferrin. *J. Clin. Invest.*, **40** (1961), 2143–2152
5. Gitlin, D., Janeway, C. A. and Farr, L. E. Studies of the metabolism of plasma proteins in nephrotic syndrome; albumin, gamma globulin and iron binding globulin. *J. Clin. Invest.*, **35** (1956), 44–56
6. Freeman, T. (1962) [131]I transferrin metabolism in human subjects. In: *Protides of the Biological Fluids*, 9th Colloquium, Bruges, 1961, 213–214
7. Awai, M. and Brown, E. B. Studies of the metabolism of ^{131}I-labelled human transferrin. *J. Lab. Clin. Med.*, **61** (1963), 363–395
8. Morgan, E. H. Factors regulating plasma total iron binding capacity in the rat and rabbit. *Quart. J. Exptl. Physiol.*, **47** (1962), 57–65
9. Cromwell, S. (1963) The metabolism of transferrin. In: *Protides of the Biological Fluids*. Proceedings of the 11th Colloquium, Bruges, H. Peeters, ed., Elsevier, Amsterdam 484–486
10. Gordon, A. H. Factors influencing plasma protein synthesis by the liver *Biochem. J.*, **90** (1964), 18P
11. Morgan, E. H. Factors affecting the synthesis of transferrin by rat tissue slices. *J. Biol. Chem.*, **244** (1969), 4193–4199
12. Lane, R. S. Changes in plasma transferrin levels following the administration of iron. *Brit. J. Haemat.*, **12** (1966), 249–258
13. Lane, R. S. Transferrin synthesis in the rat. A study using the fluorescent antibody technique. *Brit. J. Haemat.*, **15** (1968), 355–364
14. Morgan, E. H. Plasma iron binding capacity and iron stores in altered thyroid metabolism in the rat. *Quart. J. Exp. Physiol.*, **48** (1963), 176–180
15. Noyes, W. D., Bothwell, T. M. and Finch, C. A. The role of the reticulo-endothelial cell in iron metabolism. *Brit. J. Haemat.*, **6** (1960), 43–55

16. Muirhead, E. E., Halden, E. R., Mitchell, J. M., Stirman, J. A. and Jones, F. Sequestration of transferrin during iron absorption in iron deficiency. *J. Lab. Clin. Med.*, **50** (1957), 935–936
17. Weinfeld, A. Storage iron in man. *Acta Med. Scand.*, suppl. 427, **177** (1964), 1–155
18. Jarnum, S. and Lassen, N. A. Albumin and transferrin metabolism in infectious and toxic diseases. *Scand. J. Clin. Lab. Invest.*, **13** (1961), 357–368
19. O'Shea, M. J., Kershenobich, D. and Tavill, A. S. Effects of inflammation on iron and transferrin metabolism. *Brit. J. Haemat.*, **25** (1973), 707–714
20. Tavill, A. S. and Kershenobich, D. (1972). Regulation of transferrin synthesis. In: *Protides of the biological fluids*. Proceedings of the 19th Colloquium, Bruges, H. Peeters, ed., Elsevier, Amsterdam, 489–493
21. Valberg, L. S., Taylor, K. B., Witts, L. S. and Richards, W. C. D. The effect of iron deficiency on the stomach of the rat. *Brit. J. Nutr.*, **15** (1961), 473–480
22. Mancini, G., Carbonara, A. O. and Heremans, J. F. Immunochemical quantitation of antigens by single radial immunodiffusion. *Immunochemistry*, **2** (1965), 235–254
23. Drysdale, J. W. and Munro, H. N. Regulation of synthesis and turnover of ferritin in rat liver. *J. Biol. Chem.*, **241** (1966), 3630–3637
24. Fletcher, J. and Huehns, E. R. Function of transferrin. *Nature*, **218** (1968), 1211–1214

Discussion

GORDON

I would like to underline that transferrin is an especially useful and interesting plasma protein for those of us who are interested in control mechanisms because not only is there the well established iron effect which Dr Tavill has described, but a number of experiments show that transferrin may also behave as an acute phase plasma protein in the sense that its synthesis may go up after tissue damage.

TAVILL

I am sure that Dr Gordon is right when he says that transferrin may show an acute phase response under certain conditions. However, in the more commonly encountered circumstances in man, namely chronic inflammatory disease its concentration in plasma is low. Clearly its metabolic behaviour in chronic inflammation may be very different from that which occurs in response to acute inflammation.

REGOECZI

Could I ask a speculative question about the studies in iron deficient patients? Could the effects be attributable to the fact that before treatment transferrin was poorly saturated and under a different molecular conformation than during the second half of the study when the plasma transferrin was relatively highly saturated?

TAVILL

I should have stated that the radioiodine labelled transferrin was approximately 30% saturated. When it is injected into a patient with perhaps only a 10% saturation of circulating transferrin it would seem likely that it will rapidly donate its trace content of iron to the bone marrow and reach a similar saturation as the endogenous transferrin. We would maintain that any change in degradation observed following iron repletion reflects the behaviour of the endogenous protein. It is difficult to speculate on the role of iron saturation in degradation when the primary event as we have demonstrated is inhibition of synthesis.

ROTHSCHILD

Was any effort made to determine the intracellular precursor specific activity in the liver slice isotopic studies?

TAVILL

No. In the experiments that I described[1] we relied upon observations of two other reference proteins, albumin and α_1-acid glycoprotein to provide an indication of whether changes in precursor specific activity were making a major contribution. If we saw changes in incorporation under various experimental conditions occurring disproportionately in the three proteins, then the effects observed could not be entirely due to changes in precursor specific activity, particularly as Morgan[2] has provided evidence that albumin and transferrin have access to the same intracellular precursor amino acid pool. However, I would emphasise that we place far more reliance upon the non-isotopic perfusion system for quantification of protein synthesis.

MILLER

I have some questions in connection with the comparison of your observations in the liver slice and liver perfusion studies. I was interested that although you were able to document a sharp increase in transferrin synthesis following partial correction of the anaemia by an intraperitoneal infusion of red cells there was no significant change in synthesis of α_1-acid glycoprotein. Have you had a chance to do a parallel study in the isolated perfused liver? Secondly, in a limited number of observations of transferrin synthesis in long term liver perfusion going up to 12 h or 24 h we have been struck by the fact that synthesis appears to drop off substantially after 6–8 hours, in contrast to continuing or even increasing synthesis of other proteins. Do you have any comparable observations and do you have any explanation of this peculiar phenomenon?

TAVILL

We have observed the same phenomenon, namely after 5 h or so, while albumin synthesis continues linearly, transferrin synthesis begins to plateau.

While I have no evidence I can only suggest that plausible hypothesis may be that the circulating iron of the perfusate contributes to a negative feedback effect. Alternatively we must not ignore the possibility that release of haemoglobin by haemolysis may contribute to hepatic iron stores following degradation. To come back to the red cell effect I would reserve judgment until we have been able to make similar observations in the perfused liver.

MILLER

Have you been able to compare absolute rates of transferrin synthesis in liver slices and the isolated perfused liver?

TAVILL

The slices function in terms of synthesis at one-fifth or less of the rates observed in the isolated perfused liver.

References

1. Tavill, A. S. and Kershenobich, D. Regulation of transferrin synthesis. In: *Protides of the biological fluids.* Proceedings of the 19th Colloquium, 1971, 489–493
2. Morgan, E. H. Factors affecting the synthesis of transferrin by rat tissue slices. *J. Biol. Chem.,* **244** (1969), 4193–4199

16

The hepatic protein synthesising response to alcohol and fasting*

M. A. ROTHSCHILD, M. ORATZ and S. S. SCHREIBER

Introduction

The liver appears to be one of the most sensitive organs to the stress of fasting showing a marked loss of hepatic protein and RNA. Associated with this loss of hepatic RNA is a marked decrease in the ability of the liver to synthesise proteins for export[1-7]. Of interest is the fact that when the isolated liver is exposed to alcohol there is an equivalent alteration in protein production for export but there is no such marked loss in hepatic RNA[8,9]. These observations suggest that while alcohol and fasting appear to influence the synthesis of proteins by the liver, particularly albumin, in an equivalent fashion, the mechanism by which the protein production is altered might not be the same. The present study was undertaken to determine how the stresses of fasting and alcohol influenced the basic factors responsible for the synthesis of serum albumin.

Experimental methods

Donors

Fed or 24-hour fasted 1.2–1.4 kg rabbits were used in all studies. The standard rabbit chow (Wayne Rabbit Ration, Allied Mills, Inc., Chicago, Ill.) consisted of 17% protein, 2% fat, and 15% fibre and the average intake was 80–120 g day^{-1}.

Perfusate

The perfusate consisted of washed rabbit red cells made up to a final haematocrit value of 25–27% with Krebs–Henseleit bicarbonate buffer containing 3 g% rabbit or bovine albumin, 0.08% glucose and amino acids as described[8].

Supported in part by the U.S. Public Health Service Grant No. AA 00959 and The Bear Foundation.

In the alcohol studies, ethyl alcohol was added to the perfusate at an initial concentration of 0.22% by volume and this level was maintained by the constant infusion of ethanol during the experimental period.

Perfusion was directed into the portal vein at a rate of 1.0–1.4 ml g liver^{-1} min^{-1}. The perfusion volume of 140–170 ml was recirculated and oxygenated by a disc oxygenator that received the output from the inferior vena cava. Bile was collected from the cannulated biliary duct[10].

Albumin synthesis

Albumin synthesis was determined by 2 independent techniques. The ^{14}C-carbonate technique described by Swick, Reeve *et al.* and McFarlane was again the basic method employed[11–14]. In at least one study in all groups, albumin synthesis was measured by the immunologic method of Mancini employing an antiserum against rabbit albumin[15]. All batches of antiserum were tested against various dilutions of rabbit serum to assure the presence of a monospecific antibody and the antiserum was found not to cross react with bovine albumin. Therefore, in the perfusions using the immunologic technique along with the carbonate method, bovine albumin was employed in place of rabbit albumin.

Analytical methods

Total protein in the perfusate was measured by a biuret method and albumin partition by a Kern microelectrophoresis unit[16,17]. Albumin was isolated from the perfusate by preparative acrylamide gel electrophoresis[18,19], as has been described, and by repeated alcohol TCA precipitations. Cold urea was included in all of the steps in the latter method to assure against the combination of high activity ^{14}C urea bound to the albumin.

RNA and DNA determinations

After perfusion, the liver was chilled by the gentle injection of 25 ml of ice-cold 0.25 M sucrose (RNase free, Schwarz/Mann, Orangeburg, N.Y.) in TKM buffer (50 mM Tris-HCl, buffer pH 7.5, 25 mM KCl–5 mM MgCl$_2$). The homogenised liver served as the source of both DNA and RNA. DNA was determined by the indole method of Ceriotti as modified by Keck[20]. RNA was determined by the method of Fleck and Begg[21].

Polysomal techniques

Isolation
Polysomes were isolated from the whole liver following perfusion employing the methods of Blobel and Potter[22].

Analysis
The polysomes obtained from the free and bound isolations were suspended in cold distilled water and 16–20 absorbance units (260 nm) were layered over a 34 ml linear sucrose gradient (0.3–1.1 M in TKM over a 2 ml cushion of 60% sucrose). These gradients were centrifuged at 25 000 rpm in a SW 27.1 rotor at 4°C for 2 hours and the resultant gradient was analysed in an ISCO ultraviolet analyser at 254 nm (Model UA-4).

Labelling
The pattern and extent of labelling of polysomal RNA was determined by the addition of ^{3}H uridine; 1.5 mCi (26Ci mmole^{-1}) in the perfusate during the 2.5 hour perfusion. The isolated factions (bound and free) were analysed as above in an ISCO ultraviolet analyser and 1 ml fractions collected. 0.1 ml was plated on Whatman 3 mm discs. The discs were treated sequentially with ice-cold 7%, trichloroacetic acid containing nonradioactive uridine for 10 minutes, washed with ice-cold 7% trichloroacetic acid, cold ethanol, ethanol–ether (1:1 v/v), ether and then dried in air. The dried discs were suspended in 5 ml 0.7% butyl-PBD in toluene and counted in a liquid scintillation counter, appropriate corrections being made for any ^{14}C contamination in the ^{3}H channel.

Results

The data concerning the effects of specific amino acids on albumin synthesis in the isolated perfused rabbit liver are shown in table 16.1. Whereas a fast for 24

Table 16.1 The effects of amino acids and alcohol on albumin synthesis

Donor	Perfusate	Albumin synthesis (mg 100 g wet liver wt^{-1} h^{-1})	Urea synthesis
Fed	Control	16	34
Fasted	Control	8	42
Fasted	Stimulating amino acids	13–20	48–230
Fasted	Non-stimulating amino acids	5–8	19–25
Fed	Alcohol 220 mg%	6	11
Fed	Alcohol 220 mg% and stimulating amino acids	9–14	15–28
Fasted	Alcohol 220 mg% with or without stimulating amino acids	4–6	16–34

Stimulating amino acids 10 mM		Non-stimulating amino acids 10 mM
arginine	proline	leucine
ornithine	threonine	valine
lysine	tryptophan	methionine
phenylalanine	alanine	histidine
	glutamine	

hours reduced albumin synthesis from the control value of 16 mg 100 g wet liver^{-1} h^{-1} to 8, the addition, not only of tryptophan at a final concentration of 10 mM, but also of arginine, ornithine, lysine, phenylalanine, glutamine, proline, alanine and threonine resulted in marked increments in albumin production. Not all amino acids were equally effective and leucine, valine, methionine and histidine were without effect in producing this increment in albumin production. All of the amino acids which resulted in a stimulus to albumin synthesis likewise produced an increment in urea synthesis. No toxic effects of these amino acids were observed for O_2 extraction, bile production, lactate : pyruvate ratios, and glucose synthesis remained unaltered regardless of the amino acid studied (alanine stimulated glucose production). Further, no lack of penetration of the amino acids into the liver cell was noted. ^{14}C amino acids were added to the perfusate with the 10 mM test amino acid and the liver-to-plasma ratio measured after correcting for hepatic extracellular space. Alcohol did not affect this transport (table 16.2).

Table 16.2. Hepatic distribution of ^{14}C amino acids

	cts g liver^{-1} / cts ml plasma^{-1}	
	Alcohol	Control
Leucine	1.13	1.16
Lysine	1.38	1.25
Valine	0.82	0.61
Methionine	0.61	—
Histidine	1.87	1.75
Ornithine	1.47	—
Phenylalanine	0.92	—

Total hepatic water ranged from 73.5–76.8% ^{14}C-sucrose space; alcohol studies 15.7%, control studies 17.2%

Hepatic DNA remained essentially constant in all studies ranging from 100 to 120 mg. Hepatic RNA-to-DNA ratios fell from a mean control value of 2.4 ± 0.1 to 1.8 ± 0.1 following a 24-hour fast and rose slightly to average 2.1 ± 0.1 following the institution of the excess amino acids. Associated with this increment in RNA there was a concomitant increase in the degree of aggregation of the endoplasmic bound polysome (table 16.3).

Exposure of the fed liver to alcohol reduced albumin synthesis from the control value of 16 mg 100 g wet liver wt^{-1} h^{-1} to 6, and urea synthesis fell to 11. When the perfusates were augmented not only with alcohol but with the stimulating amino acids previously observed—in terms of returning albumin synthesis to control levels—to be effective in fasting, albumin production in these alcohol exposed livers likewise increased from the low value of 6 to 9–14 mg 100 g wet liver wt^{-1} h^{-1}. Urea synthesis also increased. However, when

Table 16.3 Polysome aggregation

Donor	Perfusate	Bound	Free
		Polysome aggregation	
		(% of total pattern)	
Fed	Control	76	74
Fasted	Control	48	64
Fasted	Stimulating amino acids	77	79
Fasted	Non-stimulating amino acids	51	64
Fed	Alcohol	28	54
Fed	Alcohol plus stimulating amino acids	53	58
Fasted	Alcohol	10–20	10–20

the liver was derived from a fasted donor, these procedures were totally in-effective in stimulating albumin synthesis again. The degree of reaggregation of the endoplasmic membrane bound polysome again correlated grossly with the increment in albumin synthesis. For, when the fed liver was exposed to alcohol there was a reduction in the degree of aggregation of this polysome from 76 to 28% while the free polysome this time fell from 74 to 54% aggrega-tion. The addition of the stimulating amino acids returned the aggregation of the bound polysome towards normal but did not affect the minimal degree of disaggregation of the free polysome. The combination of fasting and alcohol produced marked disaggregation for both bound and free polysomes.

Tritiated uridine incorporation into RNA
In order to study the effects of fasting and alcohol on albumin synthesis, ^{3}H uridine was added to the perfusate and served as a marker for the incorporation of new labelled ribosomes along the polysome. In order to determine *de novo* RNA synthesis, it would have been necessary to be able to determine the specific activity of the precursor uridine at the site of RNA synthesis within the nucleolus. Since this was not possible, and since the available data indicate that the ribosomal subunit pool is available eventually to both bound and free polysomes alike, we determined the specific activity of the free polysome from the trisome to the heavier aggregates, and compared the specific activity ratio of this free polysome with that obtained from the bound polysome. If the disaggregation seen in fasting and on exposure to alcohol was due to a decrease in initiation of protein synthesis by a decrease in the rate of attachment of labelled ribosomes to the mRNA, a decrease in the bound-to-free polysome specific activity ratios would be seen. Whereas, if disaggregation of the bound polysome was simply due to a decrease in available ribosomes, due to a lowered total hepatic RNA, then the bound-to-free ratio of the two polysomes in terms of their specific activity should remain unchanged regardless of the degree of aggregation, because whatever ribosomes were attached to the bound poly-some must have been derived from the same labelled subunit pool. The data indicate that during 2.5 hours of exposure of the perfused liver to ^{3}H uridine,

under control conditions, the bound-to-free polysome ^{3}H uridine ratio approaches 60%. A value nearly identical with that determined by Hulse and Wettstein[23] for chicken embryo tissue culture cells. During fasting and re-stimulation of the fasted liver by amino acids, the ratio was unchanged. How-ever, upon exposure to alcohol there was a decrease in this ratio to 40% which upon the addition of excess amino acids likewise returned to normal. When the liver was exposed to both stresses of fasting and alcohol, the free polysomes specific activity remained at 0.36 and was unaffected by excess amino acids.

Discussion

Of the essential amino acids, tryptophan has been shown to be essential for the maintenance of the aggregation of the endoplasmic membrane bound poly-some in *in vitro* studies and in the isolated perfused liver[7,24,25]. Further, trypto-phan has been shown to be capable of reversing the acute effects of fasting and of acute exposure of alcohol when the liver was derived from a fed donor[26]. The control level of tryptophan was 0.05 mM in the isolated perfused liver system described in this report, and this level of tryptophan was maintained in all perfusions. In this milieu, the other amino acids, namely, arginine, orni-thine, lysine, phenylalanine, alanine, proline, glutamine and threonine were capable not only of stimulating albumin production to values which at times exceeded those found in the fed control preparation, but also resulted in significant reaggregation of the endoplasmic bound polysome. It is of interest that the free polysome not essential to the synthesis of protein for export *in vivo* is not as sensitive to the lack of exogenous amino acids as is the endoplasmic membrane bound polysome. This observation perhaps is related to the neces-sity of maintaining protein synthesis for cellular consumption at a higher priority than that of protein synthesis for export. Associated with these changes, there was found to be no change in the bound-to-free ratio of the specific activities of polysomes indicating that, even in the face of a loss of RNA and a disaggregation of the hepatic endoplasmic bound polysome, initiation with available ribosomes on the endoplasmic membrane bound RNA was continu-ing from the same labelled ribosomal subunit pool as was present when the control fed preparation was studied.

With alcohol in the perfusion, however, the results were somewhat different. While again albumin synthesis was depressed and the endoplasmic membrane bound polysome disaggregated, there was no significant change in hepatic RNA and the bound-to-free ratio of polysome specific activities was markedly depressed. This result indicates that in the acute alcohol studies, initiation along the mRNA for albumin was inhibited so that the ribosomes that were present along this polysome were derived primarily from a ribosomal subunit pool which was existing prior to the introduction of the ^{3}H label. The reversal of this phenomenon by amino acids indicates that the mechanism by which these high levels of exogenous amino acids interfere with the acute effects of alcohol must be somewhat different than in fasting.

The combination of stresses, namely fasting and exposure to alcohol, resulted in a marked decrease in aggregation not only of the bound, but also of the free polysome, a marked loss in hepatic RNA and a low bound-to-free polysome ratio. These acute effects of the combined stresses of fasting and alcohol could not be reversed by the same procedures which were effective in altering the effects of either stress applied alone[26]. The basic mechanism by which alcohol and fasting influenced the endoplasmic membrane bound polysome appear to be different, fasting possibly decreasing the available ribosomal pool, and alcohol altering the attachment of ribosomes to existing mRNA. Why the free polysome is more resistant to the effects of these two stresses is not clear and remains to be answered.

Acknowledgements

The authors wish to express their gratitude for the technical assistance given by Mr Alvin Burks, Mrs Barbara Martin and Mr Joseph Mongelli.

References

1. Kirsch, R., Frith, L., Black, E. and Hoffenberg, R. Regulation of albumin synthesis and catabolism by alteration of dietary protein. *Nature* (Lond.), **217** (1968), 578
2. Freeman, T. and Gordon, A. H. Metabolism of albumin and γ-globulin in protein deficient rats. *Clin. Sci.*, **26** (1964), 17
3. Waterlow, J. C. Observations on the mechanism of adaptation to low protein intakes. *Lancet*, **2** (1968), 1091
4. Enwonwu, C. O. and Munro, H. N. Rate of RNA turnover in rat liver in relation to intake of protein. *Arch. Biochem. Biophys.*, **138** (1970), 532
5. Sidransky, H., Staehelin, T. and Verney, E. Protein synthesis enhanced in the liver of rats force-fed a threonine-devoid diet. *Science*, **146** (1968), 766
6. Staehelin, T., Verney, E. and Sidransky, H. The influence of nutritional change on polyribosomes of the liver. *Biochim. Biophys. Acta*, **145** (1967), 105
7. Baliga, B. S., Pronczuk, A. W. and Munro, H. N. Regulation of polysome aggregation in a cell-free system through amino acid supply. *J. Mol. Biol.*, **34** (1968), 199
8. Rothschild, M. A., Oratz, M., Mongelli, J. and Schreiber, S. S. Alcohol induced depression of albumin synthesis: Reversal by tryptophan. *J. Clin. Invest.*, **50** (1971), 1812
9. Jeejeebhoy, K. N., Phillips, M. J., Bruce-Robertson, A. Ho, J. and

Sodtke, U. The acute effect of ethanol on albumin, fibrinogen and transferrin synthesis in the rat. *Biochem. J.*, **126** (1972), 1111

10. Rothschild, M. A., Oratz, M., Mongelli, J. and Schreiber, S. S. Effects of a short-term fast on albumin synthesis studied *in vivo*, in the perfused liver, and on amino acid incorporation by hepatic microsomes. *J. Clin. Invest.*, **47** (1968), 2591

11. Swick, R. W. Measurement of protein turnover in rat liver. *J. Clin. Invest.*, **231** (1958), 751

12. Reeve, E. B., Pearson, J. R. and Martz, D. C. Plasma protein synthesis in the rat liver: method for measurement of albumin formation *in vivo*. *Science*, **139** (1963), 914

13. McFarlane, A. S. Measurement of synthesis rates of liver-produced plasma proteins. *Biochem. J.*, **89** (1963), 277

14. Oratz, M., Schreiber, S. S. and Rothschild, M. A. Study of albumin synthesis in relation to urea synthesis. *Gastroenterology*, **65** (1973), 647

15. Mancini, G., Carbonara, A. O. and Heremans, J. F. Immunochemical quantitation of antigens by single radial immunodiffusion. *Immunochemistry*, **2** (1965), 235

16. Gornall, A. G., Bradawill, C. J. and David, N. M. Determination of serum proteins by means of the biuret reaction. *J. Biol. Chem.*, **177** (1949), 751

17. Rothschild, M. A., Schreiber, S. S., Oratz, M. and McGee, H. L. The effects of adrenocortical hormones on albumin metabolism studied with ^{131}I-albumin. *J. Clin. Invest.*, **37** (1958), 1229

18. Rothschild, M. A., Oratz, M., Mongelli, J. Fishman, L. and Schreiber, S. S. Amino acid regulation of albumin synthesis. *J.Nutr.*, **98** (1969), 395

19. Scheidegger, J. J. Une micro-methode de l'immuno-electrophorese *Int. Arch. Allergy Appl. Immunol.*, **7** (1955), 103

20. Keck, K. An ultra-microtechnique for the determination of deoxypentose nucleic acid. *Arch. Biochem. Biophys.*, **63** (1956), 446

21. Fleck, A. and Begg, D. The estimation of ribonucleic acid using ultraviolet absorption measurements. *Biochem. Biophys. Acta*, **108** (1965), 333

22. Blobel, G. and Potter, V. R. An estimate of the percentage of free and membrane bound polysomes interacting with messenger RNA *in vivo*. *J. Mol. Biol.*, **28** (1967), 539

23. Hulse, J. L. and Wettstein, F. O. Two separable pools of native ribosomal subunits in chick embryo tissue culture cells. *Biochim. Biophys. Acta*, **269** (1972), 265

24. Sidransky, H., Sarma, D. S. R., Bongiorno, M. and Verney, E. Effect of dietary tryptophan on hepatic polyribosomes and protein synthesis in fasted mice. *J. Biol. Chem.*, **243** (1968), 1123

25. Munro, H. N. (1970) A general survey of mechanisms regulating protein metabolism in mammals. In *Mammalian protein metabolism*, Vol. VI, H. N. Munro, ed., Academic Press, New York, pp. 3–130

26. Rothschild, M. A., Oratz, M. and Schreiber, S. S. Effect of tryptophan on the hepatotoxic effects of alcohol and carbon tetrachloride. *Trans. Assoc. Am. Physicians*, **84** (1971), 313

Discussion

ENWONWU

You mentioned that when you perfuse liver with alcohol there is ribosomal disaggregation. Is there any change in the liver content, or in the liver activity of RNA?

ROTHSCHILD

In the separation procedures that were used, the ribosomal subunits were eliminated from the linear gradients and, therefore, the initial peak on these gradients would represent the monomers. The perfusion with alcohol resulted in only a minimal decrease in total hepatic RNA. Perfusion with excess or stimulating amino acids resulted in a slight increase in RNA from the low levels seen in fasting.

MILHAUD

What would happen if you use, as a control, animals supplied with carbohydrates or lipids, instead of fasting ones? If you use fasting animals we could expect two effects at once: the first one may be the lack of suitable amino acids, and the second one a shortage of energy.

ROTHSCHILD

An excess quantity of amino acids is the most specific factor influencing polysomal reaggregation. ATP levels have to decrease by 50% or so to result in the same degree of disaggregation as that caused by fasting. However, malnutrition, namely a poor diet, or a diet low in protein is not the same model as fasting and these two models would not be expected to behave in the same fashion. What we have reported is based on a fasted model.

MILLER

Polysomal profiles are meaningful when considered with respect to the synthesis of specific proteins. You are telling us that these polyribosomes are more concerned with the retained proteins, but you emphasise studies on albumin, which is not a retained protein. One would have to isolate the specific messengers for albumin before one could say that the albumin polysome is decreased.

ROTHSCHILD

I agree, Dr Miller, that unless we can isolate the specific polysome making a specific protein, we cannot draw an exact one-to-one relationship. However, the available evidence indicates that the bound polysome population synthesises protein for export and the major protein for export synthesised by the liver is albumin. When albumin decreases, the bound polysome profile appears to be disaggregated, and when albumin synthesis is stimulated, reaggregation occurs. Until we are capable of separating the specific albumin polysomes, this type of correlation is the best we can do. And, while polysome reaggregation is not the only mechanism for control of protein synthesis, it appears to be an effective tool permitting us to follow the effects of various stresses applied acutely to the liver.

PART 4

Nutritional Aspects of Plasma Protein Metabolism

17

Protein synthesis in the perfused liver: Comparative evaluation of the influence of amino acid supply on ribosomal activity of intact and isolated perfused rat liver

C. O. ENWONWU

Introduction

Many investigators[1,2] have demonstrated that starvation or dietary protein deprivation produces an immediate fall in synthesis of albumin which returns promptly to normal or above normal levels within 24 hours of refeeding an adequate diet, and there are suggestions that the changes in albumin synthetic rate are mediated by the availability of amino acids to the hepatic cells[1]. Such observations have triggered widespread interest in the effects of amino acid supply on the cellular organelles involved in protein biosynthesis. Hepatic cytoplasmic ribosomes, as in most mammalian cells, exist either as free ribosomes or bound to membranes of the rough endoplasmic reticulum[3].

The present study was designed mainly to examine the effects of amino acid supply on ribosomal activity in hepatic cells. The proportion of cytoplasmic ribosomes actively engaged in protein biosynthesis at the time of cell fractionation was assessed by subjecting the ribosomal preparations to ionic conditions which selectively dissociated inactive ribosomes that were not stabilised by mRNA and peptidyl-tRNA[4]. This procedure was independent of the level of endogenous free RNase activity[4-6]. Findings in protein-calorie deficient rats refed adequate protein were compared to data derived from isolated perfused organs in which the perfusate was supplemented with amino acids. Use of the isolated perfused organ provided a model system which permitted a better control of the composition of the fluid perfusing the cells and thus allowed for a direct evaluation of the effects of amino acids *per se* on cellular organelles involved in liver protein biosynthesis without the complications of hormonal and nervous influences. The results indicated that within 24 hours of refeeding protein–calorie malnourished rats with adequate protein diet, there was a significant increase in the proportion of hepatic ribosomes resistant to the dissociating effect of a high $[K^+]$ medium. Associated with the latter was an equally rapid increase in the liver content of most of the dietary essential free amino acids. In contrast, ribosomal activity in perfused livers could not be enhanced by normal or supranormal amino acid supply.

Experimental procedure

Animals and diets

For studies in intact animals, male rats of the Sprague–Dawley strain (80 to 100 g body weight) procured from Simonsen Laboratory (Gilroy, California) were used. The rats were housed in suspended wire-bottom cages under conditions of controlled lighting (6 a.m.–6 p.m.) and constant temperature and humidity. These rats were fed a 0.5% protein diet *ad libitum*. The composition (g kg^{-1}) of the synthetic low protein diet was as follows: lactalbumin (5), dextrin (558.8), sucrose (279.2), salt mixture (50), vitamin mixture (5), Wesson Oil from Wesson Oil Sales Co., Fullerton, California, (90), and choline chloride (2). The salt and vitamin mixtures were as previously reported by Rogers and Harper[7]. After 9 to 12 weeks of receiving the low protein diet, the animals were weighed, and randomised into two groups. One group was immediately autopsied for biochemical and morphological studies, while the second group was divided into three subgroups and refed Purina Rat Chow (Ralston Purina Co., St. Louis, Mo.) *ad libitum* for 1, 2 or 4 days before sacrifice.

In the isolated organ perfusion studies, two groups of rats served as liver donors. In the first series of studies, livers were obtained from male rats (220–260 g) fed the 0.5% lactalbumin diet for 5 to 8 days, while in the second series, the liver donors (male rats weighing initially 80–100 g) were fed the same low protein diet for a period of about 5 weeks. All the liver donors were fasted overnight (15–18 hours) prior to removal of the liver. The blood donors which were not fasted before use, were selected either from rats (220–260 g) fed adequately with a commercial laboratory diet (Purina Rat Chow, Ralston Purina Co., St. Louis, Mo.) or from growing rats (80–100 g) fed the 0.5% lactalbumin diet for a period of 5 weeks.

Surgery and liver perfusion

Surgical removal of the liver and the perfusion were routinely started at 11.30 a.m. to minimise any possible complications arising from diurnal variation. The modified liver perfusion apparatus of Miller *et al.*[8] obtained from Metaloglass Inc., Boston, Massachusetts was used. Detailed features of the apparatus are as described by Miller[9]. Each perfusion lasted for 1 hour. The livers were perfused with heparinised whole blood (5000 units heparin per 100 ml of rat blood) diluted to about 30% haematocrit with Krebs–Ringer bicarbonate buffer (KRB). The pH of the diluted heparinised blood was usually between 7.4 and 7.6. Prior to use, the diluted heparinised blood was filtered through a sterile surgical gauze sponge moistened with KRB to remove any gross clot which might be present.

Removal of the liver was essentially as described by Ross[10], and Miller[9], and was performed under ether anaesthesia. The bile duct was cannulated with a suitable length (6–10 cm) of fine polyethylene tube (PE 10). The portal vein was cannulated with a polyethylene tube (PE 50 or PE 60) which was previously filled with heparinised KRB and sealed at the distal end with stopcock grease. The latter was a device to prevent air bubble formation upon insertion of the cannula. No outflow cannula was used. Immediately following cannulation of the vena porta, the liver was promptly excised, rinsed with saline (0.9% NaCl, 37°C), and immediately mounted on the organ dish in the perfusion chamber (37–39°C) as described by Seglen and Jervell[11]. The positioning of the isolated liver in the perfusator dish was carefully done so as to avoid undue compression of the hepatic lobes, and thus assure an even distribution of the perfusate to virtually all the liver lobes. In most of the successful experiments, interruption of the portal flow was not allowed to exceed 3 minutes. The recirculating perfusate which was equilibrated with a gas mixture of 95% oxygen–5% carbon dioxide, flowed freely out of the cut ends of the vena cava during perfusion. The perfusate flow rate through the liver was determined to vary from 0.8–1.5 ml min^{-1} g wet liver^{-1}.

The perfusate was supplemented with glucose (400 mg glucose per 100 ml of perfusate). When indicated, single amino acids (tryptophan, methionine, valine, isoleucine, or leucine) or a mixture of 11 amino acids[12] were added to the perfusate at levels 5- or 10-fold their normal plasma concentrations in adequately fed rats. These levels were computed from the published data of Clemens and Korner[13], and were added to the perfusing fluid at the start of perfusion, followed by subsequent additions after 20 and 40 minutes of perfusion, as indicated in the results. In some of the studies involving the use of livers and blood from rats fed the 0.5% protein diet for about 5 weeks, the perfusate was supplemented with hydrocortisone sodium succinate (Solu-cortef, The Upjohn Company, Kalamazoo, Michigan) at a dose level of 4 mg per 100 g of liver donor, and crystallised beef and pork insulin from Eli Lilly and Co., Indianapolis, Indiana (4 μg ml^{-1}). The rationale for the addition of these hormones was based on previous reports of reduced plasma concentrations of insulin and corticosterone in this rodent model of the syndrome of protein–calorie malnutrition[14,15]. Ultrastructural evaluation of the adrenal cortex in these protein–calorie-deficient rats has also suggested some degree of hypofunction[16].

Fractionation of liver tissue

The livers were chilled in several volumes of ice-cold 0.25 M ribonuclease-free sucrose (Mann Research Laboratories, New York, N.Y.) solution prepared in Tris–MgCl$_2$–KCl buffer, pH 7.6 (50 mM Tris-HCl, pH 7.6; 25 mM KCl and 5 mM MgCl$_2$). This buffer is referred to in this report as the low [K$^+$] medium. All subsequent operations were carried out at a temperature near 0°C unless indicated otherwise. The livers were homogenised in 3 volumes (w/v) of the 0.25 M sucrose made up in the low [K$^+$] medium, using a Potter–Elvehjem homogeniser with a close-fitting Teflon pestle. The homogenising medium contained 10% (v/v) liver postmicrosomal supernatant as a crude source of RNase enzyme inhibitor[17–19]. Following treatment of the homogenate with rabbit antihorse ferritin serum (Calbiochem, Los Angeles, California) to bind liver ferritin[20], cell debris, nuclei, mitochondria and the ferritin–antiferritin complex were sedimented by centrifugation for 10 minutes (12 000 × g average) using a refrigerated Sorvall-SS 34 rotor (Ivan Sorvall, Inc., Newton, Connecticut).

Free and membrane-bound ribosomes

To separate the free from the membrane-attached ribosomes, 4 ml of the liver postmitochondrial fluid not treated with sodium deoxycholate (DOC) was layered on a discontinuous gradient consisting of 3 ml of 1.4 M sucrose over 4 ml of 2 M sucrose, each prepared in the low [K$^+$] medium. The gradients were centrifuged for 18–20 h at 105 000 × g in a Spinco rotor 40 (4°C) to pellet most of the free ribosomes. The bottom 2 M sucrose layer containing unsedimented monomers and some dimeric ribosomes was recovered, diluted with the low [K$^+$] buffer[21], and centrifuged for 2 hours at 105 000 × g to pellet the trapped free ribosomes. The postmitochondrial fluid trapped in the interface between the 1.4 and 2.0 M sucrose solutions following isolation of the free ribosomes was carefully aspirated, treated with DOC to a final concentration of 1%, and centrifuged for 4 hours at 105 000 × g to pellet the membrane-bound ribosomes. The various ribosomal pellets were resuspended in the low [K$^+$] buffer and stored at −60°C for no longer than 1–2 weeks before use. It must be stressed at this point that the terms free and bound ribosomes as used within the context of this study were strictly operational. Secondly, the technique for isolating the free and bound ribosomes was skewed more in favour of the former since some of the membrane-bound ribosomes sedimented with the nuclei and mitochondria during the differential centrifugation for the preparation of the postmitochondrial supernatant[22].

Ribosomal sedimentation profiles

Linear sucrose density gradients (10–40%, 10–50%, w/v) were prepared in centrifuge tubes. The sucrose solutions used in preparing the gradients were made up in the low [K$^+$] buffer. These gradients were used for evaluating the sedimentation patterns of free, membrane-bound, and total (free plus bound) hepatic ribosomal populations. The 'total ribosomal patterns' were prepared directly from the postmitochondrial supernatant using the technique of

Drysdale and Munro[20]. The PMS was treated with DOC (Mann Research Laboratories, New York, N.Y.) to a final concentration of 1%, diluted 1:2 (v/v) with the low [K^+] buffer, and 0.1 ml of the diluted supernatant containing approximately 8 mg wet weight of liver tissue was layered on a linear sucrose gradient. Profiles of the free and bound ribosomal populations were obtained from the pellets purified by zonal centrifugation[22]. The gradients were spun in the Beckman SW-50.1, six-place rotor (4°C), the 10–40% gradients usually for 60 min at 48000 rpm, while the 10–50% gradients were spun for 70 min at 45000 rpm. Distribution of the ribosomes by size was evaluated by passing the contents of the tubes at a constant flow rate through a Beckman Ultraviolet Spectrometer (5 mm flow-cell), attached to a Gilford Recorder (Gilford, Model 2400; Gilford Instrument Laboratories, Inc., Oberlin, Ohio) which scanned absorbance at 254 nm. Interpretation of the sedimentation profiles was usually accomplished by measurement of the area under each peak as described by Wunner, Bell and Munro[23]. This method of quantitation, although lacking absolute accuracy in view of incomplete separation of the different peaks, has been found to be adequate as an index of the response of the polyribosomal organization to various experimental conditions[24–26]. In some cases, especially the profiles resolved in 10–40% linear gradients, a good proportion of the fast moving polysomes ($n > 2$) sedimented at the bottom of the tube, thus obviating any attempt at a precise quantitation of the contribution of the major peaks to the total ultraviolet absorbing area. Under such conditions, interpretation of the profiles was based on visual examination of the heights of the monomer and dimer peaks relative to the heights of the remaining peaks[27,28].

Dissociation of ribosomes into subunits

Total, free or bound ribosomal suspensions as well as the postmitochondrial fluid, were adjusted to the required ionic conditions (50 mM Tris, pH 7.6, 5 mM $MgCl_2$, 800 mM KCl). Since the selective dissociation of inactive ribosomes into subunits with preservation of the integrity of functionally active ribosomes is dependent on the ratio of magnesium to potassium ions[4], preliminary investigations were undertaken to identify the best experimental condition for hepatic tissue. No marked change in the proportion of undissociated ribosomes was observed in the presence of 5 or 15 mM $MgCl_2$ when the KCl level was varied from 400 to 800 mM. Following adjustment of the [K^+] content of the ribosomal fluid, the latter was incubated at 37°C for 0.5 h to convert most of the polyribosomes ($n > 2$) into monosomes[29,30]. Studies by Faber and Tamaoki[30] have demonstrated successful isolation of ribosomal subunits from L5178Y mouse lymphoma cells only after prior incubation of the polyribosomal preparations without exogenous supplemental factors at 37°C for 45 minutes, and similar findings have been noted using hepatic ribosomes from normal and diabetic rats[31]. Following incubation, the ribosomal suspension was analysed on linear sucrose gradients (10–40%, w/v) prepared in the high $[K^+]$ buffer (50 mM Tris, pH 7.6, 5 mM $MgCl_2$ and 800 mM KCl). The gradients were centrifuged in the Beckman SW-50.1 rotor (26–28°C) for 60 min at 48000 rpm. In the interpretation of the ultraviolet (254 nm) tracing, identification of the monosome (80S) was based on the sedimentation of horse spleen ferritin (Pentex Inc., Illinois) added as an internal standard[20,32]. The two peaks sedimenting slower than the monosome (80S) were empirically designated 60S and 40S, respectively, without any assumption of accuracy in the prediction of the sedimentation coefficients. Measurement of areas under the various peaks was performed with a compensating planimeter according to the procedure suggested by Martin[4], and the percentage proportion of undissociated ribosomes (not less than 80S) served as an index of the functional activity of liver ribosomes[4,31,33].

Amino acid incorporation in cell-free system

Uptake of L-[U-^{14}C] leucine (specific activity 304 mCi mM^{-1}; New England Nuclear, Boston, Massachusetts) in a cell-free system was determined using 100 μg ribosomal protein, and the ratio of pH 5 enzyme protein to ribosomal protein in the incubation mixture was 10:1. The pH 5 enzyme was prepared from postmicrosomal liver supernatant sieved through a column of Sephadex G25 (coarse) and dialysed overnight at 0°C against several changes of TKM buffer containing 0.005 M β-mercaptoethanol[34,35]. In addition to the pH 5 enzyme and ribosomes, the reaction mixture also contained in a total volume of 1.0 ml: 4 μmole GSH, 2 mmole ATP (disodium salt), 0.25 mmole GTP, 20 mmole creatine phosphate, 20 μg creatine phosphokinase (EC.2.7.3.2), 5 mM $MgCl_2$, 25 mM KCl, 50 mM Tris–HCl buffer, 0.02 mM concentrations of 19 amino acids except leucine[36], and 0.1 μCi-^{14}C-leucine. Incubation was for 1 hour at 37°C. Subsequent processing of the samples after incubation, as well as the counting of radioactivity in a liquid scintillation counter (Packard Liquid Scintillation Spectrometer, Model 3003: Packard Instrument Co., Downers Grove, Illinois) was as previously reported[37].

Analysis of free amino acids

Portions of the liver were homogenised in 3 volumes (w/v) of ice-cold 0.2 M sucrose solution. The homogenate was treated with 12% sulphosalicylic acid (w/v) to a final concentration of 4%, mixed thoroughly, and centrifuged. The precipitate was washed once with 0.1 N HCl, and the combined protein-free supernatant was adjusted to pH 2.2 with 1 N HCl. Chromatographic separation of the amino acids was performed on a Beckman Automatic Analyser (Model 121, Beckman Instruments, Inc., Palo Alto, California) using Hi-Rez, DC-1A and DC-2A resins (Pierce Chemical Company, Rockford, Illinois) in the long and short columns, respectively.

Results

Studies in intact animals

General findings

The general features of young growing rats fed a 0.5% lactalbumin-containing diet for extended intervals of 2–3 months have been described in several communications[17,18,38]. Among the consistent and prominent findings in such protein–calorie deficient rats were growth failure, marked atrophy of many visceral organs including the thymus, spleen, pancreas and the major salivary glands, hypoalbuminaemia, extensive fatty metamorphosis of the liver, and distortion in free amino acid profiles in the liver and plasma. Some of these malnourished rats developed gross oedema[17,38]. Similar observations have also been noted in protein–calorie deficient guinea pigs[39] and in nonhuman primates[35]. Refeeding the malnourished rats with adequate protein diet produced marked remission of the oedema within 2 days[38], increased liver weight and liver RNA content, with equally prompt reduction in the fractional daily catabolic rate of hepatic ribosomal RNA (figure 17.1).

Sedimentation profiles of ribosomes and the effects of refeeding

Examination of the typical sedimentation patterns of membrane-bound and free ribosomal preparations (figure 17.2) showed that prolonged dietary protein deficiency produced marked disaggregation of polyribosomes ($n > 2$) with concomitant increase in the yield of the slow sedimenting ribosomes, compared to findings in rats pair-fed a restricted amount of adequate protein diet. This observation was to some extent in conflict with the findings of other investigators[40] who have reported that in long term (30 days) protein deficiency in rats, membrane-bound polyribosomes are preferentially disaggregated while the free ribosomes show a shift from light to heavy-sized classes of polysomes. No explanation can be offered for this discrepancy except perhaps for the difference in the duration of protein deficiency, and also the fact that the other study involved feeding rats a diet completely devoid of protein. A consistent finding in the profiles of free ribosomes from the malnourished rats was the presence of a large ultraviolet absorbing peak sedimenting slower than the monomer (figure 17.2). This peak, presumably 60S[41], was suggestive of the presence of a large population of native ribosomal subunits in livers of the malnourished rats. Associated with the disaggregated liver polysomes in protein malnutrition was a decrease in the proportion of functionally active ribosomes as evaluated by their stability under ionic conditions which selectively dissociated inactive ribosomes into ribosomal subunits (figure 17.3). In the control rats, 24% of the total liver ribosomes were inactive compared to 33% in the malnourished animals (table 17.1), a finding consistent with the reduced uptake of radioactive amino acid by ribosomal preparations from the latter group[18].

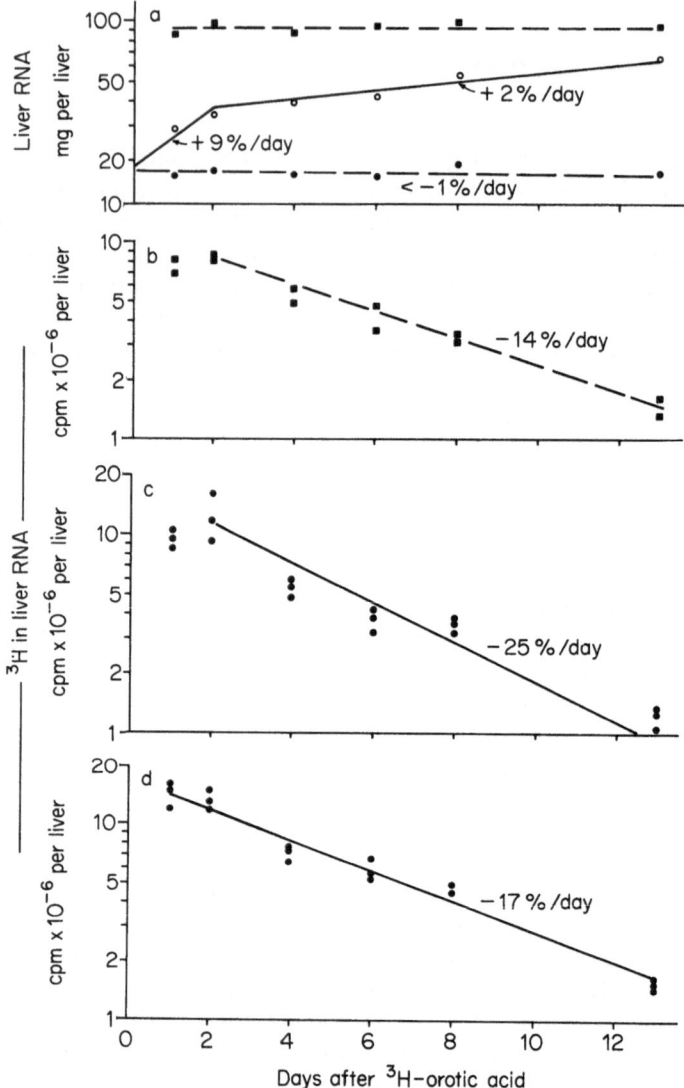

Figure 17.1. Growing rats (100–120 g) fed 0.5% lactalbumin diet for 8 weeks were each given an intraperitoneal injection of 17.5 μ Ci ^{3}H-orotic acid and randomised into two groups immediately thereafter. One group continued on the 0.5% protein diet and the second group was refed 18% lactalbumin diet *ad libitum*. Each point in figure 17.1a represents mean for two rats in adult control group and mean for three rats in the two malnourished groups. Each point in figures 1b, c, and d represents the mean of triplicate determinations for each rat. The amounts of total liver RNA (a) and of ^{3}H-RNA per liver (b, c, d) are plotted on a logarithmic scale. a: ■, control adult rats; ○, refed malnourished rats; ●, malnourished rats. b, control adult rats; c, malnourished rats; d, refed malnourished rats[18].

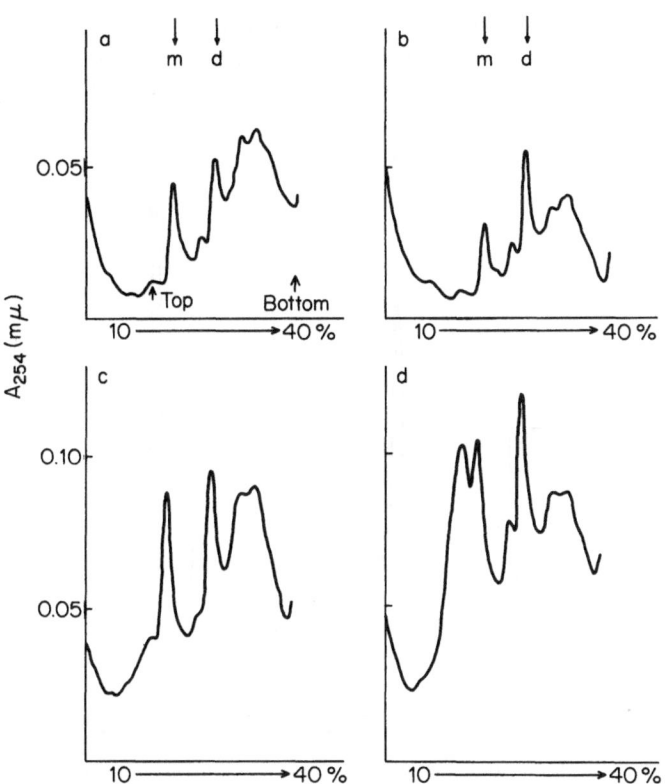

Figure 17.2. Sedimentation profiles of membrane-bound (a, b) and free (c, d) ribosomes from livers of control (a, c) and protein–calorie deficient (b, d) rats. The malnourished rats received a 0.5% protein diet for about 9 weeks while the control rats of similar age were pair-fed an adequate protein diet. Details of the separation of free and bound ribosomes on a two-step discontinuous sucrose gradient were as described in the text. Ribosomal pellet was resuspended in Tris–MgCl$_2$–KCl buffer, pH 7.6, and 0.1 ml equivalent to 8 mg wet weight of liver tissue was layered on a linear sucrose gradient (10–40%) prepared in Tris buffer of the same ionic strength. Centrifugation was for 1 h at 48000 rpm using a Beckman SW 50.1 rotor (4°C). The smaller ultraviolet absorbing areas of the bound ribosomes relative to the corresponding free ribosomes is explained by the loss of a good fraction of bound ribosomes in the nuclear pellet during preparation of the postmitochondrial supernatant fraction. m, monomer; d, dimer.

Figure 17.4 shows the absorption profiles of total hepatic ribosomes from malnourished rats (6A) refed adequate protein diet for 2 days (6B) or 4 days (6C) before autopsy. These sedimentation profiles (figure 17.4), which were representative of patterns from individual analyses of at least 3 animals in each experimental group, were isolated in low [K$^+$] medium. Each profile was prepared from 8 mg wet weight of liver tissue, and thus, the disparity between the ultraviolet absorption areas of the various groups reflected the RNA

C. O. Enwonwu

concentration in liver tissue which had previously been shown to decrease during dietary rehabilitation of malnourished rats[18]. Refeeding the protein–calorie deficient rats with adequate diet for 2 days produced some reduction

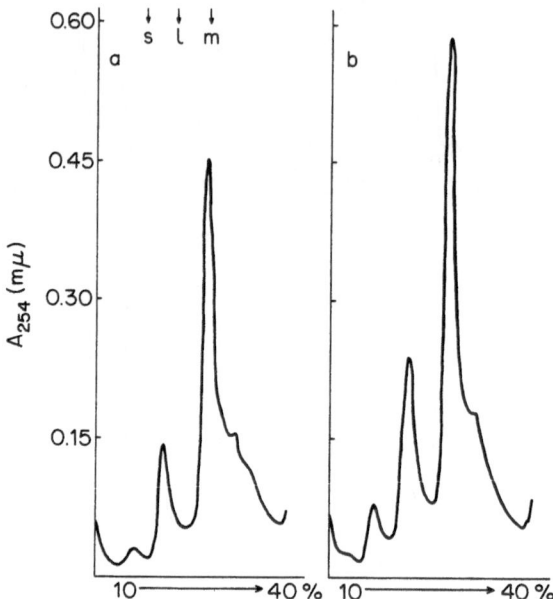

Figure 17.3. Dissociation into subunits of total liver ribosomes (bound plus free ribosomes) from control (a) and malnourished (b) rats. Ribosomes were suspended in the high $[K^+]$ buffer and incubated for 30 min at $37°C$ in the absence of supplemental factors necessary for protein synthesis. Centrifugation was in an SW 50.1 rotor at 48000 rpm for 60 min at $28°C$. m, monomer, l, large ribosomal subunit; s, small subunit.

Table 17.1. Effect of high $[K^+]$ on the stability of hepatic ribosomes from control and protein–calorie deficient rats.*

Dietary Group	Proportion of undissociated ribosomes (%)
	Total
Adequate protein	76.4 ± 0.8
0.5% protein	66.6 ± 0.7†

*Total ribosomes (free plus bound) were pelleted by centrifuging DOC-treated PMS at 50000 rpm for 2 h in a Beckman rotor Ti-50 (4°C). The pellet was resuspended in the low $[K^+]$ buffer, layered on top of 1 M sucrose solution prepared in the low $[K^+]$ buffer, and recentrifuged for 2 h at 50000 rpm. Pellet was resuspended in Tris buffer and adjusted to the required ionic strength (high K^+ medium). Incubation at 37°C for 30 min was in the *absence* of cell sap factors.

†Significantly different from corresponding control values ($p < 0.001$).

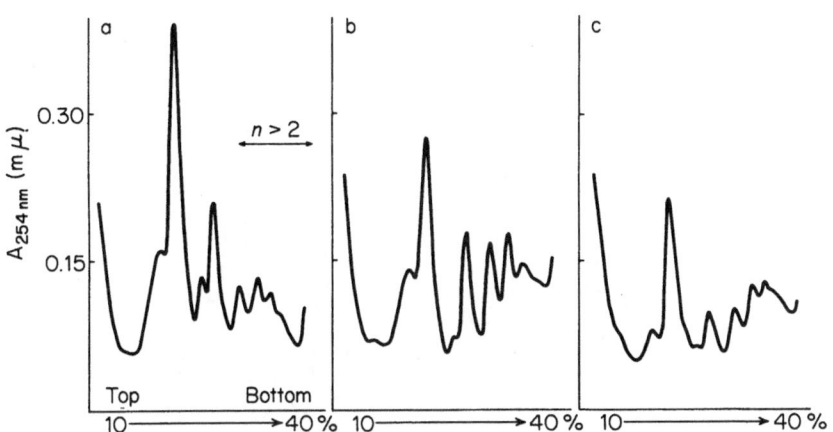

Figure 17.4. Polysome profiles of rat liver ribosomes (free plus membrane-bound). Liver from rats fed a 0.5% protein diet *ad libitum* for 9 weeks (a) and refed with Purina Rat Chow for 2 (b) or 4 days (c) were homogenised in 0.25 M sucrose solution prepared in low [K$^+$] buffer. Profiles were harvested from DOC-treated PMS. Each profile was prepared from an aliquot of PMS equivalent to 8 mg wet weight of liver tissue.

in the amount of slow sedimenting ribosomes ($n < 3$), and this feature became more marked after 4 days of refeeding.

When the ribosomal profiles were isolated in high [K$^+$] buffer following prior incubation of the PMS for 30 minutes at 37°C, a good proportion of the ribosomes sedimented slower than the monomers (figure 17.5). It must be stated that in this experiment (figure 17.5), dissociation of ribosomes was produced in the presence of autologous cell sap. The amount of the ribosomal subunits (S < 80) relative to the total ultraviolet absorbing area, varied depending on whether the malnourished rats (7A) were refed for 1 (7B), 2 (7C) or 4 days (7D) before sacrifice. It is also pertinent to point out that using PMS instead of C-ribosomes for the ribosome-dissociation studies (figure 17.5), the ratio for the absorbances of the fast (large) and slow (small) sedimenting ribosomal subunits was much lower than the theoretically expected[42] value of 2.5. No definitive explanation can be offered yet for this discrepancy. Among possible explanations are poor resolution of the subunits on the gradients, modification of the structure of ribosomal subunits during the isolation procedure, or formation[4] of artefactual 45-50S component from the 60S. Table 17.2 summarises the data obtained from several animals in each experimental group. Following 9 weeks of dietary protein–calorie deficiency, the feeding of an adequate protein diet to malnourished rats for 24 hours produced a significant increase ($p < 0.025$) in the proportion of liver ribosomes resistant to dissociation. This increase was more pronounced ($+8.5\%$) after 2 days of refeeding, and by day 4, the proportion of hepatic ribosomes resistant to the dissociating effect of high [K$^+$] buffer had returned to a value comparable to findings in animals refed for only 1 day (table 17.2).

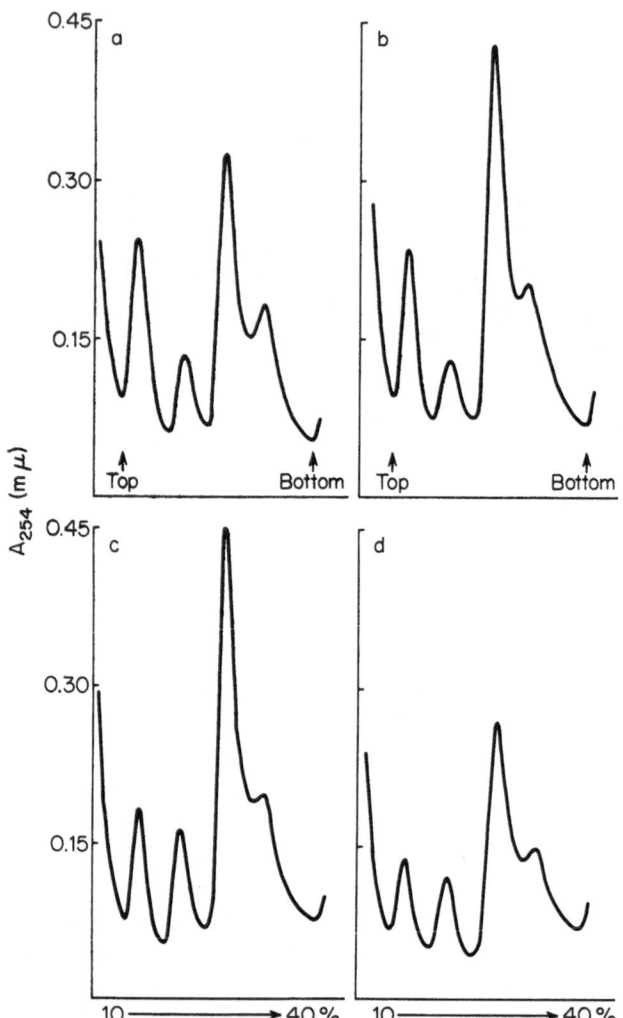

Figure 17.5. Dissociation of total ribosomes (membrane-bound and free ribosomes) from livers of rats fed a 0.5% protein diet for 9 weeks (a) and refed with Purina Rat Chow for 1 (b), 2 (c), or 4 days (d). Postmitochondrial supernatant prepared in low [K$^+$] buffer as described in the text, was adjusted to the appropriate ionic strength (50 mM Tris, pH 7.6; 800 mM KCl, 5 mM MgCl$_2$) and incubated for 30 min at 37°C. Incubation was therefore in presence of cell sap factors. An aliquot of the PMS (0.1 ml containing approximately 8 mg wet weight of liver tissue) was layered on a 10–40% sucrose gradient prepared in Tris–HCl buffer of the same high [K$^+$] content. Centrifugation was in an SW 50.1 rotor at 48 000 rpm for 60 min at 28°C.

Free amino acids in liver

Table 17.3 presents the data on liver concentrations of 17 free ninhydrin-positive substances measured in the malnourished rats, as well as the effects

Table 17.2. Resistance of ribosomes to dissociation into ribosomal subunits in high [K$^+$] medium*

Dietary group	Undissociated ribosomes† (% total profile)	P value‡
Malnourished rats—not refed (6)	61.07±0.91	—
Refed—1 day (8)	66.62±1.20	<0.025
Refed—2 days (4)	69.54±0.62	<0.001
Refed—4 days (6)	66.96±0.47	<0.001

*Rats were fed a 0.5% lactalbumin diet *ad libitum* for 9 weeks following which some of the animals were refed with adequate protein diet (Purina Rat Chow) *ad libitum* for 1, 2, or 4 days before autopsy. Dissociation of hepatic ribosomes was as described in the text. PMS rather than isolated ribosomes was used directly for the dissociation studies, and thus the ribosomes were incubated in presence of their own cell sap. Data expressed as means ±SEM. Values in parentheses represent number of animals used.

†Recovered mostly as monomers (80S) with a small dimer peak.

‡Comparison of findings in refed groups to mean value for rats not refed with Purina Rat Chow (Student *t* test).

on these substances of improving the protein quality of the diet. Refeeding the protein–calorie deficient animals with adequate diet for 4 days elicited prominent elevation (+78%) in the sum of the essential amino acids. The greatest increase (+49%) occurred within the first 24 hours of dietary rehabilitation. Unlike the pool of essential amino acids, the hepatic pool of free dietary nonessential amino acids showed an increase (+65%) during the first day of refeeding, and thereafter, returned to the level noted prior to the change in diet, thus explaining the increase in ratio of essential to nonessential amino acids noted after 2 or 4 days of refeeding (table 17.3). Accounting for most of the changes in amino acid pool during dietary rehabilitation were methionine, isoleucine, leucine, threonine, valine, tyrosine, asparagine and glutamine.

Studies in perfused livers
In studies of isolated perfused livers, several criteria were adopted for assessing the success of the perfusion. Bile flow rate was found to be rather inconsistent (0.3–0.9 ml h^{-1}), and was thus considered an unreliable index of viability of the perfused liver. All perfused livers showing blotchiness or marked swelling as determined by the dry:wet weight ratio when compared to the unperfused organ, were rejected.

Polyribosomal profiles in perfused livers
Livers of rats fed a 0.5% lactalbumin diet *ad libitum* for about 1 week showed a fairly good content of polysomes ($n>2$) even after the delay (15–20 min) in processing brought about by the surgical preparation of the organ for perfusion (figure 17.6a). Perfusion of this liver for 1 hour with diluted blood from rats fed a 0.5% protein diet for 5 weeks, and in the absence of supplemental

Table 17.3. Hepatic levels of free amino acids in malnourished rats and the immediate effects of refeeding with adequate diet§

Amino acid	Not Refed (n = 6)	Protein–Calorie-Deficient Rats		
		Refed 1 day (n = 8)	Refed 2 days (n = 4)	Refed 4 days (n = 6)
Methionine	3.06 ± 0.14	5.56 ± 0.15 (182)†	6.58 ± 0.43 (215)†	6.24 ± 0.69 (204)†
Isoleucine	12.43 ± 0.45	17.35 ± 0.61 (140)†	19.69 ± 0.09 (158)†	23.03 ± 0.75 (185)†
Leucine	15.27 ± 0.73	27.45 ± 0.91 (180)†	31.74 ± 2.94 (208)†	36.48 ± 1.64 (239)†
Threonine	63.19 ± 3.07	101.76 ± 16.52 (161)*	112.45 ± 10.25 (178)*	133.93 ± 14.04 (212)*
Valine	12.62 ± 1.19	26.45 ± 2.08 (210)†	32.60 ± 1.88 (258)†	40.81 ± 2.19 (323)†
Phenylalanine	9.20 ± 0.65	9.21 ± 0.66 (100)	9.05 ± 0.54 (98)	9.22 ± 0.43 (100)
Lysine	68.17 ± 7.78	65.69 ± 7.91 (96)	93.19 ± 9.63 (135)	107.12 ± 12.16 (157)
Histidine	71.63 ± 12.91	124.45 ± 7.84 (174)*	96.02 ± 2.97 (134)	99.28 ± 3.42 (139)†
Tyrosine	5.84 ± 0.59	11.16 ± 0.65 (191)†	12.11 ± 0.19 (207)	10.37 ± 0.67 (178)†
Aspartic acid	302.10 ± 35.77	476.45 ± 56.08 (158)	322.20 ± 17.99 (107)	363.66 ± 17.61 (120)
Serine	405.57 ± 35.12	713.94 ± 99.90 (176)	470.73 ± 178.42 (116)	296.28 ± 36.08 (73)
Asparagine-glutamine	374.48 ± 36.55	1090.74 ± 67.83 (291)†	561.68 ± 11.44 (150)*	697.00 ± 11.57 (186)†
Glutamic acid	408.92 ± 28.83	813.29 ± 93.26 (199)*	546.45 ± 51.94 (134)	329.92 ± 46.48 (81)
Glycine	299.95 ± 12.92	230.80 ± 18.86 (77)	260.33 ± 11.77 (88)	230.85 ± 17.70 (77)
Alanine	436.17 ± 38.50	455.26 ± 32.82 (104)	508.97 ± 31.94 (117)	500.95 ± 47.02 (115)
β-alanine	31.42 ± 2.77	52.93 ± 6.01 (168)	46.19 ± 6.40 (147)	38.61 ± 1.82 (123)
Phosphoethanolamine	94.60 ± 8.64	43.38 ± 1.75 (49)†	36.64 ± 1.33 (39)†	33.95 ± 1.74 (36)†
E‖	261.41	389.08 (149)	413.43 (158)	466.48 (178)
N¶	2353.21	3876.79 (165)	2753.19 (117)	2491.22 (106)
E/N (%)	11.11	10.04	15.02	18.72

§ Rats were fed a 0.5% protein diet *ad libitum* for 9 weeks when some of the animals were refed Purina Rat Chow *ad libitum* for 1, 2 or 4 days before sacrifice. Values (micromoles 100 g wet wt of liver⁻¹) represent means ± SEM. Numbers in parenthesis represent percentage of corresponding values in malnourished rats not refed with Purina Chow.

‖ Sum of dietary essential amino acids with tyrosine included.

¶ Sum of nonessential amino acids.

Symbols used

n = Number of animals examined. † = $P < 0.001$. * = $P < 0.005$. (P values between < 0.01 and < 0.05) (Means for refed groups compared to corresponding means for animals not refed with Purina Rat Chow.)

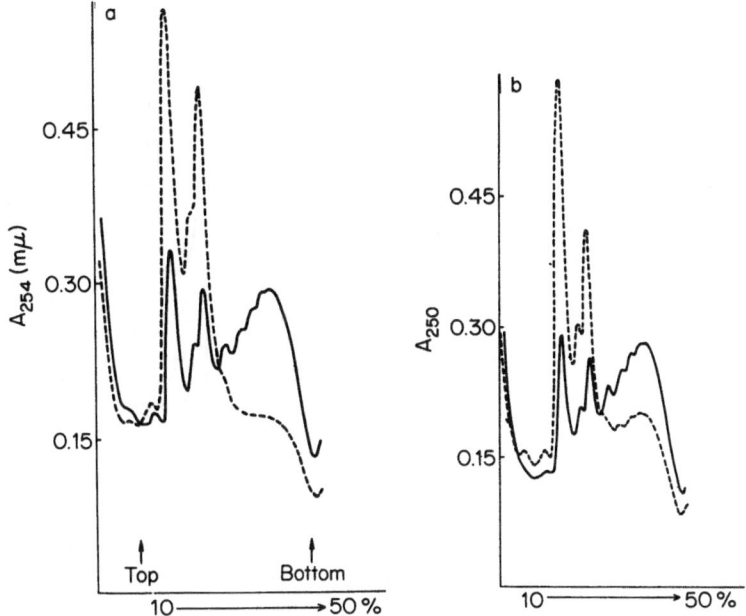

Figure 17.6. Liver ribosomal (free plus bound) profiles before (solid line) and after (dashed line) perfusion for 1 h with diluted rat blood with no supplemental factors added to the perfusate. Rats fed 0.5% protein for 5 weeks (a) or adequately fed control rats (b) served as blood donors. Liver donors were rats fed the 0.5% lactalbumin diet for about 1 week.

factors (amino acids, hormones), produced almost complete conversion of the large-sized ribosomal aggregates into slow sedimenting monomers and dimers (figure 17.6a). A similar trend was noted when the same type of liver was perfused with blood from adequately fed rats although the magnitude of polyribosomal disaggregation was slightly less pronounced (*cf.* figures 17.6a and 17.6b). These findings were in harmony with earlier reports of progressive disaggregation of liver polysomes despite the perfusion meeting many acceptable criteria of adequate liver function[43]. In our system, addition of insulin alone or in combination with cortisol tended to accelerate the disassembly of polysomes, while supplementation of the perfusate with the full complement of L-amino acids at a level five times the normal plasma concentration in rats[12] produced some stabilising effect on the polyribosomal organisation (profiles not shown). The most significant degree of stabilisation of liver polysomes following perfusion for 1 hour was observed when the perfusate was supplemented with the amino acid mixture at a level 10-fold the normal plasma concentration (figure 17.7). In the latter series of experiments, the amino acids were added to the perfusate prior to the start of the perfusion, and this was followed by two further infusions of the same amount of amino acids after 20 and 40 minutes of perfusion. Addition of tryptophan alone was without

C. O. Enwonwu

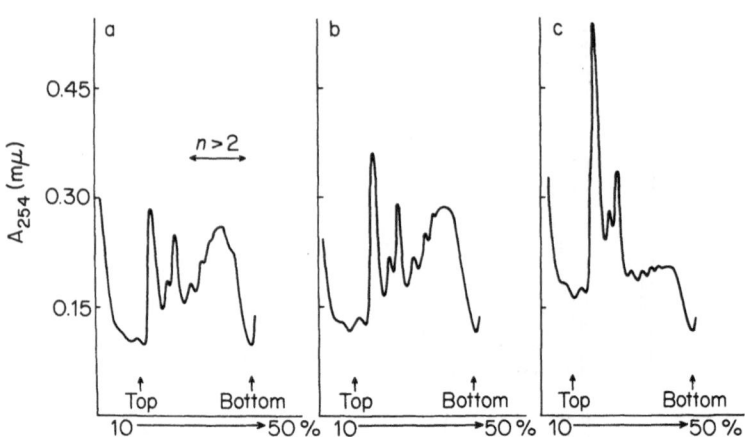

Figure 17.7. Typical polyribosomal profiles (membrane-bound and free ribosomes) of rat livers before (a) and after (b, c) perfusion for 1 h with rat blood supplemented with amino acid mixture of Jefferson and Körner[12] (b) or tryptophan alone (c) at concentrations 10 times normal plasma levels. Amino acids were added to the perfusate prior to start of perfusion and followed by two further infusions after 20 and 40 minutes. Liver donors were fed 0.5% protein for about 1 week.

much benefit (figure 17.7c), and the same was true of addition of methionine alone or in combination with tryptophan or any of the branched-chain amino acids (profiles not shown). It must be emphasised that even with supplementation of the perfusate with amino acids at ten times normal plasma levels, the yield of liver polyribosomes ($n > 2$) following 1 hour of perfusion never equalled the pre-perfusion amount from the same liver. This was noted in the sedimentation profiles of both the membrane-bound (figure 17.8) and

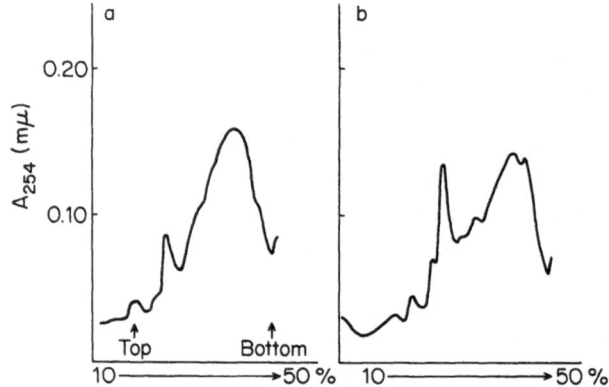

Figure 17.8. Sedimentation profiles of hepatic membrane-bound ribosomes before (a) and after (b) 1 h perfusion with blood supplemented with amino acid mixture (10 times plasma level). Liver donors were fed 0.5% protein for 1 week.

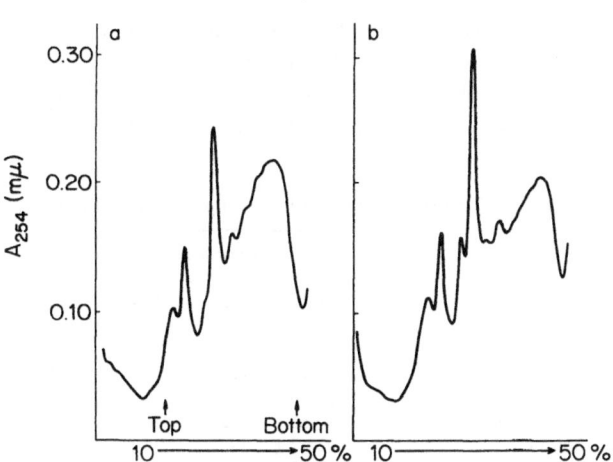

Figure 17.9. Sedimentation profiles of hepatic free ribosomes before (a) and after (b) 1 h perfusion. Other details as described in figure 17.8.

cytoplasmic-free ribosomal pools (figure 17.9), and was more pronounced when rats fed the 0.5% protein diet for about 5 weeks instead of 5 to 8 days served as liver donors. The latter findings, while in conflict with reported observations by several investigators[12,44], were nevertheless consistent with the results of Tavill *et al.*[45], showing that in the isolated perfused liver procured from normal or protein deprived rats, protein synthesis is not stimulated by supranormal amino acid supply. Table 17.4 summarises the changes in size distribution of total liver ribosomes following perfusion for 1 hour. In comparison with samples examined prior to perfusion, significant loss of polysomes ($n > 2$) occurred after perfusion in the absence of supplemental amino acids (-35%), or when the latter were added at levels five times their normal plasma con-

Table 17.4. Amount of polysomes ($n > 2$) in liver before and after perfusion*

Group	Percentage yield of polysomes ($n > 2$)†
Pre-perfusion	66.87 ± 3.26
Post-perfusion (plus amino acids)‡	61.57 ± 0.98 (92)
Post-perfusion (plus amino acids)§	53.84 ± 1.27 (81)‖
Post-perfusion (minus amino acids)	43.49 ± 2.29 (65)‖

*Total polysomes (free plus bound) were isolated from deoxycholate-treated PMS. Liver donors were fed a 0.5% protein diet for 5 days and the livers were perfused for 1 h with diluted blood from adequately fed rats. Values are expressed as means $\pm$ SE. Data were obtained from 5–7 rat livers for the various groups. Values in parentheses indicate percentage of pre-perfusion value.

†Proportion of ultraviolet absorbing area.

‡Supplemented with amino acid mixture at 10 times plasma level.

§Supplemented with amino acid mixture at 5 times plasma level.

‖Significantly different from pre-perfusion value ($P < 0.001$).

centrations (-19%). Increasing the level of amino acid mixture to 10 times the normal plasma level still produced some small but non-significant loss (-8%) in polysome yield, and this occurred regardless of the marked increase in the free amino acid pool of the perfused livers (table 17.5).

Table 17.5 Free amino acid levels in perfused rat liver*

	Experiment 1†	
Amino acid	Pre-perfusion‡ ($n = 3$)	Post-perfusion§ ($n = 2$)
Isoleucine	19.90 ± 1.41	39.21 (197)
Leucine	26.70 ± 3.01	59.35 (222)
Threonine	80.66 ± 16.42	130.59 (162)
Valine	22.42 ± 5.88	59.38 (265)
Phenylalanine	11.33 ± 0.89	24.07 (212)
Lysine	78.37 ± 23.05	128.06 (163)
Histidine	72.47 ± 11.79	63.38 (87)
Tyrosine	8.35 ± 0.21	29.72 (356)
Aspartic acid	279.33 ± 75.76	206.46 (74)
Serine	305.96 ± 87.99	348.40 (114)
Asparagine-glutamine	787.69 ± 174.10	331.18 (42)
Glutamic acid	158.41 ± 38.80	197.72 (125)
Glycine	374.81 ± 73.39	411.61 (110)
Alanine	643.29 ± 173.00	904.74 (141)
β-alanine	26.37 ± 5.50	54.71 (207)
Citrulline	12.63 ± 4.45	8.67 (69)
Taurine	152.35 ± 69.55	174.67 (115)

*Values expressed as μmole 100 g wet weight of liver^{-1}.
†Rats fed 0.5% protein diet for 5–8 days.
‡Values are expressed as means $\pm$ SD.
§Livers were perfused for 1 h with diluted blood from un-fasted well-fed animals and perfusate was supplemented with amino acids at 10 times plasma level. Values in parentheses refer to percentage of corresponding pre-perfusion levels.

Dissociation of ribosomes into subunits

The effect of a high concentration of KCl on the isolated liver ribosomes was determined by analysis of their sedimentation profiles in continuous sucrose gradients. When ribosomal preparations, incubated at 37°C for 30 minutes in the absence of factors necessary for protein synthesis, were exposed to 800 mM KCl, ribosomal subunits (presumably 40S and 60S) were formed but a large proportion of the ribosomal preparation was recovered mainly as un-dissociated 80S particles (figure 17.10). As shown in this figure, perfusion of isolated livers even in the presence of added amino acids (10-fold plasma level) reduced the proportion of total liver ribosomes resistant to the dissociating effect of a high [K$^+$] medium. This was a consistent finding regardless of whether total, bound or free ribosomes were used in the study. Data from several experiments are summarised in table 17.6 and show that perfusion in

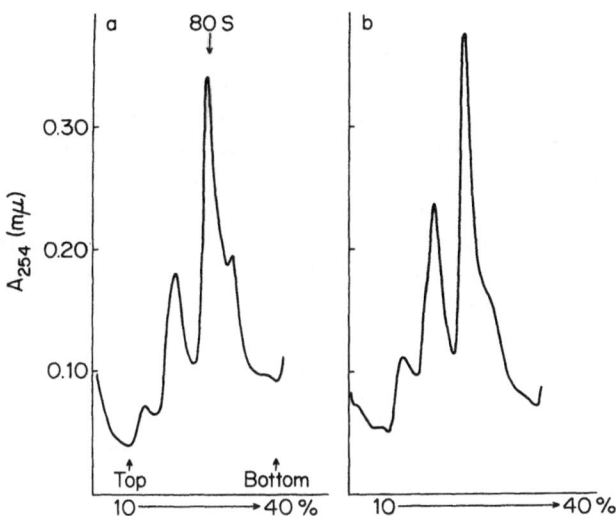

Figure 17.10. Dissociation into ribosomal subunits of rat liver ribosomes (free plus bound) before (a) and after (b) perfusion for 1 h with rat blood supplemented with amino acid mixture (10 times plasma level). Details of isolation of the profiles were as described in figure 17.3.

Table 17.6 Effect of perfusion on resistance of hepatic ribosomes to dissociation into ribosomal subunits in high [K$^+$] medium*

Group	Undissociated ribosomes‡ (% total profile)	P value§
Pre-perfusion	71.70 ± 1.79	—
Post-perfusion (minus amino acids)†	58.17 ± 2.62	< 0.001
Post-perfusion (plus amino acids)	66.78 ± 4.41	Not significant

*Livers were obtained from rats fed a 0.5 per cent protein diet for 5 to 8 days. Perfusion was for 1 h with diluted whole blood from unfasted rats fed Rat Purina Chow *ad libitum*. Total ribosomes (free and membrane-bound) were isolated by centrifuging DOC-treated PMS at 50000 rpm for 2 h. Pellet was resuspended in Tris buffer and [K$^+$] content adjusted to 800 mM. Other details are as described in Experimental Procedure. Data based on findings in 4 to 6 rats for the various groups. Values are expressed as means ± SEM.

†Amino acid mixture added to perfusate at 10 times plasma level as described in Experimental Procedure.

‡Recovered mostly as monomers (80S) with small dimer shoulder.

§Comparison of findings to mean value for pre-perfusion (Student *t* test).

the absence of supplemental amino acids was associated with the highest yield of non-functional ribosomes as exemplified by the proportion of undissociated ribosomes.

Uptake of L-^{14}C-*leucine in vitro*
Protein biosynthesis in a cell-free system was studied using equivalent amounts

of ribosomes obtained from livers before and after perfusion. Livers of adequately-fed rats served as the source of pH 5 enzyme and this was prepared from postmicrosomal supernatants following molecular sieving through Sephadex G25 and overnight dialysis against Tris buffer. Using both membrane-bound and free ribosomes, perfusion without supplemental amino acids was associated with decreased incorporation of ^{14}C-leucine (-34 to -44%) compared to pre-perfused liver samples (table 17.7). Perfusion with added amino acids markedly improved activity in an *in vitro* system although the activity never equalled the level obtained in ribosomal preparations from intact, unperfused livers (table 17.7).

Table 17.7 Incorporation of L-[U-^{14}C] leucine into protein by liver ribosomal-cell sap preparation*

Source of ribosomes	Specific radioactivity of ribosomal protein cpm mg^{-1}‡
Membrane-bound ribosomes	
Pre-perfusion	749.0 ± 49.9
Post-perfusion (plus amino acids)	598.4 ± 20.2 (79.9)†
Post-perfusion (no amino acids)	421.6 ± 29.8 (56.3)†
Free ribosomes	
Pre-perfusion	475.5 ± 60.35
Post-perfusion (plus amino acids)	426.0 ± 97.65 (89.6)
Post-perfusion (no amino acids)	312.4 ± 37.90 (65.7)†

*Data based on 3 animals for both the membrane-bound and free ribosomes. Livers were obtained from rats fed a 0.5% protein diet for 5 days. Perfusion was for 1 h with diluted whole blood from adequately fed rats. When indicated, the perfusate was supplemented with amino acid mixture (10 times plasma concentration) as indicated in the text. Values in parentheses indicate percentage of corresponding pre-perfusion values.
†Significantly different from pre-perfusion value ($P < 0.01$ or 0.025).
‡Expressed as means $\pm$ SE.

Conclusion

In the present study, on the basis of other reports[5,33], we have assumed that the extent of dissociation of hepatic ribosomes into subunits by a high [K$^+$] medium is a good measure of the number of inactive ribosomes at the time of cell fractionation. This assumption is well supported by studies[31] which show that hepatic ribosomes from diabetic rats—which are known to be less effective in protein synthesis than ribosomes from normal animals—are also more susceptible to dissociation into subunits by high concentrations of salts and elevated temperature. Similar observations have also been noted in livers of starved mice[46]. In our intact protein–calorie deficient animals, ribosomal activity (table 17.2) attained a maximum within the first 48 hours of refeeding with adequate diet, and by day 4, had returned to the level noted on day 1 of refeeding. The latter pattern of change, although difficult to explain on the basis of available data, was nevertheless consistent with the reports[2,47] that on refeeding protein-deficient rats with adequate protein diet the rate of biosynthesis of serum albumin accelerates within the first day up to or in excess of levels usually encountered in normal animals, and returns to control values on the third day.

Associated with the rapid change in ribosomal activity in the intact animals was an equally rapid alteration in liver free amino acid profile (table 17.3).

In the present study, refeeding an adequate protein diet produced within 24 hours, 49 and 65% increases in the free essential and nonessential amino acid pools respectively (table 17.3), thus suggesting that reutilization of amino acids released by catabolism of liver proteins, which is an effective adaptive mechanism during short-term deprivation[48], does not meet the needs of the cells during prolonged deprivation. Although other workers[49] have also shown a good correlation between hepatic protein synthesis in rats and alterations in the intracellular free amino acid pool, details of the precise mechanisms whereby the latter regulates the former are still not fully elucidated. Among the possible crucial points of control of protein synthesis by amino acid supply is the regulation of activities of transferases I and II (aminoacyl-tRNA binding enzyme and translocation factor).

Our studies utilised rats fed a protein-deficient diet for about one to five weeks as the main source of liver donors. In this system, perfusion with diluted rat blood from either malnourished or adequately fed rats produced extensive disaggregation of the liver ribosomes after 1 hour (figure 17.6). Supplementation of the perfusate with single amino acids such as methionine and tryptophan, or any of these two in combination with any one of the branched chain amino acids, was not capable of preventing the disassembly of the polysomes. This observation was in conflict with reports indicating specific roles for tryptophan[44,50] and methionine[44] in isolated perfused livers. It must be stated that many of the latter studies utilised livers from fed or overnight fasted animals and it is possible that under such conditions, the amino acid tryptophan might be quantitatively most limiting in the liver. Evidence against a unique role for tryptophan is provided by other studies *in vivo* in which amino acids other than tryptophan have been made most limiting[51,52]. In the present study, supplementation of the perfusate with amino acid mixture[12] at a level 10 times the normal plasma concentration, produced maximal stabilisation of the polysomal organisation after 1 hour of perfusion (table 17.4). The functional capability of the livers perfused for 1 hour with fluid adequately supplemented with amino acids, whether assessed by size of polyribosomes (table 17.4), resistance of the ribosomes to dissociation in high $[K^+]$ medium (table 17.6), or *in vitro* uptake of ^{14}C-leucine (table 17.7), never equalled the level noted in pre-perfused livers. The observed lower capacity of the perfused malnourished liver for protein synthesis is in good accord with the report[45] that synthetic rates of plasma proteins in perfused livers are about 80% of the rates noted *in vivo*. Similar observations have also been reported by Levitan and Webb[43] in connection with hydrocortisone-mediated induction of tyrosine transaminase *in vivo* and in the perfused liver.

Findings in the present study thus lead to the conclusion that at least in the perfused livers procurred from protein deficient rats, the role of amino acid supply should be viewed more in terms of reduction in the rate of disaggregation of polyribosomes rather than in the promotion of ribosomal reaggregation. This is consistent with earlier reports that reduction in the availability of amino acids produces a shift toward accumulation of monomers and subunits[23] leading to accelerated catabolism of RNA[18,26]. Data from the present study still leave unanswered the precise role of dietary amino acids in the enhanced hepatic ribosomal reaggregation and activity noted in malnourished rats, immediately following refeeding with adequate protein (figures 17.4 and 17.5).

Findings in the intact malnourished rats refed adequate protein, and in the isolated perfused livers from the protein–calorie-deficient rats, when considered together, suggest very strongly that in malnourished rat livers, transcriptional defects are probably more pronounced than the impairment of translation at the cytoplasmic level. This conclusion is consistent with the views of Tavill *et al.*[45] and the observation that the response of isolated perfused livers from 6-day fasted rats to full supplementation with amino acids and hormones was poor for albumin and α_2-(acute phase)globulin, and that partial recovery of this response was evident only if the 6-day fasted liver donors were refed for 37–47 h before sacrifice[53].

Acknowledgement

Most of the work reported in this paper was supported by United States Public Health Service Grant DE-02600-04.

References

1. Kirsch, R. E., Frith, L., Black, E. G. and Hoffenberg, R. Regulation of albumin synthesis and catabolism by alteration of dietary protein. *Nature* **217** (1968), 578

2. Morgan, E. H. and Peters, T. Jr. The biosynthesis of rat serum albumin. V. Effect of protein depletion and refeeding on albumin and transferrin synthesis. *J. Biol. Chem.*, **246** (1971), 3500

3. Fawcett, D. W. (1966) *The Cell: Its Organelles and Inclusions*. W. B. Saunders Co., Philadelphia

4. Martin, T. E. A simple general method to determine the proportion of active ribosomes in eukaryotic cells. *Exptl. Cell Res.*, **80** (1973), 496

5. Zylber, E. A. and Penman, S. The effect of high ionic strength on monomers, polyribosomes, and puromycin-treated polyribosomes. *Biochim. Biophys. Acta*, **204** (1970), 221

6. Falvey, A. K. and Staehelin, T. Structure and function of mammalian ribosomes. I. Isolation and characterisation of active liver ribosomal subunits. *J. Mol. Biol.*, **53** (1970), 1

7. Rogers, Q. R. and Harper, A. E. Amino acid diets and maximal growth in the rat. *J. Nutr.*, **87** (1965), 267

8. Miller, L. L., Bly, C. G., Watson, M. L. and Bale, W. F. The dominant role of the liver in plasma protein synthesis: a direct study of the isolated perfused rat liver with the aid of lysine-ε-C^{14}. *J. Exp. Med.*, **94** (1951), 431

9. Miller, L. L. (1973) Technique of isolated rat liver perfusion. In: *Isolated Liver Perfusion and Its Applications*, I. Bartošek, A. Guaitani and L. L. Miller, eds., Raven Press, New York, p. 11

10. Ross, B. D. (1972) *Perfusion Techniques in Biochemistry*. Clarendon Press, Oxford.

11. Seglen, P. O. and Jervell, K. F. A simple perfusion technique applied to glucocorticoid regulation of tryptophan oxygenase turnover and bile production in the isolated rat liver. Hoppe–Seyler's *Z. physiol. Chem.*, **350** (1969), 308

12. Jefferson, L. S. and Korner, A. Influence of amino acid supply on ribosomes and protein synthesis of perfused rat liver. *Biochem. J.*, **111** (1969), 703

13. Clemens, M. J. and Korner, A. Amino acid requirement for the growth-hormone stimulation of incorporation of precursors into protein and nucleic acids of liver slices. *Biochem. J.*, **119**, (1970), 629

14. Anthony, L. E. and Faloona, G. R. Plasma insulin and glucagon levels in protein-malnourished rats. *Metabolism*, **23** (1974), 303

15. Anthony, L., Geller, S. and Edozien, J. C. Liver protein synthesis in chronic protein–calorie malnutrition. *Fed. Proc.*, **28** (1968), 756

16. Enwonwu, C. O. and Jacobson, K. Relation between adrenal cortex and hepatic protein synthesis in protein–calorie-deficient rats. *J. Nutr.*, **103** (1973), 290

17. Enwonwu, C. O. Distribution of free and membrane-bound ribosomes in livers of protein–calorie-deficient rats. *Lab. Invest.*, **26** (1972), 626

18. Enwonwu, C. O. and Sreebny, L. M. Studies of hepatic lesions of experimental protein–calorie malnutrition in rats and immediate effects of refeeding an adequate protein diet. *J. Nutr.*, **101** (1971), 501

19. Enwonwu, C. O. and Sreebny, L. M. Experimental protein–calorie malnutrition in rats. Biochemical and ultrastructural studies. *Exptl. Mol. Pathol.* **12** (1970), 332

20. Drysdale, J. W. and Munro, H. N. Polysome profiles obtained from mammalian tissues by an improved procedure. *Biochim. Biophys. Acta*, **138** (1967), 616

21. Sarma, D. S. R., Reid, I. M., Verney, E. and Sidransky, H. Studies on the nature of attachment of ribosomes to membranes in liver. I. Influence of ethionine, sparsomycin, carbon tetrachloride, and puromycin on membrane-bound polyribosomal disaggregation and on detachment of membrane-bound ribosomes. *Lab. Invest.*, **27** (1972), 39

22. Blobel, G. and Potter, V. R. Studies on free and membrane-bound ribosomes in rat liver. I. Distribution as related to total cellular RNA. *J. Mol. Biol.*, **26** (1967), 293

23. Wunner, W. H., Bell, J. and Munro, H. N. The effect of feeding a tryptophan-free amino acid mixture on rat-liver polysomes and ribosomal ribonucleic acid. *Biochem. J.*, **101** (1966), 417

24. Enwonwu, C. O. Restitution of secretory proteins as reflected by changes in polyribosomal organisation in salivary glands of rats treated with isoproterenol. *Lab. Invest.*, **27** (1972) 199

25. Enwonwu, C. O. and Munro, H. N. Changes in liver polyribosome patterns following administration of hydrocortisone and actinomycin D. *Biochim. Biophys. Acta*, **238** (1971), 264

26. Enwonwu, C. O. and Munro, H. N. Rate of RNA turnover in rat liver in relation to intake of protein. *Arch. Biochem. Biophys.*, **138** (1970), 532

27. Wunner, W. H. The time sequence of RNA and protein synthesis in cellular compartments following an acute dietary challenge with amino acid mixtures. *Proc. Nutr. Soc.* (England), **26** (1967), 153

28. Ekren, T., Jervell, K. F. and Seglen, P. O. Insulin and amino acid regulation of polysomes in perfused, diabetic rat liver. *Nature New Biology*, **229** (1971), 244

29. Reader, R. W. and Stanners, C. P. On the significance of ribosome dimers in extracts of animal cells. *J. Mol. Biol.*, **28** (1967), 211

30. Faber, A. J. and Tamaoki, T. Isolation of active ribosomal subunits from L5178Y mouse lymphoma cells. *Arch. Biochem. Biophys.*, **149** (1972), 289

31. Pilkis, S. J. and Korner, A. Effect of diabetes and insulin treatment on protein synthetic activity of rat liver ribosomes. *Biochim. Biophys. Acta*, **247** (1971), 597

32. Enwonwu, C. O. Biochemical and morphologic changes in rat submandibular gland in experimental protein–calorie malnutrition. *Exptl. Mol. Pathol.*, **16** (1972), 244

33. Martin, T. E. and Wool, I. G. Active hybrid 80S particles formed from subunits of rat, rabbit and protozoan (*Tetrahymena pyriformis*) ribosomes. *J. Mol. Biol.*, **43** (1969), 151

34. Enwonwu, C. O. Alterations in ninhydrin-positive substances and cytoplasmic protein synthesis in the brains of ascorbic acid deficient guinea pigs. *J. Neurochem.*, **21** (1973), 69

35. Enwonwu, C. O., Stambaugh, R. V. and Jacobson, K. L. Protein-energy deficiency in nonhuman primates: biochemical and morphological alterations. *Am. J. Clin. Nutr.*, **26** (1973), 1287

36. Baliga, B. S., Pronczuk, A. W. and Munro, H. N. Regulation of polysome aggregation in a cell-free system through amino acid supply. *J. Mol. Biol.*, **34** (1968), 199

37. Enwonwu, C. O. and Glover, V. Alterations in cerebral protein metabolism in the progeny of protein–calorie-deficient rats. *J. Nutr.*, **103** (1973), 61

38. Edozien, J. C. Experimental kwashiorkor and marasmus. *Nature*, **220** (1968), 917

39. Enwonwu, C. O. Experimental protein–calorie malnutrition in the guinea pig and evaluation of the role of ascorbic acid status. *Lab. Invest.*, **29** (1973), 17

40. Gaetani, S., Massotti, D. and Spadoni, M. A. Studies of dietary effects on free and membrane-bound polysomes in rat liver. *J. Nutr.*, **99** (1969), 307

41. Reiss, U. and Tappel, A. L. Decreased activity in protein synthesis systems from liver of vitamin-E-deficient rats. *Biochim. Biophys. Acta*, **312** (1973), 608

42. Storb, U. and Martin, T. E. Number and activity of free and membrane-bound spleen ribosomes during the course of the immune response. *Biochim. Biophys. Acta*, **281** (1972), 406

43. Levitan, I. B. and Webb, T. E. Regulation of tyrosine transaminase in the isolated perfused rat liver. *J. Biol. Chem.*, **244** (1969), 4684

44. McGown, E., Richardson, A. G., Henderson, L. M. and Swan, P. B. Effect of amino acids on ribosome aggregation and protein synthesis in perfused rat liver. *J. Nutr.*, **103** (1973), 109

45. Tavill, A. S., East, A. G., Black, E. G., Nadkarni, D. and Hoffenberg, R. (1973) Regulatory factors in the synthesis of plasma proteins by the perfused rat liver. In: *Protein Turnover*. Ciba Foundation Symposium 9 (New Series). Elsevier, Amsterdam.

46. Norman, M., Gamulin, S. and Clark, K. The distribution of ribosomes between different functional states in livers of fed and starved mice. *Biochem. J.*, **134** (1973), 387

47. Horie, Y. and Ashida, K. Stimulation of hepatic protein synthesis in rats fed an adequate protein diet after a low protein diet. *J. Nutr.*, **101** (1971), 1319

48. Gan, J. C. and Jeffay, H. Origins and metabolism of the intracellular amino acid pools in rat liver and muscle. *Biochim. Biophys. Acta*, **148** (1967), 448

49. Wannamacher, Jr., R. W., Ribosomal RNA synthesis and function as influenced by amino acid supply and stress. *Proc. Nutr. Soc.* (England), **31** (1972), 281

50. Oratz, M., Rothschild, M. A., Burks, A., Mongelli, J. and Schreiber, S. S. (1973) The influence of amino acids and hepatotoxic agents on albumin synthesis, polysomal aggregation and RNA turnover. In: *Protein Turnover*. Ciba Foundation Symposium 9 (New Series). Elsevier. Amsterdam

51. Pronczuk, A. W., Rogers, Q. R. and Munro, H. N. Liver polysome patterns of rats fed amino acid imbalanced diets. *J. Nutr.*, **100** (1970), 1249

52. Ip, C. C. Y. and Harper, A. E. Effect of threonine supplementation on hepatic polysome patterns and protein synthesis of rats fed a threonine-deficient diet. *Biochim. Biophys. Acta*, **331** (1973), 251

53. Miller, L. L. and John, D. W. (1970) Nutritional, hormonal and temporal factors regulating net plasma protein biosynthesis in the isolated perfused rat liver. In: *Plasma Protein Metabolism: Regulation of Synthesis, Distribution and Degradation*, M. R. Rothschild and T. Waldmann, eds., Academic Press, New York, p. 207

Discussion

ROTHSCHILD

I think this points up a question which was brought up by Dr Munro. Dealing with the malnourished preparation as opposed to a starved one, I think this indicates very beautifully the reaction difference of an animal suffering *in vivo* from a whole host of interrelated factors produced by malnutrition from just the starved preparation.

It is unlikely to expect this preparation to respond as rapidly or as effectively as a short term starved preparation. I think this is a very beautiful demonstration of the correlation between the electron-microscopic pictures of polysomal disaggregation and the biochemical phenomena.

TAVILL

I note in your introduction the low protein diet on which you were able to keep your animals for quite a long period of time. It was our experience that we could keep animals alive on a 5% casein diet for not longer than about 3 weeks. And just to recapitulate our results, we to some extent agree about the following. We certainly found a great depletion of the overall polysome population, but we could not detect disaggregation as such. The problem was that our total recovery of ribosomes was extremely low in that situation.

ENWONWU

I reviewed some of the profiles which you have described and reported. It appeared to me that most of the studies were based on C-ribosomes. When you prepare C-ribosomes on a discontinuous gradient, not all the ribosomes will sediment through the 2 M sucrose solution; a lot of ribosomes remain trapped. What is happening is that you are recovering only a proportion of the total ribosomes in the supernatant, and most of the ribosomes that are trapped are actually the monosomes and disomes which are the best indicators of the changes in polysome profile. When we use C-ribosomes, we may get profiles which could be very difficult to interpret, but if we use the total ribosomes, the pictures are completely different. This is my guess as to the possible reason of the differences between your findings and ours. Another question which you raise is: How long one could keep the animals on a low protein diet? As a matter of fact, we routinely killed our animals after about two months, but I know some of the animals have survived up to $3\frac{1}{4}$ months on an 0.5% lactalbumin diet. In fact, one of my graduate students has succeeded in keeping his animals for about 4 months.

TAVILL

Just to come back on one point. We are certainly using the C-ribosome fraction, but we were able to detect monosomes and disomes which occur in response to glucagon treatment. So we were able to detect them, although we recognise the fact that one does lose quite a bit in this discontinuous gradient system.

JAMES

Just a comment. Of course, the rat is one of the most difficult animals, and the differences between the two groups might relate to the age at which you decided to put the animals on a diet, rather than to the actual nature of the protein itself.

Secondly, I think there may well be changes in total RNA and what one is really measuring is the fractional synthetic rate, if one can speak in such terms. Our data using constant infusion certainly suggest that in protein depleted animals there is a normal fractional synthetic rate for endogenous liver proteins; actually it tends to be raised. This I think would fit in with some of the data that have been presented in this session.

18

The use of a continuous infusion of ^{15}N for the study of human nitrogen metabolism

T. P. STEIN, M. J. LESKIW, H. W. WALLACE
and W. S. BLAKEMORE

Investigation of the decay curve

Noninvasive techniques are necessary for studying human protein metabolism. Because it is nonradioactive, ^{15}N has frequently been used as a tracer in such experiments. Infusing a ^{15}N-labelled amino acid for a suitable period of time can lead to an approximate isotopic steady state in certain nitrogen pools. In the absence of isotope re-entry, a true steady state should result. Usually there is some re-entry of isotope, so only an approximate isotopic steady state is observed. Methods based on the measurement of an isotopic steady state give the most reliable values for the overall rate of human protein synthesis[1,2]. A disadvantage of this approach as compared to the single injection procedure ('pulse label') is that it is not amenable to a compartmental analysis. The advantages of such an approach where only limited sampling is possible have been summarised by Dietchy and Wilson[3] in their review of cholesterol metabolism. Given a valid model and a reliable method of determining the compartmental parameters, it is possible to assess pharmacological, physiological and nutritional effects on cholesterol (nitrogen) metabolism. For these reasons it seemed worthwhile to develop a reliable compartmental method for the study of human N metabolism.

In the pulse label method the size of the metabolic N pool and its rate of turnover are calculated from the specific activity of the urinary urea during the first 2 hours after a single injection of a ^{15}N-labelled amino acid, usually glycine. The pulse labelling method depends on several assumptions, which other authors[1,4-7] have discussed in detail elsewhere. They are (i) that the metabolic N pool is homogeneous; (ii) that labelled material disappears from the pool according to a first-order reaction; and (iii) that mixing and nitrogen interchange rates are not rate limiting so that the specific activity of the urea does indeed represent that of the mixed amino acids in the metabolic N pool. The first two assumptions are undoubtedly oversimplifications; the third has been shown by Wu *et al.*[5-7] to be wrong, thereby making the interpretation of such experiments ambiguous.

Our approach requires that the rapidly turning over metabolic N pool attain a steady state by the time the ^{15}N amino acid infusion is terminated. This is the crucial assumption of our procedure. When the ^{15}N infusion is stopped, the ^{15}N decays out of the metabolic N pool into the tissue and excretory N pools via k_u and k_s (defined below). By determination of enough ^{15}N enrichment values at different times, the ^{15}N excretion curve can be accurately defined and the parameters of the Rittenberg–San Pietro scheme determined.

Our analysis is based on the Rittenberg–San Pietro 3-pool model as modified by Olesen *et al.*[8] to allow for a 'blind' tissue pool (figure 18.1). The three measurable nitrogen pools are the tissue protein pool (T), the metabolic nitrogen pool (P), and the excretory pool (U). We define λ_t^T, λ_t^P and λ_t^U as the amount of excess ^{15}N in pools T, P, and U at time t; $t = 0$ is the time at which the ^{15}N infusion is terminated.

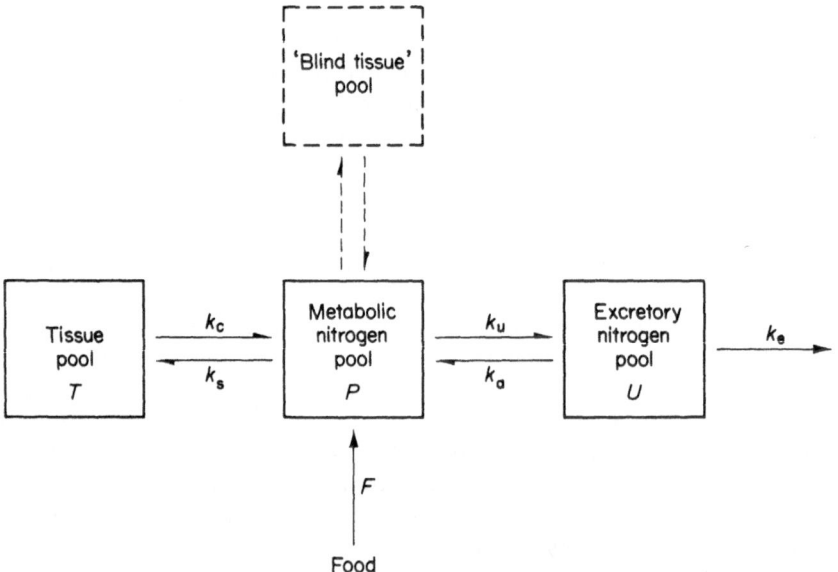

Figure 18.1. The Rittenberg–San Pietro scheme as modified by Olesen *et al.*[8] to include a 'blind tissue' pool. This consists of the relatively non-labile proteins of the body. *T* is the tissue pool that can be measured by the ^{15}N technique and consists of the labile proteins. *P* is the metabolic nitrogen pool. *U* is the excretory N pool consisting of urea and ammonia. k_s, k_c, k_u, k_e, and k_a are the rate constants for protein synthesis, tissue protein catabolism, urea plus ammonia synthesis from the metabolic N pool to the excretory N pool, the rate of urea plus ammonia excretion from the pool N and the rate of urea N recycling from pool *U* to pool *P* respectively. *F* is the rate of input of dietary N.

Let k, k_c, k_u, k_e, and k_a be the rate constants for new protein synthesis, protein breakdown, excretory nitrogen synthesis, excretory nitrogen excretion in the urine, and urea recycling by the intestinal fauna, respectively.

The following differential equations apply to the Rittenberg–San Pietro scheme.

$$\frac{d\lambda_t^T}{dt} = k_s \lambda_t^P - k_c \lambda_t^T \tag{1}$$

$$\frac{d\lambda_t^P}{dt} = k_c \lambda_t^T - (k_s + k_u)\lambda_t^T + k_a \lambda_t^T \tag{2}$$

$$\frac{d\lambda_t^U}{dt} = k_u \lambda_t^P - (k_e + k_a)\lambda_t^U \tag{3}$$

Shipley and Clarke[9] give the method for solving these three differential equations

$$\lambda_t^U = Ae^{-g_1 t} + Be^{-g_2 t} + Ce^{-g_3 t} \tag{4}$$

Since we are measuring specific activity (SA), the constants in the above equation are altered proportionately, as we assume that SA is proportional to λ^T, λ^P, and λ^U. Therefore,

$$SA \times \lambda_t^U = ae^{-g_1 t} + be^{-g_2 t} + ce^{-g_3 t} \tag{5}$$

Computer fitting of the experimental specific activity data to a three-exponent curve gives g_1, g_2, g_3, a, b, and c. Two other pieces of information are required for a complete solution of the scheme.

[15]N decay curve analysis provides no direct measure of the recycling of urea nitrogen through excretion into the gastrointestinal tract, leading to hydrolysis of the urea by intestinal microorganisms and subsequent reabsorption of the ammonia released. The evidence for a high degree (approximately 25%) of recycling comes mainly from studies in which the nitrogen load is relatively low[10,11]. Therefore, the effect may not be as marked in our studies as in others. If we assume that the bulk of the excretory pool N is a urea, then the ratio of urea hydrolysed to urea excreted can be approximated to k_a/k_e. In the absence of adequate data on normal subjects, we assume a value of $k_a/k_e = 0$. Under approximate isotopic steady-state conditions the total amount of isotope entering pool P is equal to the amount leaving this pool: the amount of isotope infused is given by

$$(k_u + k_s)\lambda_t^P \tag{6}$$

and the amount of isotope excreted via the urine is given by

$$k_u \lambda_t^P \tag{7}$$

Dividing equation 7 by 6 we obtain

$$\frac{^{15}\text{N recovered in urine}}{\text{total }^{15}\text{N infused}} = \frac{k_u}{k_u + k_s} \tag{8}$$

Having numerical values for k_a/k_e and $k_u/(k_u + k_s)$ permits a complete solution of the previous equations.

Pool sizes are calculated as follows:

By putting $t = 0$ in equation (4), it reduces to

$$\text{SA}^U_{t=0} = a + b + c \tag{9}$$

and the isotopic enrichment of pool U at $t = 0$ is known. The specific activity (excess ^{15}N) is given by the following equation, from which U can be found.

$$(1 + k_a/k_e)(\lambda^U_{t=48})/U = \text{SA}^U_{t=0} \tag{10}$$

where $(1 + k_a/k_e)$ is the hydrolysis correction and $\lambda^U_{t=48}$ is the amount of excess ^{15}N recovered in the urine up to 48 hours after termination of the infusion.

For a steady state in pool U

$$(k_e + k_a)U = k_u P \tag{11}$$

from which pool P can be calculated. The rate of protein synthesis is given by $k_s P$, which can be determined since k_s and P are both known.

For a subject in N balance, the steady state argument can be applied to pool T

$$k_s P = k_c T \tag{12}$$

and a value of T can be found.

We have completed two studies, one on a 42-year-old male and the other on a 15-year-old female. Figure 18.2 shows a urinary ^{15}N decay curve for one of the studies and the fit to a one-, two-, and three-exponential equation. The calculated rate of protein synthesis for the 42-year-old male is 25.9 mg N kg body wt^{-1} h^{-1}, and that for the 16-year-old female is 16.0 mg N kg body wt^{-1} h^{-1}.

Eight assumptions are made in the calculation:

(1). After administration, ^{15}N is rapidly distributed throughout the metabolic pool in the same pattern as unlabelled N.

(2). The specific activity of the urinary N is proportional to the amount of ^{15}N excreted.

(3). The overall processes of the model can be approximately described by first-order kinetics.

(4). The use of total urinary ^{15}N specific activity rather than fractionating the urine into ammonia and urea and analysing them separately.

(5). Pool sizes and rates are constant during the experiment. After the isotope infusion has proceeded for 6 to 8 hours, an approximate isotopic steady state is attained in the metabolic N pool.

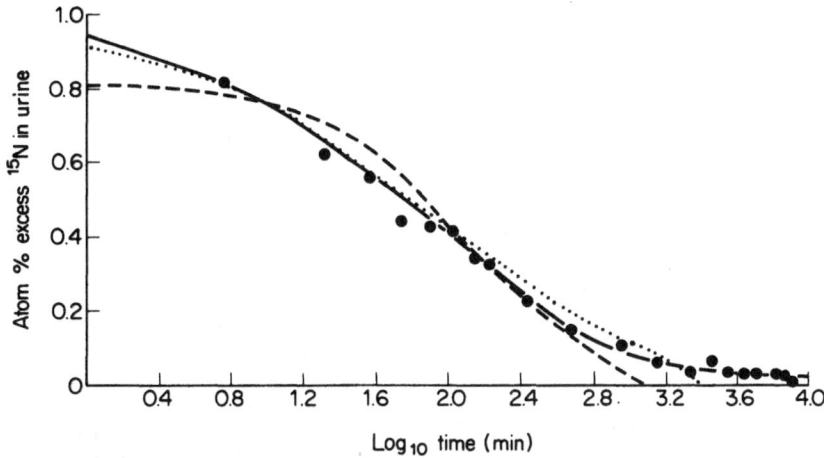

Figure 18.2. Plot of the urinary ^{15}N decay curve for study 1 (42-year-old male). SA vs. $\log_{10}$ time (min). ● = experimental; ——— = theoretical values calculated according to equation 5. The dotted line shows the fit of the data to a two-pool model (two exponents) and the dashed line to a one-pool model (one exponent).

(6). The magnitude of the urea recycling correction.

(7). Recycling of ^{15}N from protein back to the metabolic N pool.

(8). Attainment of an approximate isotopic steady state in ^{15}N in the metabolic N pool by the time the infusion is terminated.

Most of these assumptions have been discussed elsewhere by others. Only assumptions 6, 7, and 8 will be discussed here. The crucial assumption in our procedure is that the ^{15}N has attained an approximate isotopic steady state in the metabolic N pool by the time the isotope infusion is terminated. We define an approximate isotopic steady state as follows: an approximate isotopic steady state 'is assumed to exist in a given pool during an infusion of isotope when the rate of increase of isotopic enrichment is much smaller than the rate at which isotope enters and leaves that pool.' If there is re-entry of isotope, a true steady state will not be attained until all interconnected pools have equilibrated with isotope.

We have four experimental results that support our contention that after 6 to 8 hours of ^{15}N-glycine infusion an approximate isotopic steady state has been achieved in the metabolic N pool.

1. A normal adult drank 100 mg of 95% ^{15}N-glycine every hour for 11 hours following an overnight fast (figure 18.3). The plasma α-amino N approached a plateau after about 6 hours. Waterlow[12] infused ^{14}C-lysine for 30 hours into human subjects and observed a plateau in ^{14}C activity after about 10 hours. These two results may not be strictly comparable because one is concerned with carbon and the other with nitrogen.

2. Glycine can be 'trapped' in the metabolic N pool by benzoic acid. The excreted hippuric acid provides some information about the glycine in the

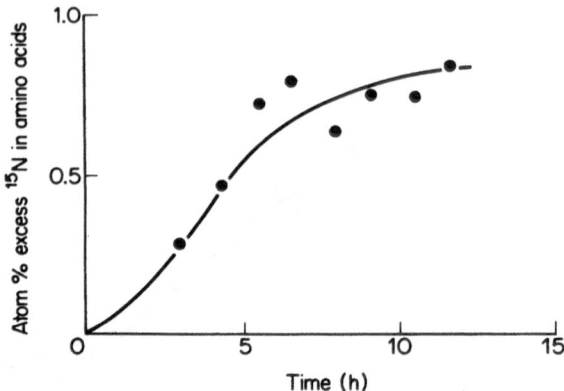

Figure 18.3. The appearance of a plateau in the plasma α-amino N. The subject, a 33-year-old male drank 100 mg of 95% ^{15}N glycine in 10 ml of water every hour for 11 hours following an overnight fast. No food was consumed while the the glycine study was in progress.

metabolic N pool. During the infusion of ^{15}N-glycine a subject drank 600 mg of sodium benzoate. The hippuric acid was isolated from the urine, and its ^{15}N specific activity was determined. The hippuric acid profile showed a rapid rise and then a flattening out after about 4 hours (figure 18.4). We interpret this to mean that the rate of influx of ^{15}N into that part of the liver metabolic N pool where glycine conjugation occurs approximately equals the rate of isotope leaving the pool via k_u and k_s. Liver glycine originates from three sources —dietary, tissue catabolism, and *de novo* synthesis. Since glycine conjugation is a rapid process, mixing of glycine in the metabolic pool may not be complete before the glycine is conjugated. Therefore, although a steady state is obtained in the hippuric acid, the actual value of the plateau may reflect a very complex situation.

3. The decay of the hippuric acid specific activity following the termination of the infusion allows the calculation of a value for $(k_u + k_s)$, the rate constant for the emptying of the metabolic pool with respect to glycine/hippuric acid. Plotting these data out as a first-order plot gave $(k_u + k_s) = 0.58$ h^{-1}. This rapid decay provides additional support for the concept of a rapidly turning over metabolic N pool.

4. For a subject, the rate of protein synthesis was determined by the less ambiguous method of Picou and Taylor-Roberts[2]. Although the two studies were some months apart, the value of 20.4 mg N kg body wt^{-1} h^{-1} is in reasonable agreement with the value obtained from decay curve analysis of 25.9 mg N kg body wt^{-1} h^{-1}. If there was something seriously amiss with our method, agreement with the Picou and Taylor-Roberts procedure would not be expected.

At the end of the 8-hour infusion the urea curve shows no sign of flattening

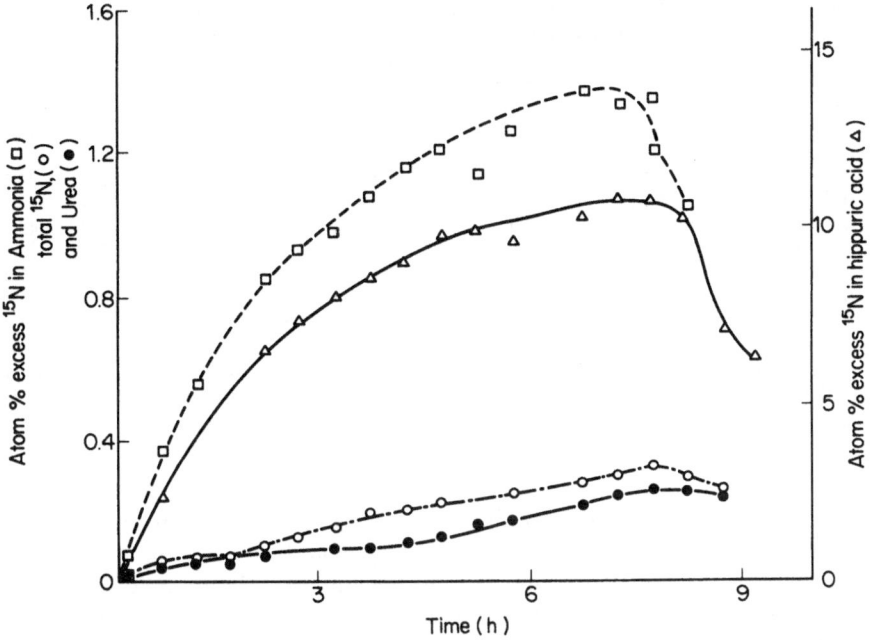

Figure 18.4. Urinary ^{15}N excretion profiles during an 8 hour infusion of ^{15}N glycine (95%) at a rate of 37.4 mg ^{15}N h^{-1}. Food was consumed at hourly intervals in the form of equal aliquots of a milkshake. Total nutrition during the 8 hour infusion was 500 kcal and 5 g N.

out (figure 18.4). This is predicted by the model, which postulates that the small, rapidly turning over metabolic nitrogen pool will rapidly become relatively highly enriched. Urea synthesised from this pool will also be relatively highly enriched. It will enter and mix with the relatively large, unlabelled urea pool, increasing its enrichment slightly. A steady state in isotope in the urea pool will not be obtained until the isotopic enrichments of the metabolic N and urea pools are equal. Since the metabolic N pool is small and turning over rapidly, it will reach an approximate isotopic steady state much sooner than the more slowly turning over urea pool. These experimental observations are consistent with the assumption of an approximate steady state in the metabolic N pool. It is important to point out that the method of calculation only requires an approximate isotopic steady state in the metabolic N pool, not a detailed description.

As previously pointed out, urea recycling cannot be measured in this type of single isotope study. We were interested to see how urea recycling could affect calculated values for the rate of protein synthesis and the various pool parameters. We therefore solved equations 1–12 for varying values of k_a/k_e (table 18.1). $k_a/k_e \times 100$ approximates the percentage of urea recycled. The table shows that the pool parameters are in fact very dependent on the amount of

Table 18.1 The effect of varying k_a/k_e, the ratio of urea N recycled to urea N excreted on the Rittenberg–San Pietro pool parameters and the rate of protein synthesis for the data on the 42-year-old male.*

$\dfrac{k_a}{k_e}$	T	P	U	k_s	k_u $\times 10$	k_e $\times 10$	k_c $\times 10^2$	k_a $\times 10^2$	$T_{t_{\frac{1}{2}}}$	$P_{t_{\frac{1}{2}}}$	$U_{t_{\frac{1}{2}}}$	PS
0	53.6	1.26	5.08	1.56	9.17	2.27	3.67	0	18.9	0.28	3.05	26.2
0.1	59.2	1.56	5.59	1.63	8.29	2.21	4.34	2.12	16.0	0.28	2.97	34.2
0.2	63.8	1.92	6.10	1.69	7.56	1.98	5.10	3.97	13.6	0.28	2.91	43.3
0.4	69.9	2.69	7.12	1.78	6.42	1.73	6.86	6.93	10.1	0.29	2.88	63.9
0.6	72.0	3.53	8.14	1.85	5.57	1.51	9.07	9.07	7.64	0.29	2.86	87.1
0.8	72.0	4.48	9.15	1.89	4.90	1.33	11.7	10.7	5.92	0.29	2.89	113
1.0	71.8	5.51	10.2	1.92	4.36	1.18	14.8	11.8	4.67	0.29	2.94	141

*T, P and U are the tissue pool, metabolic N and excretory N pool sizes in g N. k_s, k_u, k_e, k_c and k_a are the rate constants for figure 18.1 in h^{-1}. $T_{t_{\frac{1}{2}}}$, $U_{t_{\frac{1}{2}}}$ and $P_{t_{\frac{1}{2}}}$ are the half-lives, in hours for pools T, P and U, respectively. PS is the rate of protein synthesis in mg N kg body $wt^{-1} h^{-1}$.

urea recycling. Measurement of the amount of urea recycled in protein synthesis determination experiments, where the calculated rate is dependent on the specific activity of the urinary urea, would increase the accuracy of such determinations. It is, therefore, important that the method be able to handle this effect if and when a reliable value of the percentage of urea recycled is published or the method is expanded to include a direct measurement of urea recycling.

The reason for infusing the isotope over a period of hours is to make sure that most of the isotope has equilibrated with the metabolic N pool. This avoids the possibility of either the rates of glycine mixing or nitrogen interchange being rate controlling. Carrying out the infusion for a longer period of time would further decrease the mixing and nitrogen interchange contributions. The reason that we did not do so is that increasing the infusion time might make protein recycling a more significant factor in the interpretation of the results[13].

The calculated excretory N pool, which consists of ammonia and urea N, reflects contributions from both. Using equation 14, an approximate value for the size of the urea pool can be calculated. Assuming that the amount of ammonia in the excretory pool is negligible, the size of the urea pool can be found, since under these conditions all the ^{15}N in the pool is assumed to be urea ($\gamma^U_{t=0}$). Analysis of the urinary urea for ^{15}N at the time the infusion was terminated gives a value for the specific activity of the urea pool at $t = 0$, and Equation 10 can be solved with $k_a/k_e = 0$. The calculated values are 11.9 g N (male) and 4.1 g N (female). These values are within the range to be expected for normals.

Had the urine been analysed only for urea ^{15}N, the Rittenberg–San Pietro model would probably have given reasonable values for the size and half-life of the urea pool. The effect of ammonia, which decays much faster than the urea pool, is to give a size and half-life for the excretory N pool that is somewhat smaller than the urea pool proper.

One of the problems of human N pool size determination from ^{15}N urinary decay curve analysis is the relatively small size of the calculated tissue protein pool compared with the large amount of protein in the body. Pulse label experiments give values of about 100 g N and 60 hours for a 60 kg male[14]. Our values are smaller (59.2 g N and 16 hours). Olesen *et al.*[8] postulated the existence of a blind tissue pool to account for the discrepancy. According to this modified version of the Rittenberg–San Pietro model, the ^{15}N-labelled protein falls into two classes. The 'blind tissue pool' includes such proteins as collagen and other proteins with a half-life much greater than the duration of the experiment. [15]N incorporated into these proteins is lost from the system being studied because none of this ^{15}N-labelled protein will break down during the period in which urine collections are made. The second protein pool consists of proteins whose half-lives are shorter, or of the same order of magnitude, as the time of the experiment. Some ^{15}N from these proteins will re-enter the metabolic N pool during the course of the experiment, and a fraction of this will appear in the excretory N pool.

Analysing decay curves from an assumption of an approximate ^{15}N steady state in the metabolic N pool gives values that may not be absolutely correct but should be reasonably accurate, since the method is internally consistent.

Investigation of the uptake profiles

During the infusion of ^{15}N we noticed that the urinary ^{15}N excretion profile was very sensitive to perturbation. Figure 18.5 shows what happens when a

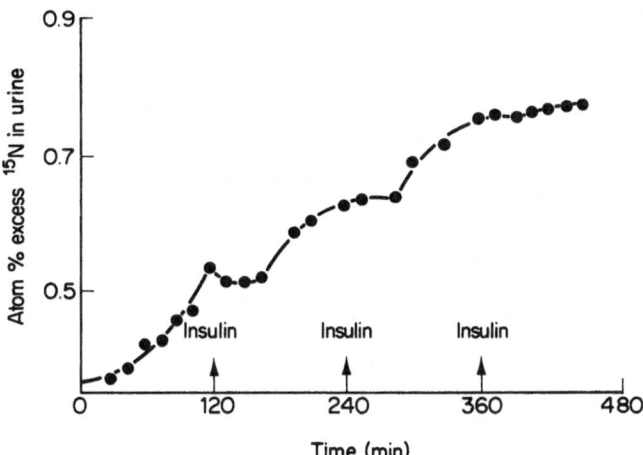

Figure 18.5. Effect of insulin on the urinary ^{15}N excretion. The subject was a 72-year-old male diabetic who had been undergoing intravenous hyperalimentation for the 3-week period prior to the study. For 8 hours he received 16.1 mg [15]N (as glycine). At $t = 120$ min, 1 unit of insulin was given, at $t = 240$, 2 units and at $t = 360$, 3 units.

diabetic's requirement of 10 units of regular insulin per 8 hours is given in four separate pulses. It is apparent that insulin has a profound effect. Obviously this result reflects a very complex situation, so before trying to interpret this result we decided to study a simple effect which would give a known result. We selected the effect of a meal on the urinary ^{15}N excretion profiles. 1.5 g of 95% ^{15}N-glycine was infused for 8 hours. Three hours after the start of the infusion the subject consumed a normal meal within 30 minutes (figure 18.6). If no meal is consumed, the curves are smooth (figure 18.4), thus leading us to the following interpretation of figure 18.5.

For the most part, the urinary amino acid profiles reflected the ammonia profile. This is probably because both originate from the arterial blood perfusing the kidneys. The ammonia profile continues to rise for about an hour after the meal, reflecting the time required for gastric digestion and emptying. Appreciable amounts of food N then begin to enter the bloodstream. As long as this exogenous N is being absorbed, the ammonia precursor pool is being diluted. When the dietary N absorption is almost complete, the pool ^{15}N concentration increases again with the influx of ^{15}N, and the ammonia specific activity increases concomitantly. From a mean of several studies we concluded

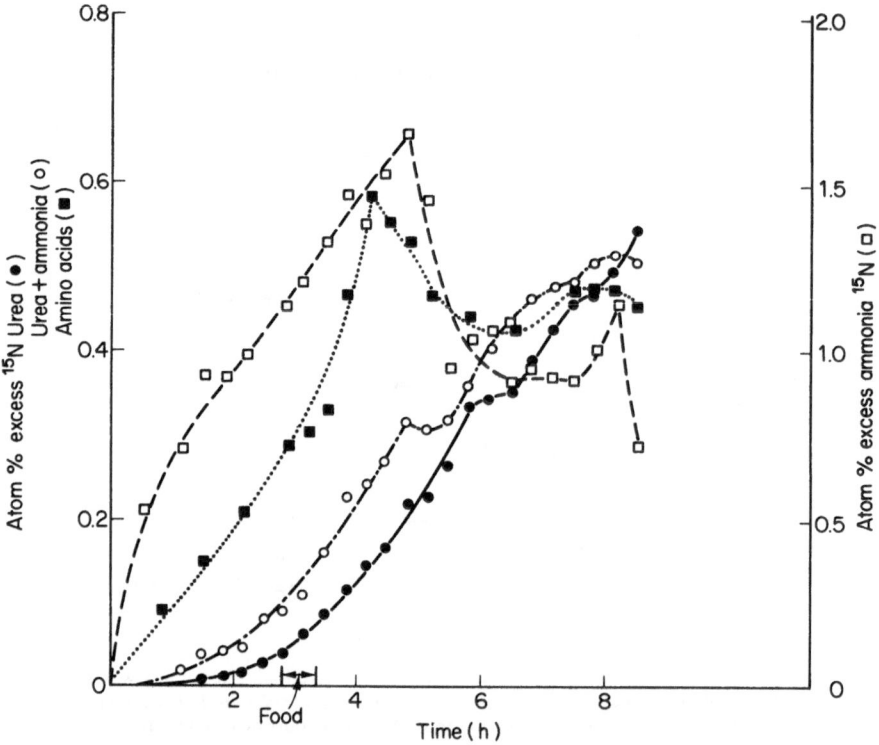

Figure 18.6. The effects of a meal on the urinary ^{15}N excretion profiles.

that the time of nongastric digestion ranged from 1.5 to 3.8 hours. The total time for digestion and absorption was about 1 hour longer (2.5–4.8 h).

Figures 18.7 and 18.8 show the effects of 8 units of insulin (regular) on a 54-year-old woman whose diet was being supplemented by infusion at a constant rate of amino acids in 2.5% dextrose. For 8 hours ^{15}N-labelled glycine was

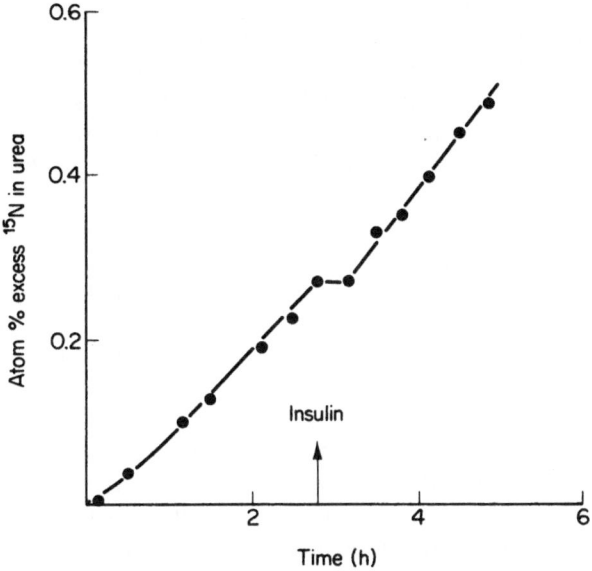

Figure 18.7. The effects of 8 units of insulin on the SA of the excreted urinary urea. ^{15}N glycine was infused.

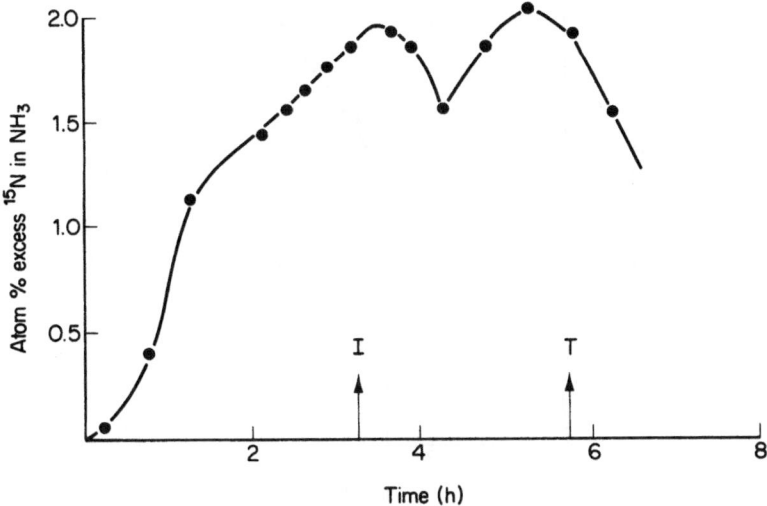

Figure 18.8. The urinary ammonia excretion profile for figure 18.6. I = insulin and T = termination of infusion.

added to the infusate at a rate of 34 mg ^{15}N h^{-1}. Urine samples were collected as frequently as possible, and the urinary urea, ammonia, and amino acids were isolated by standard techniques.

Interpretation was not possible until answers to the following questions were obtained:

1. *Is the body being overloaded with glycine during the ^{15}N infusion?*
In the average adult the glycine flux is 1 gm kg body wt^{-1} day^{-1} = 7.8 mg N kg body wt^{-1} h^{-1} for a 70 kg person[15,16]. 35 mg ^{15}N is infused per hour. For a 70 kg person, this corresponds to 0.5 mg N kg body wt^{-1} h^{-1}. The ^{15}N-labelled amino acid therefore contributes less than 10% to the total glycine flux.

2. *Is the effect peculiar to glycine?*
That the effect is not peculiar to glycine is demonstrated by figures 18.9 and 18.10. ^{15}NH$_4$Cl was the ^{15}N carrier in these studies. Apart from the much higher urinary ammonia enrichments the curves are the same.

3. *In the context of these experiments, what is the relationship between the various nitrogen precursor pools and the observed specific activity of the urinary urea and ammonia?*
We assume that the SA of the urinary urea represents the SA of the urea pool at the midpoint time between collections. When ^{15}N labelled urea was infused we observed no change in the slope of the urea ^{15}N excretion profile following the injection of insulin. While it is known that in rat liver perfusion experi-

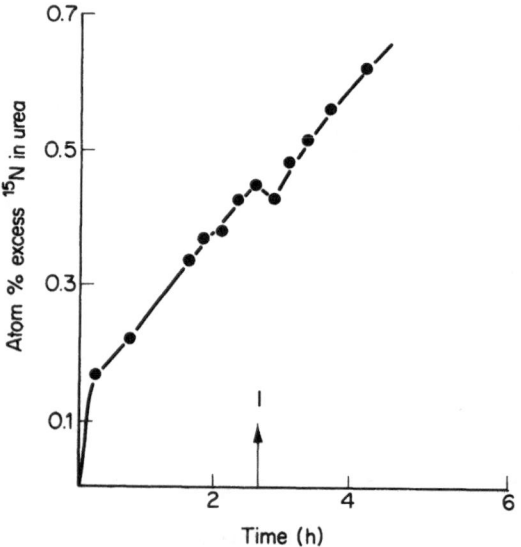

Figure 18.9. The effect of 8 units of insulin (regular) on the urinary urea SA. when ^{15}NH$_4$Cl is used as the ^{15}N carrier. I = insulin.

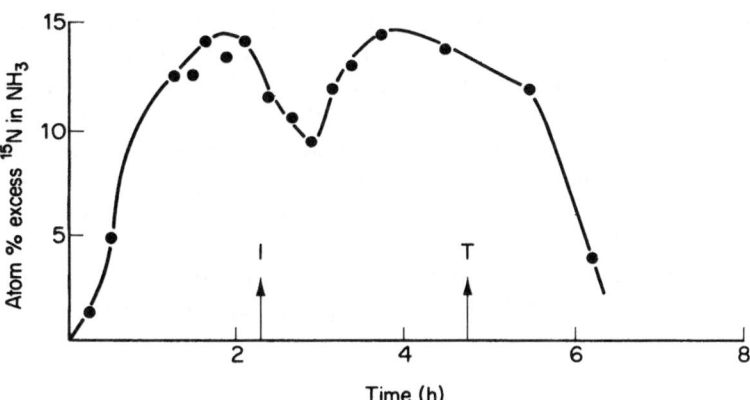

Figure 18.10. The urinary ammonia profile corresponding to figure 18.8.
I = insulin and T = termination of infusion.

ments insulin does inhibit urea synthesis by inhibiting liver protein catabolism this result suggests that any inhibition of urea synthesis is too small for us to observe. One explanation of our results is that insulin changes the origin of the urea and ammonia precursor pools. Felig[18] pointed out that while insulin increases the uptake of certain amino acids by muscle, no effect is observed on the net A–V difference across muscle for alanine. He suggested that this was because insulin stimulated peripheral alanine formation. 'The tendency for insulin to increase muscle uptake of circulating alanine is thus counterbalanced by augmented intra-cellular production and release.' Alanine originating from muscle will have a lower SA than alanine originating from gut, liver or kidney. According to Pitts, alanine accounts for about 5.7% of the ammonia N. Thus a decrease in the circulating alanine ^{15}N SA could account for the effect on the urinary NH_3 SA. However, glutamine, like alanine also functions as an N carrier from the periphery to the splachnic regions. Experiments are currently in progress to see how insulin effects the SA of circulating glutamine.

Insulin decreases hepatic gluco-neogenesis by decreasing alanine uptake[18], and inhibits hepatic proteolysis[17]. This should lead to a decrease in the size of the urea precursor pool in the liver, and therefore, a decrease in the rate of urea synthesis and a decrease in the SA of the newly synthesised urea. In fact, we only observed the latter. These results suggest that the major effect of insulin *in vivo* in man is on the origin of the urea precursor amino N rather than on the actual rate of synthesis. In rats the effect of insulin on urea synthesis has been hard to prove because the effect is small[17].

Interaction of muscle with blood amino acids

A continuous infusion of ^{15}N into the body provides a method of investigating the interaction of the blood amino acids with a given tissue. We used $^{15}NH_4Cl$

as the ^{15}N carrier. One gram of ^{15}NH$_4$Cl (95+ atom%) was infused at a constant rate over a period of 6 hours into healthy adult volunteers. At 4.5, 5, and 5.5 hours, radial blood and brachial venous blood samples were taken in heparinised syringes. The α-amino N from the plasma and blood cells, respectively, was isolated by standard methods and analysed for ^{15}N. For the ten sets of blood samples analysed, the arterial α-amino N specific activities ranged between 0.5 and 1.3 atom % excess, and the corresponding venous values ranged between 0.3 and 1.1 atom % excess.

Administration of ^{15}N-labelled ammonia leads to the labelling of the non-essential amino acids only[1]. Defining the percentage change as

$$\frac{\text{SA}_{\text{arterial}} - \text{SA}_{\text{venous}}}{\text{SA}_{\text{arterial}}} \times 100$$

the mean percentage change observed was $18.4 \pm 4.2\%$. Felig et al.[20] in A–V difference studies observed an output of amino acids from limb muscle at rest in the region of 5–10%. This is much less than that found by looking at ^{15}N. It suggests that (i) investigation of the interaction of amino acids with tissues requires more than net balance data, and (ii) fractionation of isotopically labelled plasma amino acids and subsequent analysis for isotope could provide a useful procedure for defining the interaction between blood and tissue amino acid pools. These experiments are currently in progress.

Acknowledgements

We wish to thank the U.S. Public Health Service for financial support (Grant AM-16658).

References

1. Waterlow, J. C. (1969) Protein nutrition and metabolism in the whole animal. In *Mammalian Protein Metabolism*, H. N. Munro, ed., Academic Press, New York, Vol. 3, p. 325–390
2. Picou, D. and Taylor-Roberts, T. Nitrogen turnover in infants receiving different amounts of dietary protein. *Clin. Sci.*, **36** (1969), 283–296
3. Dietschy, J. and Wilson, J. D. Regulation of cholesterol metabolism. *New Eng. J. Med.*, **282** (1970), 1128–1138
4. Tschudy, D. P., Bacchus, H., Weisman, S., Watkin, D. M. and White, J. Studies on the effect of dietary protein and caloric levels on the kinetics of nitrogen metabolism using ^{15}N-L-aspartic acid. *J. Clin. Invest.*, **38** (1959), 892–901
5. Wu, H. and Bishop, C. W. Pattern of ^{15}N excretion in man following the administration of ^{15}N labelled glycine. *J. Appl. Physiol.*, **14** (1959), 1–5

6. Wu, H. and Sendroy, J. Jr. Pattern of ^{15}N excretion in man following the administration of ^{15}N labelled glycine. *J. Appl. Physiol.*, **14** (1959), 6–10

7. Wu, H., Sendroy, J. Jr. and Bishop, C. W. Interpretation of urinary ^{15}N excretion data following administration of ^{15}N labelled glycine. *J. Appl. Physiol.*, **14** (1959), 11–15

8. Olesen, K., Heilskov, N. C. S. and Schonheyder, S. The excretion of ^{15}N in urine after administration of ^{15}N glycine. *Biochim. Biophys. Acta*, **15** (1954), 95–107

9. Shipley, R. A. and Clark, R. E. (1972) *Tracer Methods for In Vivo Kinetics*, Academic Press, New York and London, p. 215 et seq.

10. Walser, M. and Bodenlos, L. J. Urea metabolism in man. *J. Clin. Invest.*, **38** (1959), 1617–1626

11. Walser, M. Urea metabolism in chronic renal failure. *J. Clin. Invest.*, **53** (1974), 1385–1392

12. Waterlow, J. C. Lysine turnover in man measured by intravenous infusion of L-[14]U-C lysine. *Clin. Sci.*, **33** (1967), 507–515

13. Aub, M. R. and Waterlow, J. C. Analysis of a five compartment system with continuous infusion and its application to the study of amino acid turnover. *J. Theor. Biol.*, **26** (1970), 243–250

14. San Pietro, A. and Rittenberg, D. A study of the rate of protein synthesis in humans. I Measurement of urea pool and urea space. II. Measurement of the metabolic pool and rate of protein synthesis. *J. Biol. Chem.*, **201** (1953), 445–456 and 457–473

15. Watts, R. W. E. and Crawhall, J. C. The first glycine metabolic pool in man. *Biochem. J.*, **73** (1959), 277–282

16. Nyhan, W. L. and Childs, B. Hyperglycinemia. V. The miscible pool and turnover of glycine and the formation of serine. *J. Clin. Invest.*, **43** (1964), 2404–2411

17. Mortimore, G. E. and Mondon, C. E. Inhibition by insulin of valine turnover in liver. *J. Biol. Chem.*, **245** (1970), 2375–2383

18. Felig, P. The glucose–alanine cycle. *Metabolism*, **22** (1973), 179–207

19. Fürst, P., Jonsson, A., Josephson, B. and Vinnars, E. Distribution in muscle and liver vein protein of ^{15}N administered as ammonium acetate to man. *J. Appl. Physiol.*, **29** (1970), 307–312

20. Felig, P., Wahren, J. and Raf, L. Evidence of inter-organ amino acid transport by blood cells in humans. *Proc. Natl. Acad. Sci. U.S.A.*, **70** (1973), 1775–1779

Discussion

REGOECZI

I would like to go back to the Rittenberg–San Pietro scheme (figure 18.1 in the text). You said that you were concerned about the possible effect of urea recycling on the calculated rate of protein synthesis. By urea recycling I take it you mean the hydrolysis of urea by the bacteria in the gut to ammonia, and the ammonia then re-entering the metabolic N pool. How did you calculate the values in table 18.1? What data did you use?

STEIN

We took the experimental data for the study from a 42-year-old male, and with the aid of a computer solved equations 1–12 for various amounts of urea hydrolysed to urea excreted (k_a/k_e). We did this because we were interested to see whether the well documented phenomenon of urea hydrolysis could affect the calculation of the various pool parameters and the rate of protein synthesis.

This type of experiment depends on the validity of the scheme (figure 18.1) and measurements of the specific activity of urinary urea. The theoretical values shown in the table prove that making an allowance for urea recycling does change the calculated values. It might, therefore, be important in such experiments to measure the degree of urea recycling.

REGOECZI

Nine years ago, when I was working on this problem with Dr McFarlane in his laboratory, we did some double-labelling experiments with ^{15}N and ^{14}C labelled urea in rabbits. We found that ^{15}N labelled urea had a much longer half-life than ^{14}C labelled urea.

TAVILL

Could I add a comment to the discussion of the problem of quantification of urea recycling? When figures of 25 to 30% are quoted, they refer to the failure to recover in the urine all the urea which is synthesised. Only 70–80% may be recoverable[1,2]. The fate of the unrecovered urea seems to be decided by the intestinal flora which hydrolyse it to CO_2 and NH_3. True recycling denotes the re-utilisation of that ammonia for anabolic purposes, and it is the defficiency of that process which needs to be defined.

STEIN

I think that this question of urea recycling is a major problem in measuring the rate of protein synthesis with [15]N. The method depends on defining the excretory pathway accurately and then finding the synthetic rate by difference[3]. If there is a significant amount of urea recycling, then correction should be made for this. Possibly this could be done by comparing the urea pool size calculated from decay curve analysis with the value obtained from the BUN. I think that the unknown magnitude of the urea recycled is the major source of error in this type of experiment. Like Aub and Waterlow[4], we did some theoretical calculations on the effect of re-entry of ^{15}N into the metabolic N pool from the protein pool and found it not to be significant.

JAMES

I want to make two points: (i) In steady-state methods it is important to attain a good plateau. (ii) Your evidence for an approximate isotopic steady state in the metabolic N pool is based mainly on products from the liver amino N pool, e.g. hippuric acid. Having established that the liver metabolic N pool is in an approximate steady state, you then derive from that a value for the total body protein turnover. This value is necessarily dependent on the discrepancy between the weighted mean precursor specific activity in the body, and that in the hepatic pool. The hepatic pool specific activity is in turn dependent on the relative rate of turnover of protein within the liver and the exchange rate of amino acids between the liver and the tissues via the blood. I think, therefore, that values for whole body protein synthesis rates based on measurements of urea ^{15}N specific activity may overestimate the contribution of the liver and give rather higher values than if ^{14}C were used as the tracer. We have some animal data that suggest that this might be the case.

STEIN

Figure 18.3 shows that we also observed an approximate isotopic steady state in the plasma α-amino N, as well as in the liver. This observation suggests that most of the body amino N pools have equilibrated with the isotope by the time the infusion is terminated. We are currently using your method of determining the rate of protein synthesis in the rat using ^{15}N glycine instead of [14]C. Our preliminary values are in good agreement with your ^{14}C published results.

References

1. Walser, M. and Bodenlos, L. J. Urea metabolism in man. *J. Clin. Invest.*, **38** (1959), 1617–1626
2. Walser, M. Urea metabolism in chronic renal failure. *J. Clin. Invest.*, **53** (1974), 1385–1392
3. San Pietro, A. and Rittenberg, D. A study of the rate of protein synthesis in humans. I. Measurement of urea pool and urea space. II. Measurement of the metabolic pool and rate of protein synthesis. *J. Biol. Chem.*, **201** (1953), 445–456 and 457–473
4. Aub, M. R. and Waterlow, J. C. Analysis of a fine compartment system with continuous infusion and its application to the study of amino acid turnover. *J. Theor. Biol.*, **26** (1970), 243–250

19

Albumin depletion in uremic patients on conservative management*

R. BIANCHI, G. MARIANI, A. PILO, M. G. TONI and F. CARMASSI

Introduction

The purpose of this work was to investigate how a long-term low protein diet could affect albumin metabolism of patients with chronic uremia, on the assumption that this protein reflects to some extent the behaviour of the whole body proteins[1-3]. In fact, prolonged low protein diet in these patients as the only type of conservative management, raises the problem as to whether a dangerous protein malnutrition could ensue[4].

Albumin metabolism has been assessed by the evaluation of protein catabolism and distribution in 62 uremic patients, and by the direct measurement of albumin synthesis in 10 of them.

Material and methods

Patients
The 62 uremics submitted to the study were all under steady-state conditions, as judged by the constancy of body weight, serum creatinine and protein concentrations throughout the whole experimental period. All patients followed a Giovannetti diet[5] adapted to the individual renal function, in every case containing not less than 20 g of high biological value proteins per day, supplemented by vitamins and, where necessary, salts.

On the basis of the duration of the diet, they were divided into two groups: 35 subjects on dietary treatment for periods ranging from 6 to 30 days (first group), and 27 subjects following the diet from 6 months to 5 years (second group).

*Work partly supported by Public Health Service U.S.A., Contract No. PH-43-68-707.

Serum creatinine in the first group ranged from 1.7 to 17 mg 100 ml⁻¹; patients of the second group had creatinine levels ranging from 2.3 to 21.3 mg 100 ml⁻¹ (see tables 19.1 and 19.2).

Experimental procedure

Starting from 3 days before the experiment, 10 drops of Lugol's solution were administered twice a day, to prevent iodide uptake by the thyroid gland.

For catabolism and distribution studies, intravenous injections of ^{131}I-Human Serum albumin (^{131}I-HSA, electrolytically labelled[6]) and of [125]I-

Table 19.1 Albumin metabolism in uremic patients on low protein diet from 6 to 30 days.

Case No.	Sex	Age (years)	Blood creatinine (mg 100 ml⁻¹)	Time on diet (days)	Body weight (kg)	Plasma volume (ml kg⁻¹)	Serum albumin concentration (g 100 ml⁻¹)	Intravascular albumin mass (g kg⁻¹)	Extravascular albumin mass (g kg⁻¹)	Total albumin mass (g kg⁻¹)	Extra/intravascular albumin ratio	Fractional catabolic rate (% day⁻¹)	Absolute catabolic rate (mg kg⁻¹ day⁻¹)	Fractional synthetic rate (% day⁻¹)	Absolute synthetic rate (mm kg⁻¹ day⁻¹)	Albuminuria (g day⁻¹)
138	M	62	10.8	17	63.6	40.6	3.22	1.31	1.10	2.41	0.84	12.0	157	—	—	—
145	F	69	7.5	32	64.0	43.6	4.40	1.90	2.18	4.08	1.15	11.7	223	—	—	—
146	M	24	4.3	19	59.0	52.6	4.74	2.50	2.37	4.87	0.95	10.6	264	—	—	—
148	M	27	6.9	13	59.5	43.0	4.50	1.96	1.49	3.45	0.76	12.6	246	—	—	1.56
150	F	53	6.0	18	68.5	38.5	3.64	1.40	1.47	2.87	1.05	9.3	131	—	—	0.16
151	M	29	4.9	10	58.0	51.1	3.70	1.89	1.55	3.44	0.82	11.6	220	—	—	1.00
158	M	52	4.0	28	49.0	54.6	2.87	1.57	1.32	2.89	0.84	9.6	152	—	—	0.95
160	M	38	13.0	28	57.0	30.2	4.00	1.21	1.03	2.24	0.85	9.3	112	—	—	0.50
166	M	65	10.6	18	71.0	32.0	3.10	0.99	0.59	1.58	0.60	8.4	83	—	—	0.18
167	M	32	2.8	16	59.0	41.1	3.22	1.32	1.27	2.59	0.96	11.8	156	—	—	—
169	F	47	4.9	14	77.0	35.6	3.43	1.22	0.98	2.20	0.80	15.1	185	—	—	0.83
171	M	37	7.3	18	64.0	58.0	3.41	1.98	1.18	3.16	0.60	8.8	174	—	—	0.96
173	M	18	1.7	fd	64.0	45.0	1.70	0.76	0.30	1.06	0.39	17.8	136	—	—	29.40
174	M	38	3.9	22	61.0	43.5	2.77	1.21	0.40	1.61	0.33	11.0	134	—	—	6.95
175	M	25	5.3	17	60.0	49.8	3.02	1.50	1.34	2.84	0.89	8.9	133	—	—	0.22
182	M	59	11.0	15	54.0	59.2	3.18	1.88	1.39	3.27	0.74	7.8	147	—	—	0.88
184	M	65	3.2	30	77.0	46.9	3.20	1.50	1.91	3.41	1.27	10.7	161	—	—	1.12
185	M	29	8.5	16	62.0	55.3	3.80	2.10	2.50	4.60	1.19	13.1	275	—	—	1.22
189	M	28	7.4	13	44.0	55.1	3.43	1.93	1.60	3.53	0.83	7.8	150	—	—	—
006	M	38	10.6	29	60.0	38.5	3.70	1.43	0.75	2.18	0.53	15.2	218	15.0	214	—
030	M	40	10.3	26	72.0	44.1	3.34	1.47	1.34	2.81	0.91	9.2	135	8.4	124	—
032	M	63	4.5	17	59.0	41.9	3.80	1.59	1.93	3.52	1.21	14.1	222	14.1	222	—
037	M	50	17.0	6	62.0	85.2	3.00	2.55	—	—	—	—	—	19.0	484	—
041	F	57	8.3	26	57.0	48.9	3.48	1.70	1.39	3.09	0.82	9.0	152	9.2	156	—
207	M	23	5.3	10	54.0	43.5	4.00	1.74	—	—	—	9.1	158	—	—	0.51
055	M	28	8.9	10	65.0	60.2	3.66	2.20	1.30	3.50	0.59	4.8	106	—	—	0.31
219	M	44	9.3	10	67.0	39.0	3.30	1.29	1.60	2.89	1.24	8.0	103	—	—	0.43
221	M	50	5.0	10	72.5	39.1	3.20	1.25	1.50	2.75	1.20	9.4	117	—	—	1.09
225	M	30	3.5	10	57.0	54.1	3.50	1.89	1.82	3.71	0.96	13.5	255	—	—	0.97
228	M	31	9.5	10	85.0	45.1	3.88	1.75	2.01	3.76	1.15	12.9	226	—	—	0.45
236	M	51	2.2	fd	69.0	56.3	3.70	2.08	0.75	2.83	0.36	17.0	354	—	—	3.45
247	M	69	2.2	fd	62.0	34.0	2.86	0.97	0.46	1.43	0.47	15.8	153	—	—	13.01
249	M	31	7.9	10	75.0	32.9	3.40	1.12	1.09	2.21	0.97	7.0	78	—	—	1.76
252	M	35	2.6	10	84.0	40.4	3.50	1.41	0.47	1.88	0.33	9.2	130	—	—	20.90
265	M	18	2.3	fd	58.5	51.5	3.00	1.54	0.40	1.94	0.26	12.3	189	—	—	15.54

fd = free protein diet.

Table 19.2 Albumin metabolism in uremic patients on low protein diet from 6 months to 5 years.

Case No.	Sex	Age (years)	Blood creatinine (mg 100 ml^{-1})	Time on diet (years)	Body weight (kg)	Plasma volume (ml kg^{-1})	Serum albumin concentration (g 100 ml^{-1})	Intravascular albumin mass (g kg^{-1})	Extravascular albumin mass (g kg^{-1})	Total albumin mass (g kg^{-1})	Extra/intravascular albumin mass ratio	Fractional catabolic rate (% day^{-1})	Absolute catabolic rate (mg kg^{-1} day^{-1})	Fractional synthetic rate (% day^{-1})	Absolute synthetic rate (mg kg^{-1} day^{-1})	Albuminuria (g day^{-1})
147	F	33	13.6	1	49.0	47.8	3.80	1.82	2.12	3.94	1.17	8.0	146	—	—	—
187	M	56	12.0	5	113.0	41.4	4.00	1.66	1.11	2.77	0.67	6.9	114	—	—	0.65
188	M	38	6.8	0.5	63.0	48.8	3.74	1.91	0.92	2.83	0.48	11.1	211	—	—	0.87
190	F	61	8.0	4	45.0	37.6	3.43	1.29	2.14	3.43	1.66	11.4	147	—	—	0.63
191	M	47	21.0	4	55.5	49.2	4.01	1.97	1.22	3.19	0.62	7.2	142	—	—	0.06
192	M	39	8.3	1	64.0	49.3	3.37	1.62	1.07	2.69	0.66	10.9	176	—	—	0.05
008	F	33	14.2	3	54.0	61.4	4.00	2.45	—	—	—	13.1	320	13.0	318	—
195	M	60	15.0	0.6	73.0	45.1	3.46	1.56	0.86	2.42	0.55	8.6	134	—	—	0.30
034	M	49	15.6	2	61.5	58.9	3.38	1.99	1.47	3.46	0.74	9.3	185	9.6	191	—
036	M	41	14.3	2	75.0	44.9	3.33	1.49	—	—	—	8.7	130	8.5	127	—
040	F	36	21.3	1	66.0	51.7	3.60	1.86	—	—	—	—	—	11.2	208	—
042	F	32	5.8	2.4	56.0	44.8	3.38	1.51	1.12	2.63	0.74	10.6	160	11.0	166	—
043	M	56	10.7	3	55.5	51.6	3.95	2.04	1.49	3.53	0.73	8.3	169	—	—	—
051	M	55	9.3	0.5	76.5	37.1	4.00	1.48	1.65	3.13	1.11	7.8	115	—	—	0.41
053	M	51	7.1	1	60.0	44.4	3.26	1.45	1.23	2.68	0.85	10.0	145	—	—	0.22
054	M	41	2.1	1.25	74.0	36.1	3.90	1.40	1.48	2.88	1.05	10.7	150	—	—	0.38
209	M	47	2.3	5	77.0	34.9	3.46	1.21	1.53	2.74	1.27	13.8	167	—	—	0.40
214	M	63	4.7	4.2	67.0	50.0	3.32	1.68	1.89	3.57	1.13	8.6	145	—	—	0.29
216	M	60	9.0	4	64.0	48.7	3.21	1.56	1.49	3.05	0.95	12.7	198	—	—	0.15
218	M	51	2.4	2.25	58.0	40.9	3.70	1.51	1.48	2.99	0.98	14.0	211	—	—	0.17
220	M	32	4.5	0.75	82.0	39.1	3.70	1.45	1.95	3.40	1.35	6.0	87	—	—	0.59
223	M	42	2.4	2	73.0	43.3	4.10	1.78	1.46	3.24	0.82	10.1	180	—	—	0.36
231	M	57	9.9	4	61.0	52.3	3.15	1.65	—	—	—	8.7	144	—	—	1.31
237	M	62	9.4	3	69.5	54.5	3.35	1.82	1.57	3.39	0.86	9.8	178	—	—	1.01
248	M	31	4.5	0.5	56.0	49.2	3.87	1.91	1.01	2.92	0.53	14.0	267	—	—	0.21
251	M	55	7.3	1	41.0	67.5	3.62	2.44	2.05	4.49	0.84	6.3	154	—	—	1.20
271	M	61	13.6	3	66.5	38.2	3.66	1.40	2.04	3.44	1.46	6.0	84	—	—	0.41

iodide were performed simultaneously (doses were about 100 and 50 μCi, respectively)

Heparinised venous blood samples were taken after injection at 6 minutes, 1, 3, 6, 12, 24, 36 and 48 h, and thereafter once daily until the end of the experiment (6–8 days in most of the cases). This procedure helped to define the plasma disappearance curves of both ^{131}I-HSA and ^{125}I-iodide.

Throughout the experiment urine samples were collected at 24-hour intervals. Protein-bound ^{131}I activity in urines was measured on the eluates from IRA 400 Amberlite ion-exchange columns.

From case no. 216 on, separation of inorganic ^{131}I-iodide arising from ^{131}I-HSA catabolism was achieved in each plasma sample by gel filtration on Sephadex G-10.

All iodide radioactivity gamma counts were measured (both in plasma and urine) by a well-type scintillation counter connected to a double channel spectrometer unit.

In the 10 patients in whom albumin synthesis was independently measured, sodium ^{14}C-carbonate (about 120 μCi) additionally was injected intravenously, simultaneously with the tracers cited above; the specific activities in plasma of urea carbon and albumin guanido carbon (thus *in vivo* labelled by the liver) were measured by the xanthydrol technique[7].

Theoretical approaches

The theoretical models and experimental validations of the methods employed to measure albumin metabolism have been already described in detail elsewhere[8-14]; we give here a brief account of the theoretical approaches.

Albumin synthesis

Production rate of the protein is directly determined by two tracers: ^{14}C-carbonate and radioiodinated human serum albumin. This last label permits calculation of the kinetics of newly synthesised albumin without the need for unwarranted assumptions about protein compartments[15].

A monocompartmental model for urea kinetics avoids the use of a third tracer for complete urea system description; the influence of this hypothesis on computed albumin synthetic rates is negligible for the values from the fifth hour on[10].

Albumin catabolism

Degradation rate is measured by the two-tracer method[8,11-14], which is based on the assumption that the behaviour of the iodide released from albumin breakdown is the same as that of the free, directly injected iodide. This fact enables us to take into account early distribution and removal rates of the tracer released from protein degradation; in fact, iodide accumulates in the body fluids as the renal excretion falls. The two-tracer method becomes then mandatory for short-term studies of albumin metabolism in uremic patients, in whom iodide removal by routes other than kidney takes place to an extent which is even comparable with the rate of renal excretion[8].

From case no. 216 on, we used this method also in the version based on plasma data only, which implies separation of very small amounts of non-protein-bound from protein-bound label by gel filtration on Sephadex G-10, and which was extensively described at the last meeting in London[14].

Figure 19.1 schematises the principles of the two-tracer technique. In both versions the input from labelled albumin to the iodide system is the tracer released from protein breakdown, that is the fractional catabolic rate of intravascular albumin, FCR, multiplied by plasma activity at the same time, $P(t)$.

When considering urine activities, the transfer function of the iodide system is the cumulative excretion of free ^{125}I-iodide, $UI(t)$; in the case of plasma sampling only, the iodide transit time distribution function is the plasma time course activity of the same tracer, $I(t)$. In the first case, output of the albumin

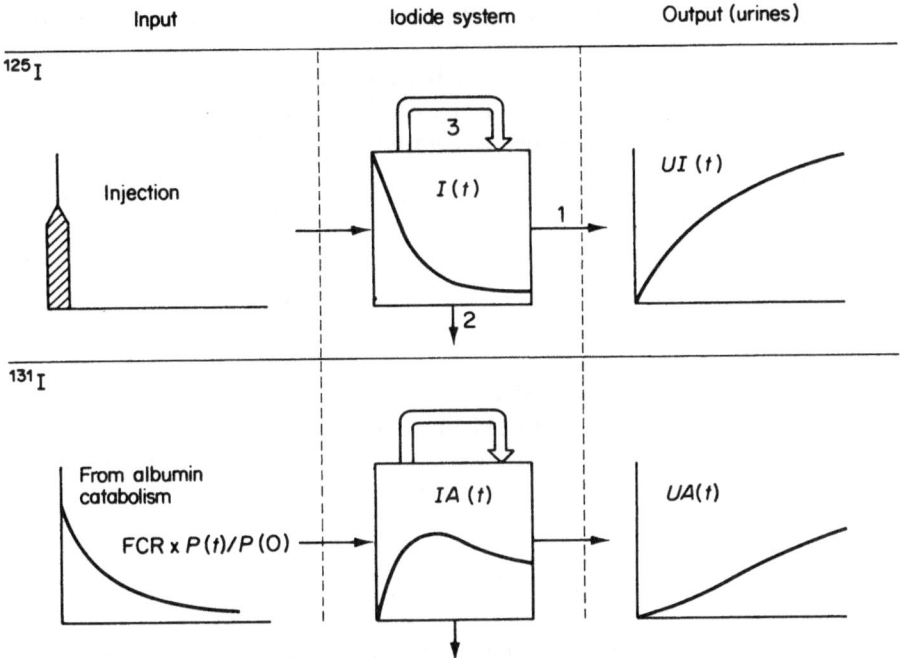

Figure 19.1. Theoretical approach of the two-tracer method for albumin catabolism running by plasma data only and by plasma and urine data (see text).

released label is the cumulative urine excretion of ^{131}I-iodide, $UA(t)$, while in the second it is represented by the ^{131}I-iodide arising in plasma from albumin catabolism, $IA(t)$.

Reminding the general relation existing between input, output and transfer function of a given system*, we can equate in each case the output to the convolution product of the input by the respective transfer function, that is

$$UA(t) = \text{FCR} \times P(t)/P(0) \ast UI(t)$$

for urine activities, and

$$IA(t) = \text{FCR} \times P(t)/P(0) \ast I(t)$$

for plasma data only.

In both equations all experimental data appear except fractional catabolism, which, being the only unknown, can therefore be computed by the least-squares method.

Albumin distribution
Intravascular albumin mass was computed by the distribution volume of the

*$O = I \ast G$, where I = input, O = output, and G = transfer function of a given linear system, and the symbol $\ast$ means convolution product operation.

labelled protein at zero time, $P(t)/P(0)$, multiplied by serum albumin concentration.

The distribution ratio of albumin between its extravascular and intravascular pools was derived by means of the equilibrium time method[16], extravascular activity, $E(t)$, being computed according to the following relation:

$$E(t) = D - P(t) - \text{FCR} \times \int_0^t P(t) \, dt$$

where D is the injected dose.

All the computations involved in analysing experimental data have been performed on a digital computer IBM 360, using the FORTRAN IV language.

Significance of all the differences between metabolic results obtained in the two groups was judged by Student's t test.

Results

The results obtained are reported in detail in tables 19.1 and 19.2, and schematically represented in figure 19.2; we report thereafter the mean values with the respective Standard Deviations.

Serum albumin concentration was slightly, but significantly, reduced in both groups (3.44 ± 0.55 g 100 ml^{-1}, with $P < 0.001$, and 3.62 ± 0.29 g 100 ml^{-1}, $P < 0.005$, respectively, in the first and second group, in comparison with the mean normal value of 4 ± 0.5 g 100 ml^{-1}), without significant difference between them.

Intravascular albumin pool (12.6 ± 0.35 g kg body weight^{-1} in normal subjects) was found to be within the normal range either in the short-term or the long-term diet patients (1.60 ± 0.42 g kg^{-1} and 1.70 ± 0.31 g kg^{-1}, respectively), thanks to the slightly enlarged plasma volume in both groups (46.6 ± 10.6 ml kg^{-1}, with $P < 0.1$, and 47 ± 7.9 ml kg^{-1}, $P < 0.05$, respectively, in comparison with 42 ± 11.5 ml kg^{-1} in normal subjects).

The ratios between extravascular and intravascular pools (EAM/IAM) and the extravascular albumin mass (EAM) (respectively, 1.5 ± 0.5 and 2.60 ± 0.5 g kg^{-1} in normals) were markedly reduced in both groups (with $P < 0.001$ in all the respects), without significant difference between them (EAM/IAM = 0.78 ± 0.30 and 0.92 ± 0.31; EAM = 1.30 ± 0.58 g kg^{-1} and 1.49 ± 0.39 g kg^{-1}, respectively in the first and second group).

Total albumin mass was likewise reduced in the two groups (2.67 ± 0.88 g kg^{-1} and 3.16 ± 0.47 g kg^{-1}, respectively, with $P < 0.001$, as compared with 4.22 ± 0.33 g kg^{-1} in normals), irrespective of the period the patients were on the diet.

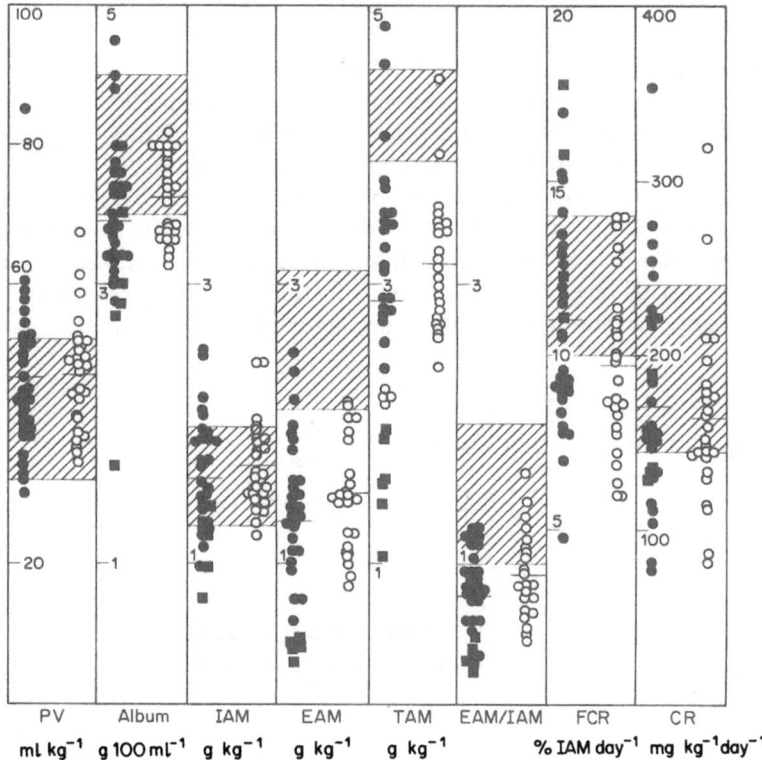

Figure 19.2. Turnover results obtained in all patients. Shaded areas represent normal ranges, while horizontal continuous lines indicate the mean values within each group. Squares stand for patients with chronic renal failure and heavy proteinuria.

Both fractional (FCR) and absolute (CR) catabolic rates of albumin were slightly reduced in the patients studied, with some statistical differences. FCR = $11 \pm 3\%$ day^{-1}, with $P < 0.1$, and $9.7 \pm 2.4\%$ day^{-1}, $P < 0.005$; CR = 171.6 ± 60.8 mg kg^{-1} day^{-1}, $P < 0.1$, and 163.8 ± 50.4 mg kg^{-1} day^{-1}, $P < 0.025$, respectively, in the first and second group, normal mean values being $12 \pm 2\%$ day^{-1} and 194 ± 49 mg kg^{-1} day^{-1}, respectively.

The values of the synthesis rate obtained by direct measure were in close agreement with the respective catabolic results in all the patients in whom they were independently determined (see tables 19.1 and 19.2).

Discussion

When considering the results obtained in the present study, which complete previous observations made on the same kind of patients[17,18], the following conclusions can be drawn concerning the effects of dietary treatment on albumin metabolism in subjects with chronic renal failure.

First, the very limited usefulness of serum albumin concentration as a reliable index of the actual body albumin depletion[19-24] is here confirmed. In fact, very different values of extra- to intravascular pool ratio correspond to similar levels of serum albumin concentration, which appears within the normal range in most of the subjects (see figure 19.3).

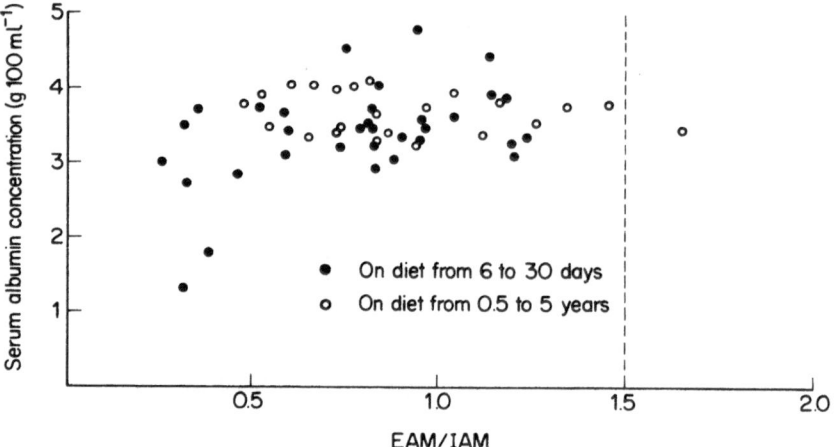

Figure 19.3. Vertical dashed line represents mean normal value of the extra/intravascular pool ratio (EAM/IAM).

Second, none of the turnover parameters revealed in the patients of the second group (on the low protein diet from 6 months to 5 years) differed significantly from the values observed in the first group (short-term diet). Therefore, although a marked protein depletion exists in chronic uremia, it appears that dietary treatment is not directly responsible for it; moreover, the results obtained in the long-term diet group alone strongly support such a conclusion. In fact, no statistically significant variations of total albumin stores were found in relation to the duration of the low protein diet. This conclusion is also supported by the results obtained in a patient submitted to the study 18 days, 6 months and 1 year, respectively, after the beginning of the diet (see figure 19.4).

It is to be noted that total albumin pool is even less reduced in the long-term diet group than in patients of the first group (see Results), despite the lower degrees of residual renal function in the former group (see blood creatinine levels, tables 19.1 and 19.2).

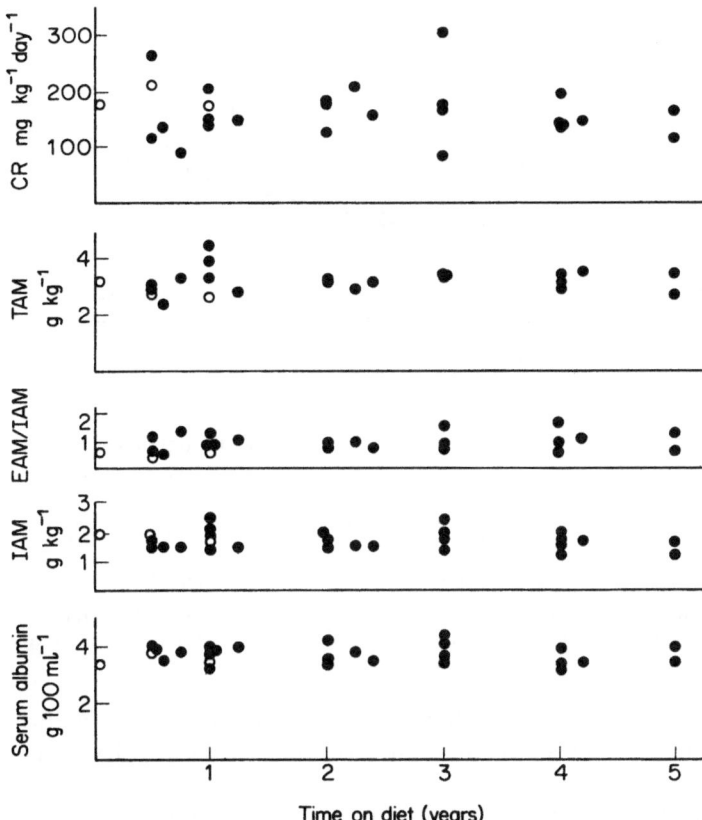

Figure 19.4. Results obtained in the long-term low protein diet group. Open circles represent repeated turnover studies in the same patient at different periods after the start of dietary treatment (see text).

Furthermore, the observed protein depletion is associated with adaptive mechanisms preventing a further decrease of the albumin stores, and maintaining sufficient levels of intravascular albumin. We observe, indeed, that plasma albumin depletion occurs almost solely at the expense of its extravascular fraction (see Results).

Moreover, the decline of serum albumin is buffered also by a reduction of catabolism, a statistically significant positive correlation existing between absolute catabolic rate and total albumin mass (see figure 19.5). Our data confirm previous observations made also in different kinds of patients[11,17–20,22–25].

Steady-state conditions throughout the study are confirmed not only by the stability of the hematochemical and chemical signs, but also by the close agreement between synthesis and catabolism values obtained in all the subjects in whom they were independently measured.

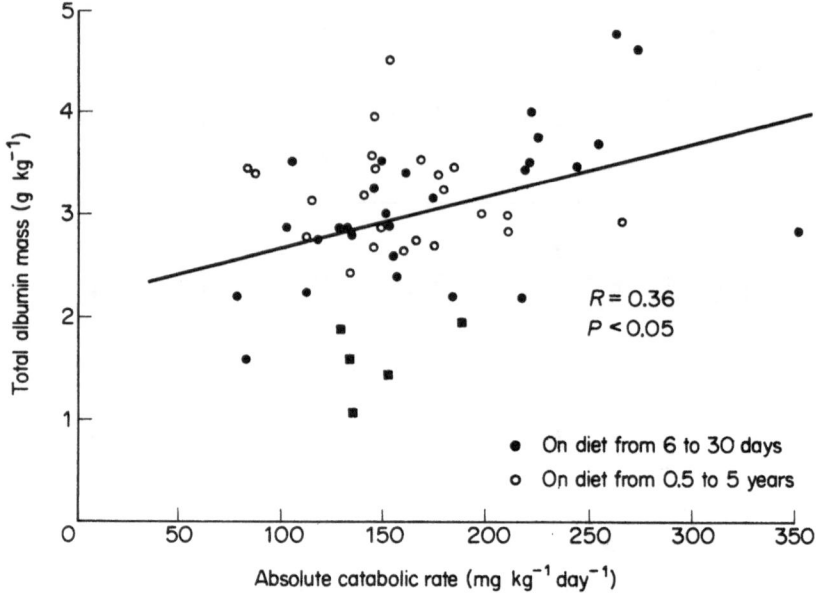

Figure 19.5. Linear regression line ($y = 2.16 + 0.005x$) is plotted together with all experimental points. Squares stand for patients with chronic renal failure and heavy proteinuria.

All these findings suggest that in chronic uremia a new equilibrium state is attained in which body albumin turnover is reset at a level significantly lower than normal.

A possible explanation for this finding is that the readjustment could follow rapid transient changes of albumin turnover in the early phase of uremia, during which catabolism exceeds production. This has been recently demonstrated in a group of uremic patients undergoing hemo- or peritoneal dialysis treatment[26], and is reported elsewhere in this volume.

The extent of the depletion would then depend on the absolute difference between production and catabolism, and on the duration of the unbalanced state. The lack of an inverted compensatory turnover pattern to restore the previous metabolic state still remains unexplained in chronic uremia.

References

1. Shemin, D., Rittenberg, D. Some interrelationships in general nitrogen metabolism. *J. Biol. Chem.*, **153** (1944), 401–421
2. Solomon, G., Tarver, H. The effect of diet on the rate of loss of labelled amino acids from tissue proteins. *J. Biol. Chem.*, **195** (1952), 447–452
3. Garrow, J. S. The effect of protein depletion on the distribution of protein synthesis in the dog. *J. Clin. Invest.*, **38** (1952), 1241–1250

4. Wang, F. Conservative management of chronic renal failure. In *Med. Clin. North Am.*, **55** (1971), 137–154

5. Giovannetti, S., Maggiore, Q. A low nitrogen diet with proteins of high biological value for severe chronic uremia. *Lancet*, **1** (1964), 1000–1003

6. Rosa, U., Pennisi, F., Bianchi, R., Federighi, G., Donato, L. Chemical and biological effects of iodination on human albumin. *Biochim. Biophys. Acta* **133** (1967), 486–498

7. Bianchi, R., Mariani, G. An improved biochemical method for the determination of synthesis rate of liver produced plasma proteins by ^{14}C-carbonate. *J. Nucl. Biol. Med.*, **13** (1969), 167–171

8. Donato, L., Vitek, F., Bianchi, R., Federighi, G. A double tracer method for metabolic studies with iodinated proteins or polypeptides in presence of a relatively slow excretion of iodide. *J. Nucl. Biol. Med.*, **11** (1967), 1–8

9. Bianchi, R., Donato, L., Mancini, P., Mariani, G. Pilo, A. and Vitek, F. (1970) Models for protein synthesis studies. In *Plasma Protein Metabolism*, M. A. Rothschild and T. A. Waldmann, eds., Academic Press, New York, pp. 25–49

10. Bianchi, R., Mariani, G. and Pilo, A. Albumin synthesis measurement by means of an improved two tracer method in patients with chronic renal failure. Effects of low protein diet. *J. Nucl. Biol. Med.*, **14** (1970), 136–144

11. Bianchi, R., Mariani, G. and Pilo, A. (1972) Albumin metabolism in uremic patients on low protein diet. In *Uremia*, R. Kluthe, G. M. Berlyne and B. T. Burton, eds., Georg Thieme Verlag, Stuttgart, 206–219

12. Bianchi, R., Mariani, G. and Pilo, A. Short-term measurement of plasma protein catabolism in man. In *Radioaktive Isotope in Klinik und Forschung*, 10th ed., K. Fellinger and R. Höfer, eds., Urban and Scharzenberg, München, pp. 94–101

13. Bianchi, R., Mariani, G., Pilo, A. and Toni, M. G. Short-term determination of serum albumin catabolism in man from plasma data only *J. Nucl. Biol. Med.*, **17** (1973), 117–118

14. Bianchi, R., Mariani, G., Pilo, A., Toni, M. G. and Donato, L. (1973) Short-term determination of plasma protein turnover by a two-tracer technique using plasma only or plasma and urine data. In *Protein Turnover* ASP, Amsterdam, pp. 47–72

15. Jones, E. A., Craigie, A., Tavill, A. S., Simon, W. and Rosenoer, V. M. Urea kinetics and the direct measurement of the synthetic rate of albumin utilising ^{14}C-carbonate. *Clin. Sci.*, **35** (1968), 553–564

16. Pearson, J. D., Veall, N. and Vetter, H. A practical method for plasma albumin turnover studies. *Strahlentherapie*, **38** (1956), 290

17. Bianchi, R., Mariani, G., Pilo, A. and Donato, L. (1972) Albumin metabolism in patients with chronic renal failure on low protein balanced diet. In *Protides of the Biological Fluids*, 19th ed., H. Peeters, ed., Pergamon Press, Oxford, pp. 483–487

18. Bianchi, R., Federighi, G., Giagnoni, P., Giordani, R., Navalesi, R.

and Donato, L. (1968) Patterns of serum albumin metabolism in renal diseases. In *Physiology and Pathophysiology of Plasma Protein Metabolism*, G. Birke, R. Norberg and L. O. Plantin, eds., Pergamon Press, Oxford, pp. 195–211

19. Picou, D. and Waterlow, J. C. The effect of malnutrition on the metabolism of plasma albumin. *Clin. Sci.*, **22** (1962), 459–468
20. Hoffenberg, R., Black, E. and Brock, J. F. Albumin and gamma globulin tracer studies in protein depletion states. *J. Clin. Invest.*, **45** (1966), 143–152
21. Whitehead, R. G., Frood, J. D. L. and Poskitt, E. M. E. Value of serum albumin measurements in nutritional surveys. A reappraisal. *Lancet*, **2** (1971), 287–289
22. Cohen, S. and Hansen, J. D. L. Metabolism of albumin and gamma-globulin in kwashiorkor. *Clin. Sci.*, **23** (1962), 351–359
23. James, W. P. T. and Hay, A. M. Albumin metabolism. Effect of the nutritional state and the dietary protein intake. *J.Clin. Invest.*, **47** (1968), 1958–1972
24. Kirsch, R., Frith, L., Black, E. and Hoffenberg, R. Regulation of albumin synthesis and catabolism by alteration of dietary protein. *Nature*, **217** (1968), 578–579
25. Bianchi, R., Mariani, G., Pilo, A. and Toni, M. G. Serum albumin turnover in liver cirrhosis. *J. Nucl. Biol. Med.*, **18** (1974), 20–29
26. Mariani, G., Bianchi, R., Pilo, A., Palla, R., Toni, M. G. and Fusani, L. Albumin catabolism measurement by a two tracer technique in uremic patients during a single dialytic treatment. *Europe. J. Clin. Invest.*, 1974 **4** (1974), 435

Discussion

ROSSING

I am most interested in the redistribution of albumin from the extravascular to the intravascular compartment, this being well known also from malnourished people. My question is whether you have used your technique for measuring the fractional transfer rate (FTR). The reduction of extravascular compartment is it due to reduction of FTR or to increase of lymphatic or extravascular return rate?

BIANCHI

Unfortunately, I cannot give any information to this point, since we have not yet analysed our data in this respect. However, when trying to explain reduction of the extravascular albumin pool by changes of the respective output and/or input rates, one must keep in mind that the present series of observations relate to uremic patients in metabolic steady-state, while possible transfer changes would have occurred in the early phase of uremia.

REEVE

It has been shown by Arthur Guyton and co-workers[1] that with a low plasma albumin level, as after plasmapheresis, lymph flow is very much increased so that the return rate of extravascular protein might be much increased.

BIANCHI

I think that this would be a very interesting point to assess.

HOFFENBERG

I was a little puzzled with the fact that the serum albumin concentration in the intravascular compartment did not fall in patients who were taking less than 20 g of protein per day, because I think that experimentally this would be below the level of protein intake required to maintain serum albumin mass at a normal level. Does this imply that there is something special about uremic patients that they are able to compensate for that?

Certainly, when I had human subjects taking 20 g or less of proteins per day this serum protein pattern changed a lot.

BIANCHI

Actually, 18 g was the least value of protein intake daily, but many patients took 30–45 g, according to the level of their renal function. However, in one case we have performed the measurement of albumin turnover in the same patient at the start of the diet, after 6 months and after 1 year, and we did not find any significant difference in serum albumin concentration and the other turnover and distribution parameters.

ROTHSCHILD

I am curious about this too, and I wonder how much the recycling of the urea nitrogen plays a role in the overall nitrogen balance, perhaps permitting an adequate nitrogen source at some level, thereby saving whatever amino acids might be needed for protein production.

BIANCHI

Yes, this is a hypothesis that Giordano made some years ago, of the possibility of some recycling of nitrogen from urea for protein synthesis[2]. More evidence for this point has been recently given by Richards, who proved that in chronic uremia retained urea nitrogen is recycled to the liver as ammonia. A proportion of this nitrogen seems to be reutilised for synthesis of non-essential amino acids and, if their carbon skeletons are supplied, even for synthesis of essential amino acids[3].

TAVILL

We have some recent and as yet unpublished data just on this point. We have been administering ^{15}N labelled urea to uremic patients, and looking for enrichment of albumin with ^{15}N.

Our findings are that recycling of urea is appreciably increased in uremic patients. Since the total body urea mass in many of these patients is increased to 80 g or more there is a very much higher utilisation of urea nitrogen for albumin synthesis. However, when we calculated the efficiency of incorporation of nitrogen released by hydrolysis from urea in the gut, we found it to be very inefficient in both uremic and normal subjects. Although the efficiency was somewhat higher in uremics, I do not think that it could account for the maintenance of the albumin synthetic rate, if dietary amino acids were not already adequate[4].

BIANCHI

I would simply recall that the Giovannetti diet we use in our uremic patients supplies more than the minimum daily requirement of essential amino acids, and we believe its effectiveness in maintaining a relatively satisfactory metabolic state resides just in this.

References

1. Guyton, A. personal communication
2. Giordano, C. Use of exogenous and endogenous urea for protein synthesis in normal and uremic subjects. *J. Lab. Clin. Med.*, **62** (1963), 231
3. Richards, P. Protein metabolism in uremia. *Nephron*, **14** (1975), 135
4. Varcoe, R., Halliday, D., Carson, E. R., Richards, P. and Tavill, A. S. (1975) Efficiency of utilization of urea nitrogen for albumin synthesis by chronically uraemic and normal man. *Clin. Sci. and Mol. Med.*, In press.

20

Nutritional aspects of plasma protein metabolism: the relevance of protein turnover rates during malnutrition and its remission in man

W. P. T. JAMES, P. M. SENDER and
J. C. WATERLOW

It has been known for years that many children with protein-energy malnutrition (PEM) have reduced serum albumin concentrations. Gopalan[1] has suggested that the difference between children with marasmus and those with kwashiorkor is not the result of differences in diet but in the capacities of children to adapt to a reduced intake of energy and protein; the marasmic state is thus the result of well maintained physiological mechanisms of adaptation with conservation of albumin mass, whereas in kwashiorkor there is a pathological breakdown in adaptive mechanisms with a fall in the body's albumin content. Whether these differences represent the result of dietary differences or variations in adaptive capacities has still to be resolved[2,3] but it is clear that kwashiorkor is particularly likely to occur in areas where the staple food has a low protein content, for example in Uganda. Most children in developing countries have to adapt to a reduced intake of food but some diets may demand a greater degree of adaptation if albumin is to be conserved. A better understanding of the pathophysiology of albumin metabolism is an important aspect of the work needed to resolve these problems.

Early experimental studies by Hoffenberg et al.[4] showed that when normal adults were fed for 3–6 weeks on a low protein diet of less than 10 g each day a small reduction in serum albumin concentration was accompanied by a marked fall in the catabolic rate of albumin. Other studies on South African children[5] and in Jamaica[6] had shown that malnourished children had albumin catabolic rates which were about half those found on recovery. In the South African study the children were fed a low protein diet whereas the Jamaican studies were conducted with malnourished children consuming 4 g protein

kg^{-1} day^{-1} (table 20.1). This suggested that the catabolic rate of albumin might reflect the nutritional state rather than the intake of protein[7].

More detailed studies in adults by Hoffenberg[8] and later in children[9] showed that the concept of an albumin catabolic rate adjusting slowly to changes in nutritional state needed modifying. Within a week in adults, and 3–4 days in children on a low protein diet, there was a steady fall in the catabolic rate. A malnourished child, fed a high protein diet, might have a normal catabolic rate within 2–3 weeks despite remaining 30% below his correct weight. Thus albumin catabolism was more dependent on the dietary protein intake than the nutritional state and seemed to change in response to changes in albumin mass.

Table 20.1. Albumin catabolism: its relationship to nutritional state and dietary protein intake

	Nutritional State	Number Studied	Protein Intake $(mg\,kg^{-1}\,day^{-1})$	Albumin Catabolism $(mg\,kg^{-1}\,day^{-1})$
Children				
Cohen and Hansen (1962)[5]	Malnourished	7	0.5–1.3	106
James and Hay (1968)[9]	,,	9	0.73	170
Picou and Waterlow (1962)[6]	,,	6	4.0	231
James and Hay	,,	9	4.3	165
Cohen and Hansen	Recovered	7	4–7	245
Picou and Waterlow	,,	6	4	317
James and Hay	,,	9	3.4	219
Adults				
Hoffenberg *et al.* (1966)[8]	Malnourished	18	1.2	87
	Controls	41	1.2	148

No direct measurements of the synthesis rates of albumin have been made in man to test his responsiveness to changes in protein intake. These measurements, involving the repeated use of ^{14}C bicarbonate would have been difficult to justify in our studies on children. We therefore resorted to a computer analysis, first introduced by Matthews[10], which was designed to overcome the problems which arise when the albumin pools are not in a steady state. The model used was a simple two-pool system. Children were fed sequentially on different protein intakes.

Whole body counting, plasma volume and serum albumin measurements permitted an accurate measurement of whole body activity a month after the single injection of ^{131}I-albumin and a division of the remaining whole body activity into intravascular and extravascular albumin activities. Computer simulation of the distribution of isotope in these pools was compared with the observed values for isotope distribution and loss. It was found empirically that a 10% change in any of the rate constants gave curves which were clearly different from the measured activities so that although the model was an oversimplification of the true state of metabolism, the rate constants could be derived fairly accurately and the generated curves fitted the experimental points very closely indeed. The derived values were used to generate curves to match a week's observations following a change of diet. We expected these computer-generated curves to be very different from the experimental values in the early part of the week as the rate constants adjusted slowly to the altered diet. In practice, however, the change in the rates of synthesis and distribution seemed to occur with sufficient rapidity for the experimental values 24 hours after the change of diet to reflect the rates of distribution and loss observed for the rest of the week. We concluded therefore that synthesis rates changed rapidly in marked contrast to the slow alteration in the rate of albumin catabolism.

Several interesting features emerged from this analysis. First, despite the known fatty infiltration of the liver and the tendency to hepatic failure in malnutrition, these children with PEM had a capacity to synthesise albumin which was completely normal if the supply of dietary protein was adequate (table 20.2). This matched the clinical observations of rapid regeneration of albumin in treated hypoproteinaemic children. Accumulation of albumin would clearly be helped by the combination of a normal or raised synthetic rate and the low catabolic rate.

Table 20.2. The importance of the nutritional state in buffering the effects of low protein feeding on albumin synthesis[9]

Protein intake ($g\ kg^{-1}\ day^{-1}$)	Albumin synthesis ($mg\ kg^{-1}\ day^{-1}$)		fall in synthesis (%)
	3.3–5.0	0.7–1.2	
Malnourished	233 ± 11.2	101 ± 14.6	57
Recovered	222 ± 11.6	148 ± 6.9	33

If all our observations on malnourished children are combined, then it can be seen that albumin synthesis in the malnourished children is clearly related to the protein intake (figure 20.1). All the protein intakes were at or above the intakes known to keep children in nitrogen balance[11]. Thus marked changes in albumin synthesis could occur when protein intakes were considered adequate on the basis of nitrogen balance. These observations parallel those of Whitehead *et al.* who in epidemiological studies have recently shown that albumin levels may fall slowly at a time when growth (and presumably nitrogen retention) are continuing[12]. Neither growth nor nitrogen retention are therefore indices of 'optimal' nutrition since albumin metabolism may already have adapted with changes in synthesis and catabolism and a small reduction in serum albumin concentration. The data in figure 20.1 suggest that as protein intakes fall close to requirement levels then marked changes in albumin synthesis will occur and small reductions in intake below the requirement value, e.g. 0.7 from protein per kg for a year-old child, may lead to falls in synthesis too great for albumin catabolism to compensate.

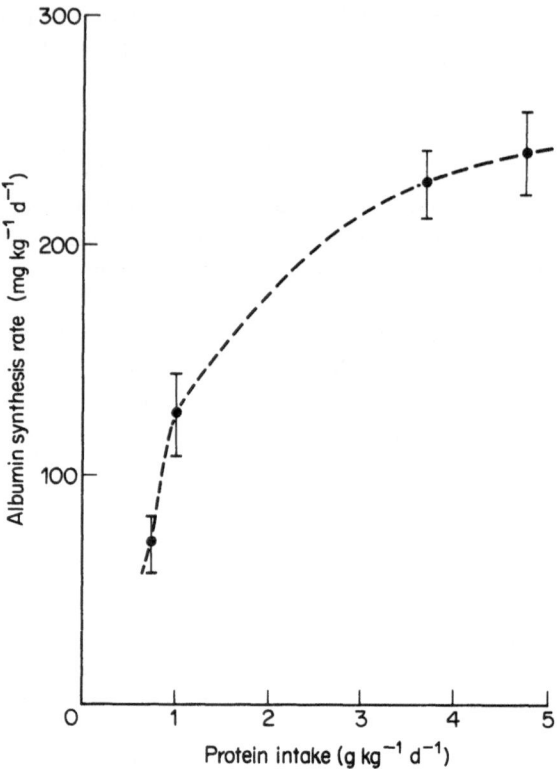

Figure 20.1. The effect of different levels of protein intake on the synthetic rate of albumin synthesis in malnourished children[9].

Our data also showed that the malnourished child is peculiarly sensitive to dietary changes in protein. On low protein feeding a well nourished child's synthesis rate falls by only a third but the fall in the malnourished child's synthesis was significantly greater, i.e. 57% $p < 0.01$ (table 20.2). This buffering effect of a good nutritional state has great practical importance in clinical medicine and in public health.

The greater capacity of a healthy child to maintain albumin synthesis could relate either to a more efficient mechanism for conserving nitrogen or to the greater availability of amino acids from other body proteins in the well nourished child. Nitrogen conservation, however, has been shown to be more, not less, efficient in the malnourished child[11]. It is probable that the recovered child with his much greater protein mass is able to divert amino acids from less important tissues, e.g. muscle, to the liver for albumin synthesis. As the dietary inflow of protein is reduced the liver must derive its amino acids for the synthesis of export proteins either from its own hepatic proteins or by a supply from elsewhere, e.g. muscle. We have suggested[13] that liver differs from muscle in modulating its mass primarily by changes in protein breakdown while muscle mass is influenced usually by rapid changes in protein synthesis. If liver protein is broken down at a faster rate in states of protein deficiency, this may contribute some amino acids to albumin synthesis. The major supply of amino acids must, however, come from extrahepatic sources. Whole body analyses have shown that in malnutrition the mass of non-collagen protein may be reduced to only a half of its normal value with a very marked reduction in muscle mass[14]. With a small muscle mass the flux of amino acids to the liver may well be impaired in malnutrition and lead to a much greater sensitivity of albumin synthesis to the dietary inflow of protein (figure 20.1). The small release of amino acids from muscle appears in part to be an adaptation to prolonged dietary restriction[15], and associated with a slowing of the breakdown rate of muscle protein[16] but the quantitative reduction in muscle protein available for breakdown must be a major factor.

In starvation, amino acid release from muscle may be greater than on a protein deficient diet because plasma insulin levels are lower—feeding energy will stimulate insulin release and tend to limit the mobilisation of muscle protein and increase the peripheral distribution of dietary amino acids[17]. Our studies were all conducted with isocaloric diets: insulin stimulation during low protein feeding was presumably maintained and perhaps limited albumin synthesis. Studies on the effects of energy restriction might have shown much smaller effects on albumin turnover.

Finally, our observations in children supported those of Hoffenberg *et al.*[8] in adults in suggesting that extravascular albumin mass was mobilised when synthesis rates fell. These changes in extravascular ^{131}I albumin were a necessary consequence of changes in albumin synthesis if the intravascular mass was to be preserved. No detectable change in serum albumin concentration occurred and no consistent trend in plasma volumes was observed which could have reduced the validity of the analysis. We concluded that the fall in extravascular activity probably did reflect changes in the distribution of albumin (table 20.3). This then suggested that the exchange of albumin between plasma and tissues is under some control which is independent of either the rates of albumin synthesis or catabolism and independent of the serum albumin concentration.

Table 20.3. Net transfer rates of albumin into extravascular tissues in malnourished and recovered children*

| Dietary intake | Intravascular albumin pool accumulating in tissues per day (%) | | | |
	High	Low	Low	High
Time from beginning of study (day)	0–10	11–17	18–24	18–24
Malnourished	+3.2	−5.7	−4.1	+2.9
Recovered	+0.6	−3.8	−1.5	+2.8

*Calculated from reference 9.

Decreases in extravascular albumin seemed to be an immediate effect which buffered circulating albumin levels before catabolic rates could adjust (table 20.3). Animal studies by Sellers *et al.*[18] and by Katz *et al.*[19] have also shown that in hypoalbuminaemic rats either on a protein deficient diet or with nephrosis, the extravascular albumin mass is also depleted and may be selectively lost (table 20.4). The depletion of extravascular albumin from individual tissues in protein depletion seemed to be related to the protein turnover rates of the tissue studied. Thus, gut has probably the highest turnover rate in the body, and muscle one of the lowest: depletion of extravascular albumin was greatest in gut and least in muscle. Whether the turnover of both tissue proteins and albumin within the tissue is under the same control needs further work but a relationship seems to exist between the rate of tissue protein breakdown and tissue albumin depletion in severe protein deficiency.

Table 20.4. Distribution of albumin in protein depletion*

	Control (%)
Total body albumin	55.2
Blood	61.7
Gut	29.0
Skin	47.3
Viscera	52.2
Muscle and Bone	77.3

*Recalculated from data in reference 18.

We might expect these changes in albumin turnover to be accompanied by alterations in body protein turnover. Picou and Taylor-Roberts[20] measured total turnover with ^{15}N glycine introduced for such studies by Sprinson and Rittenberg[21]. Picou, however, employed Waterlow's technique[22] of infusing the isotope rather than giving it as a single injection. This both simplified the calculations of turnover and enabled a check to be made on label recycling from protein; recycling is minimal if constant labelling of the urea pool is achieved. The results of these studies are summarised in table 20.5.

We have recently used two isotopes U-^{14}C-L-tyrosine[23] and 1-^{14}C-L-leucine

Table 20.5. The turnover rate of nitrogen in children when malnourished and after recovery[20]

	Number	Nitrogen flux (mgN kg^{-1} h^{-1})	(mgN h^{-1})
Malnourished	5	87.6 ± 14.1	425.6 ± 82.6
Recovered	5	55.8 ± 5.5	427.6 ± 41.4
		Mean ± SEM	

to measure plasma flux of amino acids in adults. We have observed a marked fall in flux on semi-starvation so that after three weeks there is a fall of more than 50% in tyrosine flux. If this is related to protein turnover rates by assuming a tyrosine content of protein of 3% and a leucine content of 8% then marked changes in turnover occur, consistent with a steady adaptation in protein breakdown rates. This is shown in table 20.6. Whilst these observations reflect the response to semi-starvation, not protein depletion, they do fit in with the results of Kerr *et al.*[24] on reduced amino acid oxidation in malnutrition. The turnover data are also consistent with the reduced excretion in malnutrition of 3-methyl-histidine which is a non-reutilisable derivative from myofibrillar protein breakdown[16]. They also match the decreased efflux of amino acids from muscle in prolonged starvation[15] and the fall in 3-methyl histidine output[25].

Table 20.6. The fall in plasma flux of amino acids in semi-starvation

Length of semi-starvation (day)	3	20
Amino acid studied	Leucine*	Tyrosine†
Amino acid flux % control	87.9±4.4	44.5±4.7
	(Mean ± SEM. Five patients in each group)	

*O'Keefe, Sender and James (unpublished)
†Reference 23.

There may be several reasons for the high turnover rates found in the ^{15}N glycine studies. Changes in body composition were marked with a great loss of muscle mass. Since muscle mass, despite its lower turnover rate than liver, accounts for a considerable proportion of the body protein turnover, we would expect the total rate in absolute terms to have fallen sharply in the malnourished children. Changes in body composition cannot account for these observations therefore. This is emphasised by the recent report of Steffee *et al.*[26] who found increased turnover rates in normal adults fed on a moderately low protein intake for 3–6 weeks and assessed with ^{15}N glycine. In these adult cases compositional changes could not have been substantial.

Two main reasons may be given for the change in turnover on low protein diets when measured with ^{15}N glycine. First the ^{15}N excretion will reflect the hepatic labelling of ^{15}N and this may become dominated by changes in hepatic protein turnover if exchange rates between plasma and liver do not alter also. Thus increased compartmentation of the hepatic pool may occur in nutritional stress when protein turnover in the liver changes with increases in both fractional synthesis and breakdown (Garlick, Millward and James, unpublished).

Secondly, increasing transamination in protein deficiency may also make available tracer ^{15}N to more amino acid pools with the possibility of sequestration of label. Increased transamination is known to occur in malnutrition with high transaminase levels in both liver and muscle in animals fed a reduced protein intake[27,28].

Our animal studies on tissue turnover rates on protein free diets or in starvation all show a fall in turnover and reduced brain, heart, muscle and kidney synthesis and breakdown rates for protein. This, together with our ^{14}C studies in adults and the reduced excretion of 3-methyl histidine in malnutrition, favours the view that body protein turnover falls, not rises, in malnutrition. If we accept this interpretation, more methodological studies with ^{15}N are clearly needed.

In conclusion, it seems reasonable to suppose that the steady fall in albumin catabolism which follows the feeding of a low protein diet occurs in parallel with changes in total protein turnover. Although individual tissue proteins have their individual turnover rates, it seems possible that the same mechanisms which control protein breakdown rates in tissues are also responsible for the control of albumin catabolism.

Acknowledgement

We thank Dr A. Coward for his helpful comments on extravascular albumin mass.

References

1. Gopalan, C. Kwashiorkor and marasmus-evolution and distinguishing features. (1968). In *Calorie Deficiencies and Protein Deficiencies*, McCance, R. A. and Widdowson, E. M., eds., J. A. Churchill, Ltd., London, p. 49 ·
2. Whitehead, R. G. and Alleyne, G. A. O. Pathophysiological factors of importance in protein–calorie malnutrition. *Brit. Med. Bull.*, **27** (1972), 72
3. Rao, K. S. J. Evolution of kwashiorkor and marasmus. *Lancet*, **i** (1974), 709
4. Hoffenberg, R., Saunders, S., Linder, G. C., Black, E. and Brock, J. F. (1962) [131]I albumin metabolism in human adults after experimental protein depletion and repletion. In *Protein Metabolism*, F. Gross, ed., Springer-Verlag, Berlin, p. 314
5. Cohen, S. and Hansen, J. D. L. Metabolism of albumin and γ-globulin in kwashiorkor. *Clin. Sci.*, **23** (1962), 351
6. Picou, D. and Waterlow, J. C. The effect of malnutrition on the metabolism of plasma albumin. *Clin. Sci.*, **22** (1962), 459
7. Waterlow, J. C. Protein malnutrition and albumin breakdown. *Lancet*, **2** (1962), 1279
8. Hoffenberg, R., Black, E. and Brock, J. F. Albumin and γ-globulin tracer studies in protein depletion states. *J. Clin. Invest.*, **45** (1966), 143

9. James, W. P. T. and Hay, A. M. Albumin metabolism: effect of the nutritional state and the dietary protein intake. *J. Clin. Invest.*, **47** (1968), 1958

10. Matthews, C. M. E. (1965) Application of an analogue computer to analysis of experiments with ^{131}I labelled plasma proteins when pools are not in dynamic equilibrium. In *Radioaktive Isotope in Klinik und Forschung*, K. Fellinger and R. Hofer, eds., Urban and Schwarzenberg, Munich, p. 240

11. Chan, H. and Waterlow, J. C. The protein requirement of infants at the age of about one year. *Brit. J. Nutr.*, **20** (1966), 775

12. Whitehead, R. G., Frood, J. D. L. and Poskitt, E. M. E. Value of serum albumin measurements in nutritional surveys. *Lancet*, **ii** (1971), 288

13. Garlick, P. J., Millward, D. J. and James, W. P. T. The diurnal response of muscle and liver protein synthesis *in vivo* in meal-fed rats. *Biochem. J.*, **136** (1973), 935

14. Garrow, J. S., Fletcher, K. and Halliday, D. Body composition in severe infantile malnutrition. *J. Clin. Invest.*, **44** (1965), 417

15. Pozefsky, T., Felig, P., Tobin, J. Amino acid balance across tissues of the forearm in postabsorptive man. Effects of insulin at two dose levels. *J. Clin. Invest.*, **48** (1969), 2273

16. Rao, B. S. and Nagabhushan, V. S. Urinary excretion of 3-methylhistidine in children suffering from protein–calorie malnutrition. *Life Sciences*, **12** (1973), 205

17. Whitehead, R. G. (1971) Metabolic changes in protein–calorie malnutrition and the assessment of nutritional status. In *Proceedings of the International Congress of Pediatrics*, Vol. II. Nutritional and Gastroenterology, Vienna, p. 231

18. Sellers, A. L., Katz, J., Bonorris, G. and Okuyama, S. Determination of extravascular albumin in the rat. *J. Lab. Clin. Med.*, **68** (1966), 177

19. Katz, J., Bonorris, G., Golden, S. and Sellers, A. L. Extravascular albumin mass and exchange in rat tissues. *Clin. Sci.*, **39** (1970), 705

20. Picou, D. and Taylor-Roberts, T. The measurement of total protein synthesis and catabolism and nitrogen turnover in infants in different nutritional states and receiving different amounts of protein. *Clin. Sci.*, **36** (1969), 283

21. Sprinson, D. B. and Rittenberg, D. The rate of utilisation of ammonia for protein synthesis. *J. Biol. Chem.*, **180** (1949), 707

22. Waterlow, J. C. Lysine turnover in man measured by intravenous infusion of L-U-^{14}C-lysine. *Clin. Sci.*, **33** (1971), 507

23. James, W. P. T., Garlick, P. J. and Sender, P. M. Studies of protein metabolism in man with infusions of [^{14}C] tyrosine. *Clin. Sci. Mol. Med.*, **46** (1974), 8

24. Kerr, D. S., Stevens, M. C. G., Robinson, H. M. and Picou, D. (1973) Hypoglycaemia and the regulation of fasting glucose metabolism in malnutrition. In *Endocrine Aspects of Malnutrition*, Gardner, L. I. and

Amacher, P. eds., Kroc Found. Sympos. No. 1, p. 313

25. Young, V. R., Haverberg, L. N., Bilmazes, C. and Munro, H. N. Potential use of 3-methyl histidine excretion as an index of progressive reduction in muscle protein catabolism during starvation. *Metabolism*, **22** (1973), 1429

26. Steffee, W. P., Pencharz, P. V., Goldsmith, R. S., Anderson, C. F. and Young, V. R. Protein intake and total body protein turnover in adult subjects. *Fed. Proc.*, **32** (1973), 916

27. Mimura, T., Yamada, C. and Swendseid, M. E. Influence of dietary protein levels and hydrocortisone administration on the branched chain amino acid transaminase activity in rat tissues. *J. Nutr.*, **95** (1968), 493

28. Ichihara, A., Noda, C. and Ogawa, K. Control of leucine metabolism with special reference to branched chain amino and transaminase isozymes. *Adv. Enzyme Regn.* **11** (1973), 155

Discussion

REEVE

I want to put forward an hypothesis. I think of albumin as a constitutive protein, in the sense that there is a certain amount of messenger RNA for albumin which is being made continuously. The main regulation of synthesis is in the amounts of amino acids that get to the messenger RNA, whereas proteins such as fibrinogen and some of the acute phase proteins are non-constitutive proteins in the sense that they have intrahepatic regulatory mechanisms that depend not just on translation but on transcription too.

JAMES

I would agree that there certainly appear to be differences in the way intra-cellular hepatic and hepatic export proteins are regulated. All the data from perfusion work suggest that control of synthesis is important for export proteins but we believe that hepatic proteins may be controlled primarily by changes in breakdown rates.

I think you are now trying to propose different types of control for export proteins which we have not investigated.

TAVILL

I may have misunderstood you, but did you say that in protein malnutrition the turnover of muscle protein is reduced? How then does the muscle waste, in simple terms?

JAMES

The mass is dependent on the balance between synthesis and breakdown of muscle protein. Breakdown, as reflected by 3-methylhistidine excretion, is completely different from the net outflow of amino acids from a muscle. I think that what we are demonstrating is a very slow steady change in break-down rates which fall over a period of time.

The balance of movement of amino acids into or out of muscle is dependent on a movement of synthesis either above or below that breakdown level.

TAVILL

Can we get that point clear? You are saying that the muscle wasting is a result of reduced synthesis and the extreme loss of weight that patients on a low protein diet develop is a consequence of reduced synthesis, not of increased breakdown?

JAMES

Correct, in fact it may be a reduced synthesis, even in face of a reduced breakdown rate, but the fall in synthesis is greater than the fall in breakdown.

TAVILL

Yes, this is a concept that previously has been held for liver-produced export proteins, and you are extending it now to general tissue proteins.

JAMES

Correct.

STEIN

Going back to ^{15}N plateaus, first of all they are apparent and I think it is a very important fact. In 1955 Neuberger took a rabbit and waited 45 days before seeing a plateau in ^{15}N, using glycine as the ^{15}N carrier[1]. If you waited 45 days in your procedure, you would see a plateau between 35 and 45 days and during the period of 1–2 days after injection there would be a fairly steep slope. It is very much a matter of the timescale you use.

Secondly, all the ^{15}N methods for calculating the rate of human protein synthesis are only comparable with each other. They do not give absolute values. Your method is dependent on the height of the plateau in blood, which in turn is dependent on the time you decide to measure the plateau labelling.

The key thing is that nobody has really shown whether these ^{15}N methods give really valid measures of human protein synthesis.

JAMES

On the first point, we can demonstrate a plateau value (with 1-^{14}C-leucine) from a time point of 8–10 hours through to 30 hours and there is no question under such circumstances of recycling of label from labelled proteins. We are in fact effectively able to ignore re-entry. These methods when used in both normal and abnormal people give highly consistent results. Whether the result is a true measure of the absolute total body protein turnover depends primarily on the heterogeneity of the free amino acid pool, which for various reasons in the human we think is likely to be a small problem.

We are not able to say what the normal protein turnover rate is until we have done many more labelling studies with different labels, particularly ^{14}C, to see if we get the same value. So far, our values with labelled tyrosine and leucine appeared to be the same.

On your second point, very little work has been done with ^{15}N.

I think ^{15}N is probably going to be fine under normal dietary circumstances but not when we start to manipulate the protein intake, which will create problems. Do not forget that the ^{15}N can move to a lot of different labelling sites, and this may be one of the explanations for the apparently high synthetic rate values obtained from measurements over short periods of time. You in fact are having outflows of label not into protein but into other pools, into glutathione, nucleotide synthesis and so on.

ENWONWU

Dr James, you mentioned 3-methylhistidine in urine. Is its urinary excretion low at all stages of starvation, or do you pass through a phase in which the 3-methylhistidine in the blood coming from the breakdown of muscle could be moving in directions other than excretion, e.g., into the brain? There is evidence that there is a high concentration and an increase in 3-methylhistidine in the brain. Could you comment on this?

JAMES

The published data, as you know, suggest that during progressive starvation there is indeed a progressive fall without an immediate preliminary rise in 3-methylhistidine excretion[2]. Now, in relation to the direction of movement or the validity of 3-methylhistidine as an index of muscle breakdown, I think one has to wonder whether platelets with their 3-methylhistidine content contribute to urinary excretion, how much they contribute, and so on. As far as I know, there are no quantitative determinations of the relationships between 3-methylhistidine excretion and an independent measure of muscle breakdown.

MILLER

Can you relate your work on rates of breakdown in malnutrition to the observations of Flatt and Blackburn in their treatment of protein loss in post-operative surgical situations[3]? I have a feeling that certain fundamental principles operate here and may determine the extent of the protein loss in individuals who are taking moderate but insufficient calories. An individual seems to be worse off in terms of losing protein from the body protein mass if they take a modest caloric intake, i.e. 40–50% of optimum, than they are when taking very few calories as pure protein. This implies of course they have some body fat to depend on in fasting ketosis.

JAMES

We are well aware of these data and at the moment are checking the ideas of Flatt and Blackburn in conjunction with turnover studies. I think it is entirely possible that there are insulin mediated effects post-operatively which depend on insulin sensitivity as it applies to protein synthesis and fat mobilisation. You may get a situation where muscle synthesis will tend to be preserved by amino acid infusions with amino acid oxidation continuing normally.

References

1. Henriques, O. B., Henriques, S. B. and Neuberger, E. Quantitative aspects of glycine metabolism in the rabbit. *Biochem. J.*, **60** (1955), 409
2. Young, V. R., Haverberj, L. N., Bilmazes, C. and Munro, H. N. Potential use of 3-methylhistidine excretion as an index of progressive reduction in muscle protein catabolism during starvation. *Metab. Clin. Experim.*, **22** (1973), 1429
3. Blackburn, G. L., Flatt, J. P., Clowes, G. H. and O'Donnel, T. E. Peripheral intravenous feeding with isotonic amino acid solution. *Am. J. Surg.*, **125** (1973), 447

21
Regulation of synthesis of immunoglobulins

T. A. WALDMANN, S. BRODER, M. DURM and M. BLACKMAN

Turnover studies using radioiodinated purified immunoglobulin molecules have been used to define the metabolic parameters of the different classes of immunoglobulins in normal man and to define the physiological factors regulating immunoglobulin catabolism and transport[1]. In addition, as we have emphasised in previous Congresses, such studies have lead to the development of new insights into the pathophysiology of abnormalities of immunoglobulin levels and have led to the discovery of new classes of immunodeficiency disease associated with short survivals of immunoglobulin molecules[2]. Such turnover studies, however, are not of major value in the definition of the complex pathways of cellular differentiation and biosynthesis which are required for the normal immune response and for immunoglobulin biosynthesis, nor are they of value in defining the precise defects in these events that occur in patients with the different immunodeficiency diseases. We have, therefore, developed an entirely different technique to study the factors controlling immunoglobulin synthesis and secretion by lymphocytes and their daughter cells. These studies have been directed towards defining the physiological control mechanisms that regulate immunoglobulin synthesis and toward developing new insights into the nature of the defects of immunoglobulin synthesis and secretion that occur in patients with the primary immunodeficiency diseases or with the neoplastic diseases affecting lymphocytes and plasma cells.

Before considering the results of these specific studies we will introduce the features of cellular maturation and cellular interaction required for the normal immune response (figure 21.1). There is now ample evidence that cells destined to subserve immunological functions derive from primitive stem cells located in the bone marrow. Such cells migrate to central lymphoid organs where under appropriate inductive influences they differentiate into cells that can interact with antigens. One route of migration is through the thymus. At this site, cells differentiate into T-lymphocytes or T-cells which give rise to

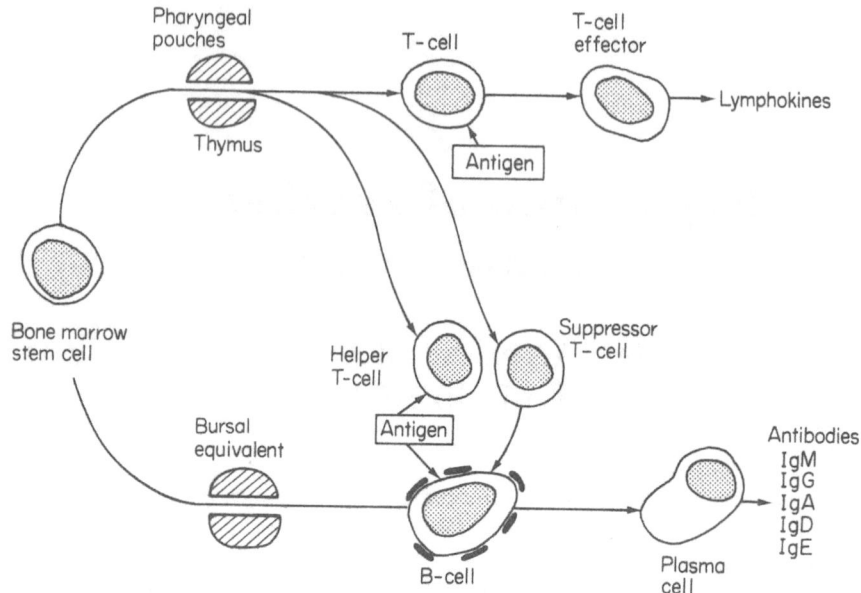

Figure 21.1. Scheme of cellular differentiation, cellular interaction and cellular
biosynthesis required for specific immune response.

populations of cells that take part in cellular immune reactions such as delayed
skin responses or transplant rejection. The functional capacity of these T-
lymphocytes can be assessed *in vitro* by a number of techniques. In one such
technique the capacity of such cells to form lymphoblasts which incorporate
thymidine into DNA when cultured in the presence of antigens or certain
mitogens termed 'T' cell mitogens such as phytohaemagglutinin, staphlococcal
filtrate and concanavalin A is assessed.

The development of antibody-producing plasma cells can also be considered
a discontinuous developmental process which can be divided conveniently
into two stages. The first stage involves the primary differentiation of stem
cells into immunocompetent B-cells or B-lymphocytes without requirement of
exogenous antigenic stimulation but with the cooperation of a central lymphoid organ, the bursa of Fabricius in birds or its yet unknown equivalent in
man. B-lymphocytes can be identified in the peripheral blood by the demonstration of immunoglobulins bound to the cell surface membrane in high
density using immunofluorescent antisera. These membrane bound immunoglobulins are the receptors through which antigens are recognised. The union
of antigen and surface antibody triggers second stage events which include
B-cell proliferation and terminal differentiation into antibody-secreting
plasma cells. This process of B-cell maturation can also be activated by certain
stimulants including pokeweed mitogen and lipopolysaccharide. All or most of
the B-cells appear to have receptors on their surface for these stimulants by

which they can be triggered into terminal differentiation and immuno-globulin synthesis. The T- and B-cell pathways of cellular differentiation interact in the course of normal humoral immune responses (figure 21.1). Specifically, many antigens require both the presence of T- and B-lymphocytes to induce a full circulating antibody response. For these antigens the T-cells serve as a helper cell and the B-lymphocyte is the cell that matures to synthesise the antibody. More recently it has been recognised that certain T-cells may act as negative regulators of B-cell maturation and inhibit this process. There is evidence for both antigen specific and antigen non-specific T-cell regulatory or suppressor effects[3].

The immunological deficiency diseases associated with hypogamma-globulinemia have different disorders with defects at distinct positions in the sequence of development of immunoglobulin producing cells. For example, patients with the disorder severe combined immunodeficiency have been identified as having a defect in the stem cell precursor of immunocompetent cells and the majority of patients with infantile x-linked agammaglobulinemia (Bruton-type agammaglobulinemia) appear to have a defect in the differentia-tion of these stem cells into B-lymphocytes. Patients with x-linked agamma-globulinemia are in general quite deficient in B-lymphocytes[4-6]. In contrast, many patients with the common variable immunodeficiency or acquired agammaglobulinemia have normal percentages of circulating B lympho-cytes[4,6,7]. Nevertheless, these patients are unable to produce normal quantities of immunoglobulin. Similarly, most patients with isolated IgA deficiency have normal proportions of IgA-bearing B-lymphocytes yet are unable to produce normal amounts of circulating and secretory IgA[5,6,8]. These observations have led to the view that both patients with common variable hypogammaglobulin-emia and those with isolated IgA deficiency have a defect in the terminal differentiation of B-lymphocytes into mature immunoglobulin secreting and synthesising cells. These observations that the most common immunodefi-ciency diseases in man are associated with an apparent defect in the differen-tiation of B-lymphocytes into immunoglobulin synthesising and secreting cells have led us to develop a technique wherein we could define the nature of this critical process and its defects in patients with immunodeficiency disorders.

Methods

To study the terminal differentiation of lymphocytes, exhaustively washed cells were cultured *in vitro* in the presence or absence of a B-cell mitogen, poke-weed mitogen. The immunoglobulin produced and secreted by such cells after 7 days was then measured with sensitive radioimmunoassay techniques specific for IgM, IgG, or IgA. In this technique supernatant cells obtained from 20 to 50 cc of heparinised blood were sedimented at 37°C. In order to remove all human serum proteins, the cells were washed 12 times with balanced salt solution containing 5% heat-inactivated foetal calf serum. The cells at a concentration of 2×10^6 lymphocytes ml^{-1} were then incubated at 37°C in 5% CO_2 in loosely capped vials in the presence or absence of pokeweed mitogen using 1640 medium supplemented with glutamine, penicillin, streptomycin and 10% foetal calf serum. At the termination of the culture period the tubes were centrifuged at 2500 rpm (1200 G) for 10 minutes. The amount of IgG, IgA, and IgM synthesised and secreted into the culture medium was then determined by double antibody radioimmunoassay of these immunoglobulins. The techniques for the production of the antisera, defining their specificity and performance of the radioimmunoassay test were essentially identical to those previously described for double antibody radioimmunoassay of IgE[9].

In a number of studies, aliquots of cells obtained on the seventh day of culture were assayed for cytoplasmic immunoglobulin molecules by direct and indirect staining using fluorescein labelled antibodies. Purified T-cells were prepared by spontaneous sheep cell rosette techniques combined with density centrifugation[10]. The T-cell preparations were shown to contain less than 1% contamination of B-cells as assessed by cell surface immunofluorescence and less than 1% macrophages as defined using a stain for non-specific esterase[11].

Results and discussion

Peripheral blood lymphocytes of normal individuals synthesised modest quantities of immunoglobulins when cultured in the absence of pokeweed mitogen. In these studies the geometric mean rates of IgG, IgA, and IgM synthesis were 212, 303, and 537 ng per 2×10^6 lymphocytes, respectively, over the 7-day culture period. Immunoglobulin production by lymphocytes was markedly stimulated when pokeweed mitogen was added at the initiation of the culture. The geometric mean synthetic rates for IgG, IgA, and IgM were 1641, 1698, and 3715 ng per 2×10^6 lymphocytes, respectively, over the 7-day culture period. This represents a 5- to 7-fold stimulation of immunoglobin synthesis in the presence of pokeweed mitogen as compared to immunoglobulin synthesis when incubated in foetal calf serum containing medium alone. An analysis of the time course of synthesis and release of immunoglobulin molecules by lymphocytes in the presence of pokeweed mitogen indicated there was very little synthesis and release of immunoglobulin by lymphocytes during the first three to four days of culture, however, after this time the concentration of immunoglobulin detectable in the culture medium increased rapidly indicating maturation from non-immunoglobulin secreting B-cells into immunoglobulin synthesising and secreting cells. The first immunoglobulin class to appear was IgM, with IgG and IgA following shortly thereafter. The immunoglobulin synthetic process could be inhibited by at least 90% by irradiation of the cells with 2000R or by the addition of puromycin (5×10^{-4} M), actinomycin D (10 μg ml^{-1}), mitomycin C (25 μg ml^{-1}), or cytosine arbinoside (2×10^{-4} M) to the culture medium. These studies, taken in conjunction with the time course studies, suggest that a critical mitosis is required before a major increase in lymphocyte immunoglobulin synthesis is possible following exposure of these cells to pokeweed mitogen.

Another finding of major significance in our understanding of the regulation of B-cell maturation is that the addition of 'T' cell mitogens such as phyto-haemagglutinin, staphylococcal filtrate or concanavalin A to a culture of normal B-lymphocytes and the B-cell stimulant pokeweed mitogen inhibited B-cell maturation and immunoglobulin synthesis. This observation suggests that activated T-cells or their products can act as non-specific inhibitory regulators of B-cell maturation and immunoglobulin synthesis.

The technique for the study of *in vitro* immunoglobulin biosynthesis and secretion was applied to the study of the peripheral blood lymphocytes of patients with a variety of immunological deficiency diseases. As noted earlier, patients with common variable hypogammaglobulinemia or acquired hypo-gammaglobulinemia have markedly reduced rates of antibody production *in vivo* yet have significant numbers of B-cells in the majority of patients. The lymphocytes of 13 patients with this disorder studied using the *in vitro* culture technique were not able to make significant quantities of IgG, IgA, or IgM. That is, in no case did any of the patients synthesise and secrete as much as 100 ng of any of the three immunoglobulin classes studied in the presence or in the absence of pokeweed mitogen during this 7-day culture period. There was also no evidence of immunoglobulin synthesis as assessed by cytoplasmic immunofluorescence in 10 of the 11 patients studied. The remaining patient had cells showing cytoplasmic immunofluorescence although no immuno-globulin was detectable in the culture medium. Thus, these patients were shown to have a disorder in the terminal differentiation of B-cells into cells which can both synthesise and secrete immunoglobulin molecules. To deter-mine whether the failure of lymphocyte immunoglobulin synthesis by these patients with hypogammaglobulinemia was due solely to an intrinsic defect in the maturation of their B-lymphocytes or whether it could be due to the pre-sence of suppressor cells, the lymphocytes from patients with common variable hypogammaglobulinemia were incubated with lymphocytes from normal individuals. In these studies, lymphocytes from the controls and lymphocytes from the patients were cultured together (2×10^6 cells from each source in 2 ml of media) in the presence of pokeweed mitogen. The synthesis of immuno-globulins from the cells of two subjects in co-culture as related to the sum of that synthesised by the cells of the individuals cultured alone was determined using the following relationship:

$$\text{synthesis of immunoglobulin (Ig) by cells in co-culture as a percentage of that synthesised individually} = \frac{\text{synthesis of Ig by cells in co-culture} \times 100}{\text{sum of Ig synthesised by cells of each individual cultured separately}}$$

The synthesis of immunoglobulins by the normal cells was reduced by 84–100% of normal when incubated with lymphocytes from 6 of the 9 hypogammaglobulinemic patients studied. The cells of the remaining 3 patients did not inhibit normal immunoglobulin synthesis in co-culture. A comparable suppression of immunoglobulin synthesis by normal lymphocytes was observed when they were co-cultured with purified thymus-derived lymphocytes (T-cells) from a hypogammaglobulinemic patient. In control studies, suppression of immunoglobulin synthesis was not significant when normal cells were co-cultured with lymphocytes from unrelated normal individuals or from patients with the Sezary syndrome. Nor were they significantly inhibited when incubated with purified T-cells from normal individuals. In addition, no suppression of immunoglobulin synthesis by normal lymphocytes was observed when they were cultured in the serum from the hypogammaglobulinemic patients. These studies suggest that in some patients the disease common variable hypogammaglobulinemia may be caused or perpetuated by an abnormality of regulatory T-cells which act to suppress B-cell maturation and antibody production. This suppression of B-cell maturation and immunoglobulin biosynthesis by T-cell suppressors represents a new previously undescribed pathogenic mechanism resulting in hypogammaglobulinemia and is the first disorder of suppressor T-cells observed in man.

We next applied the *in vitro* lymphocyte immunoglobulin biosynthesis technique to the study of patients with isolated IgA deficiency or with IgA deficiency associated with the immunodeficiency syndrome ataxia-telangiectasia. Virtually all of these patients with absence of IgA from the serum had normal numbers of B-lymphocytes that bear IgA molecules on their cell membrane. Thus, these patients again appear to have a defect in the terminal differentiation of IgA bearing B cells into IgA secreting plasma cells. In the *in vitro* culture system with pokeweed mitogen the lymphocytes of these patients secreted virtually no IgA into the medium. However, in contrast to the patients with common variable hypogammaglobulinemia 8 of 9 such patients studied had cells with IgA in their cytoplasm, as assessed by cytoplasmic immunofluorescent studies, following 7 days of *in vitro* culture in the presence of pokeweed mitogen. These studies suggest that limited IgA synthesis, at least, is possible in response to pokeweed mitogen in such patients but that secretion of IgA from these cells is abnormal. In general, such patients did not have circulating or cellular inhibitors of B-cell maturation nor could they be made to synthesise and secrete IgA into the media using a variety of techniques. There was one interesting exception to this latter generalisation. In this patient there was a severe defect in thymic function as well as the defect in IgA production. The cells of this patient did not make significant quantities of IgA when cultured with pokeweed mitogen alone. However, when T-cells from a patient with the Sezary syndrome, which did not in themselves synthesise IgA, were added to this culture system, the cells of the patient with thymic and IgA deficiency produced large quantities of IgA in the presence of pokeweed mitogen indicating that an interaction of T- and B-cells might be required for the terminal differentiation and secretion of IgA by appropriate B-cells.

The technique for the study of the terminal differentiation of B-cells into immunoglobulin synthesising and secreting cells has most recently been applied to the study of peripheral blood lymphocytes of patients with malignancy of the T- or B-lymphocyte systems. The peripheral blood lymphocytes of patients with the Sezary syndrome, a T-cell leukemia, synthesised essentially no immunoglobulin when cultured *in vitro*. Similarly, the peripheral blood lymphocytes of patients with chronic lymphocytic leukemia, a leukemia of the B-cell system, synthesised low quantities of IgM and markedly reduced quantities of IgG and IgA when cultured in the presence of pokeweed mitogen. In addition, the peripheral blood lymphocytes of patients with multiple myeloma synthesised abnormally low quantities of polyclonal immunoglobulin, when studied with this system. The synthesis of immunoglobulins by normal lymphocytes was inhibited when incubated with lymphocytes from 3 of 5 patients with myeloma and hypogammaglobulinemia studied. This suggests the possibility that the polyclonal hypogammaglobulinemia associated with myeloma may be produced by a mechanism similar to that associated with common variable hypogammaglobulinemia; that is the inhibition of B-cell immunoglobulin synthesis by a circulating suppressor cell.

Summary

The physiological and pathophysiological factors regulating immunoglobulin biosynthesis were analysed using a new technique which permits the study of the terminal differentiation of B-lymphocytes into immunoglobulin synthesising and secreting cells. The peripheral blood lymphocytes from normal individuals had geometric mean synthetic rates of 1641 ng per 2×10^6 cells for IgG, 1698 for IgA, and 3715 for IgM when cultured for 7 days in the presence of pokeweed mitogen. This synthetic process could be abrogated by irradiation of cells or by the addition of cytosine arabinoside, mitomycin C, actinomycin D, or puromycin to the culture media suggesting that a critical mitosis is required for the maturation of B-cells into immunoglobulin synthesising and secreting cells. The addition of T-cell mitogens such as phytohaemagglutinin, staphylococcal filtrate or concanavalin A to the media with the normal lymphocytes and the B-cell stimulant pokeweed mitogen prevented B-cell maturation and immunoglobulin synthesis. This observation adds major support to the view that T-cell regulators of B-cells exist and that following activation they may be of importance in controlling the specific immune response.

The lymphocytes from patients with common variable immunodeficiency did not synthesise and secrete significant quantities of any class of immunoglobulin in the presence of pokeweed mitogen during the 7-day culture period. In addition, when lymphocytes from 6 of 9 such hypogammaglobulinemic patients were incubated with normal lymphocytes and pokeweed mitogen, the synthesis and secretion of immunoglobulin by normal cells was inhibited. This observation suggests that in some patients the disease common variable hypogammaglobulinemia may be caused or perpetuated by an abnormality of regulatory T-cells which act to suppress B-cell maturation and antibody production. Patients with isolated IgA deficiency or with IgA deficiency associated with ataxia-telangiectasia have cytoplasmic IgA as demonstrated by immunofluorescent studies following incubation *in vitro* with pokeweed mitogen but do not secrete IgA into the media. This suggests that such patients either have a quantitative defect in the amount of IgA they synthesise or have a defect in the secretion of synthesised IgA from their cells. Polyclonal immunoglobulin synthesis was reduced when lymphocytes from patients with the Sezary syndrome, a T-cell leukemia; chronic lymphocytic leukemia, a B-cell leukemia; or multiple myeloma were studied using this assay. Circulating suppressor cells were shown to be a factor in the reduced polyclonal immunoglobulin synthesis of patients with multiple myeloma.

References

1. Waldmann, T. A. and Strober, W. *Progr. Allergy.* **13** (1969), 1–110
2. Strober, W., Blaese, R. M. and Waldmann, T. A. (1970) In *Plasma Protein Metabolism*, Rothschild, M. A. and Waldmann, T., eds., Academic Press, New York, pp. 287–305
3. Gershon, R. K. (1974) In *Contemporary Topics in Immunobiology*, Cooper, M. D. and Warner, N. L., eds. Volume 3, pp. 1–40
4. Grey, H. M., Rabellino, E. and Pirofsky, B. *J. Clin. Invest.*, **50** (1971), 2368–2375
5. Preud'homme, J. L. and Seligmann, M. *Lancet*, **1** (1972), 442
6. Gajl-Peczalska, K. J., Park, B. H., Biggar, W. D. and Good, R. A. *J. Clin. Invest.*, **52** (1973), 919–928
7. Cooper, M. D., Lawton, A. R. and Bockman, D. E. *Lancet*, **2** (1971), 791–795
8. Lawton, A. R., Royal, S. A., Self, K. S. and Cooper, M. D. *J. Lab. Clin. Med.*, **80** (1972), 26–33
9. Waldmann, T. A., Polmar, S. H., Balestra, S. T., Jost, M. C., Bruce, R. M. and Terry, W. D. *J. Immunol.*, **109** (1972), 304–310
10. Wybran, J., Chantler, S. and Fudenberg, H. H. *Lancet*, **1** (1973), 126–129
11. Yam, L. T., Li, C. Y. and Crosby, W. H. *Amer. J. Clin. Path.*, **55** (1971), 283–290

Discussion

BIRGER JENSEN

In kidney transplantation irradiation of the peripheral blood has been used to kill the relatively short-lived T-cells. Would it be possible to kill those suppressing T-cells in your system by means of irradiation, and thus get the B-cells to secrete immunoglobulins?

WALDMANN

The radiosensitivity of suppressor T-cells is as follows: prior to activation of the T-cell suppressors the precursors are quite radiosensitive, but once activated they are quite radioresistant.

Since the B-cells are quite radiosensitive prior to their activation to become plasma cells, one cannot use irradiation to eliminate the activated T-cell suppression of B-cells. However, we are working at other approaches to eliminate the suppressor T-cells. When the suppressor T-cells are eliminated *in vitro* from cultures of lymphocytes from patients with common variable hypogammaglobulinemia, the remaining lymphocytes of these patients can be activated to make immunoglobulins normally.

TAVILL

Do the patients with paraproteinemia show evidence of polyclonal immunoglobulins inside their cytoplasm by immunofluorescent studies? In other words, do they have a secretion defect like the IgA deficient patients?

WALDMANN

We have studied cytoplasmic immunofluorescence in the lymphocytes of patients with chronic lymphocytic leukemia and those with the Sezary syndrome. The cells of these patients do not have immunoglobulins in their cytoplasm. We do not have similar studies in patients with myeloma and paraproteinemia. However, we have shown that patients with myeloma may have activated suppressor cells that inhibit polyclonal immunoglobulin synthesis by normal lymphocytes in culture.

HOFFENBERG

Dr Rosenoer, I believe that you stated earlier that you felt that the results of turnover studies of immunoglobulin metabolism published so far are of no value except in patients with urinary or gastro-intestinal protein loss. I wonder if you would like to discuss your views on this issue.

ROSENOER

What I said was that if you have a protein which is synthesised but does not totally pass into the vascular compartment on its way to destruction, then an injection of a labelled protein into the intravascular compartment and assessment of its disappearance curve may be very misleading in the assessment of that protein's catabolic and synthetic rate in the whole organism. I think there is evidence to suggest that a number of immunoglobulins are in fact synthesised and destroyed without passing through the intravascular compartment.

WALDMANN

I agree in part with what Dr Rosenoer has said. That is, there are immunoglobulins especially IgA that are synthesised and catabolised without ever passing through the intravascular compartment. However, with radioiodin-ated turnover studies one can measure the quantity of a protein that is both synthesised and delivered into the intravascular compartment as well as that fraction of the compartment which is catabolised.

Thus, with these turnover studies one can differentiate between patients with hypogammaglobulinemia due to deficient synthesis and delivery of immunoglobulins into the intravascular compartment and patients who have hypogammaglobulinemia due to excessively short survival time or sojourn time of the immunoglobulins in this compartment. By this mechanism one has been able to discover a new class of diseases with low total circulating immuno-globulin levels. That is, the class of diseases with disorders of endogenous catab-olism of these immunoglobulins, a disorder that could not have been discovered without the use of iodinated immunoglobulin turnover studies. Similarly, with elevated immunoglobulin levels one can differentiate between patients with increased immunoglobulin synthesis and delivery into the intravascular compartment and patients who have disorders which result in a very prolonged immunoglobulin survival and a reduced fractional catabolic rate of these proteins. In this fashion the elevated serum levels of immunoglobulin light-chains in patients with renal disease were shown to be due to a prolonged survival of these immunoglobulin fragments, rather than due to an increased synthesis and delivery of these fragments to the intravascular compartment. Thus, although, one cannot use iodinated immunoglobulin turnover studies to get exact estimates of total body synthetic rates of these proteins, they can be used to discover new pathogenetic mechanisms for disordered serum immunoglobulin levels.

TAVILL

Could I ask Dr Waldmann about his views on the defect in the IgA deficiency syndrome? Are this syndrome and α-antitrypsin deficiency the only situations where a defect in secretion can be incriminated?

WALDMANN

Failure of secretion of synthesised immunoglobulin molecules may occur in a number of conditions. It has been described in rare patients with common variable hypogammaglobulinemia[1]. In addition, it has been noted in trans-plantable myeloma tumors of mice[2]. Finally, as we have noted earlier, it is a potential defect in patients who have isolated IgA deficiency. We know that the addition of terminal polysaccharides to the polypeptide chain is required for the secretion of immunoglobulin molecules.

We are investigating the possibility that patients with isolated IgA deficiency can make the polypeptide chain for the molecule, but are unable to add the required polysaccharide residues.

References

1. Geha, R. S., Schneeberger, E., Merler, E. and Rosen, F. S. Heterogeneity

of 'acquired' or common variable agammaglobulinemia. *New Engl. J. Med.*, **291** (1974), 1

2. Sherr, C. J. and Uhr, J. W. Immunoglobulin synthesis and secretion. VI. Synthesis and intracellular transport of immunoglobulin in nonsecretory lymphoma cells. *J. Exp. Med.*, **133** (1971), 901

22
Thrombosis and metabolic studies of fibrinogen, prothrombin, plasminogen and tissue thromboplastins*

Y. TAKEDA, H. GONMORI, T. R. PARKHILL
and N. KOBAYASHI

Introduction

An important feature of thrombosis is that fibrin is contained in all three kinds of thrombi regardless of their types. Generally, three factors are claimed to be responsible for the formation of thrombus. These are injury to vascular walls, retardation of blood flow, and alteration of blood constituents, and they usually coexist in thrombosis to a variable degree. Therefore, the solution of the problems must come from careful studies of the three factors mentioned above. We too have been studying the problems of thrombosis, particularly, the alteration of blood constituents in thrombosis, with the aim of devising a simple and effective means for prevention, diagnosis and treatment of the disease. In this paper we would like to discuss metabolic studies of fibrinogen, prothrombin, plasminogen and tissue thromboplastins in relation to thrombosis, based mainly on our contributions in the past 10 years.

Fibrinogen metabolism under normal conditions
Fibrinogen has a molecular weight of about $350\,000$[1]. It forms fibrin with a release of fibrinopeptides A and B when acted upon by thrombin[2]. Since thrombi contain greater or smaller amounts of fibrin within their structure, it is of considerable importance to understand fibrinogen metabolism. First, it was necessary to study fibrinogen metabolism under normal conditions. One important question is whether or not fibrin is generated during normal metabolism of fibrinogen. Our tracer data were analysed by a two-compartment model in which fibrinogen was assumed directly catabolised without the

*Supported by Research Grant HL-11686 from the National Heart and Lung Institute.

formation of fibrin[3]. The fact that this model closely predicted and described the *in vivo* behaviour of intravenously injected ^{125}I-fibrinogen suggested that a major portion of fibrinogen at least is directly catabolised without the formation of fibrin[4]. To further substantiate this thesis, studies were also carried out in hemophilia A patients as described later. We also performed studies of fibrinogen metabolism in 10 healthy females[5], using ^{125}I-fibrinogen. These results and further analyses showed that in healthy females the plasma fibrinogen concentration, the amount of plasma and interstitial fibrinogen were significantly lower $(0.001 > p)$, but that the transcapillary flux and catabolic (synthetic) rate were not significantly different $(0.9 > p > 0.8)$ compared with normal males. It was also found that the plasma half-life of ^{125}I-fibrinogen was significantly shorter in females $(0.02 > p > 0.01)$.

Fibrinogen metabolism under pathological conditions

Up to the present, fibrinogen metabolism has been studied in a number of disease states such as hemophilia A, chronic glomerulonephritis, rheumatoid arthritis, essential hypertension and myocardial infarction[6-10]. The main purpose of these studies was to gain further insight into the pathogenesis of thrombosis by studying fibrinogen metabolism in hypercoagulable and hypocoagulable states, or in conditions in which plasma fibrinogen concentration is increased. The main finding in hemophilia A patients[6] was that the plasma half-life of ^{125}I-fibrinogen was not longer than in healthy subjects. In hemophilia A in which Factor VIII is deficient, fibrin formation would be diminished if it were formed under normal conditions. Therefore, if a major portion of fibrinogen were catabolised after fibrin is formed, a longer half-life of plasma ^{125}I-fibrinogen would be expected in hemophilia A patients. The finding that the plasma half-life of fibrinogen in hemophilia A patients was not longer than that in healthy subjects thus supports the thesis that a major portion of fibrinogen is directly catabolised without the formation of fibrin in healthy subjects. The main findings in chronic glomerulonephritis[7] and rheumatoid arthritis[8] were that fractional transcapillary transfer rate of plasma fibrinogen (j_1) and fractional catabolic rate (j_3) were not altered compared to normal values, but that because of the increased amount of plasma fibrinogen $(\bar{x})$, the transcapillary and catabolic fluxes $(j_1\bar{x}$, and $j_3\bar{x})$ were increased. The increased plasma fibrinogen appeared to be secondary to the tissue damage. We also studied fibrinogen metabolism in 15 female patients with essential hypertension[9]. The average behaviour of plasma ^{125}I-fibrinogen and the plasma half-life of ^{125}I-fibrinogen showed no significant difference $(0.1 > p > 0.05)$ compared with healthy females. However, the amount of plasma fibrinogen was significantly increased $(0.001 > p)$ over normal values. The amount of interstitial fibrinogen was significantly reduced $(0.001 > p)$. On the other hand, the fractional transcapillary rate, j_1 in patients with essential hypertension was greatly decreased $(0.001 > p)$ compared with the normal value in healthy females, but the fractional catabolic rate was not appreciably different from normal. On balance, the transcapillary flux of fibrinogen showed no significant difference $(0.3 > p > 0.2)$ compared with normal, but the catabolic synthetic rate showed a marked increase $(0.001 > p)$. Further analyses of the data showed that the fractional transcapillary transfer rate was closely correlated $(0.001 > p)$ with diastolic blood pressure (P) with the regression equation of $j_1 = \exp(-0.0285P)$ (figure 22.1). Close correlations $(0.001 > p)$ were also found between $\bar{x}$ and j_1 with the regression equation of $\bar{x} = 226 \exp(-0.938j_1)$ (figure 22.2) and between $\bar{y}$ and j_1 with the regression

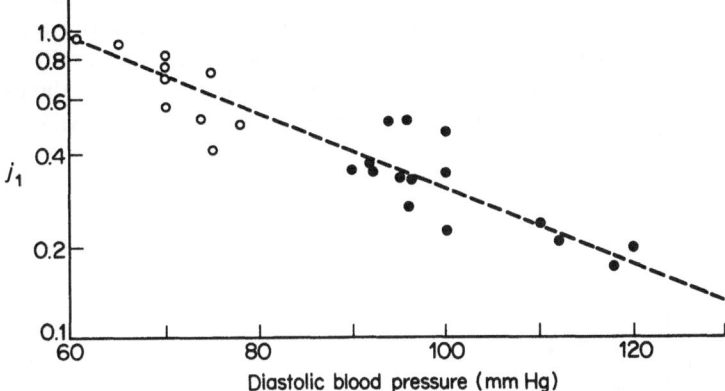

Figure 22.1. Correlation between fractional transcapillary transfer rate of plasma fibrinogen $(j_1$ per day) and diastolic blood pressure $(P$ mmHg). The open dots are the data in healthy female subjects and the solid dots are those obtained in patients. Analysis showed significant correlation between j_1 and P with a correlation coefficient of -0.838 $(p < .001)$. The regression equation was
$$j_1 = \exp(-0.0285P).$$

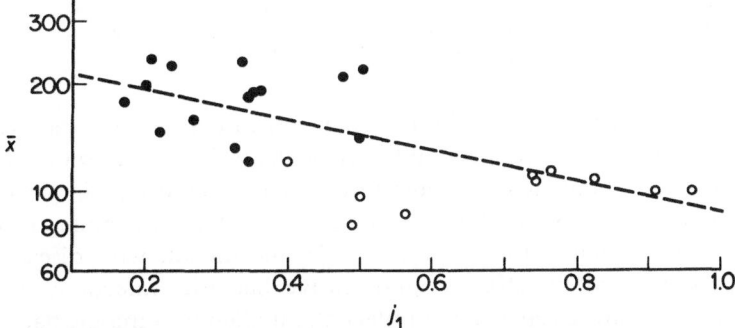

Figure 22.2. Correlation between fractional transcapillary transfer rate of plasma fibrinogen $(j_1$ per day) and plasma fibrinogen $(\bar{x}$ mg per kg body weight). The open dots are the data in healthy female subjects and the solid dots are those obtained in the patients. Analysis showed significant correlation between j_1 and $\bar{x}$ with a correlation coefficient of -0.655 $(p < .001)$. The regression equation was $\bar{x} = 226 \exp(-0.938j_1)$.

equation of $\bar{y} = 12.6 \exp(0.624 j_1)$ (figure 22.3). These findings were interpreted as follows: the observed increased $\bar{x}$ and decreased $\bar{y}$ are due to the decreased fractional transcapillary transfer rate (j_1) which was probably caused by vasoconstriction or the increased muscle tone of the precapillaries and the increased $j_3\bar{x}$ was due to the increased amount of $\bar{x}$. We also studied fibrinogen metabolism in 10 male patients with myocardial infarction[10]. The tracer data in these patients were essentially the same as those in healthy male subjects, but because of the marked increase in the amount of plasma fibrinogen ($\bar{x}$), $j_1\bar{x}$ and $j_3\bar{x}$ were also greatly increased. The question is whether or not these increases were primary or secondary to myocardial damage. It is extremely difficult to measure plasma fibrinogen concentration in healthy individuals every day or every hour until the onset of the disease. Therefore, this question is not completely solved, but it seems reasonable to assume that the observed increases of $\bar{x}$ and consequently of $j_3\bar{x}$ are secondary to myocardial damage.

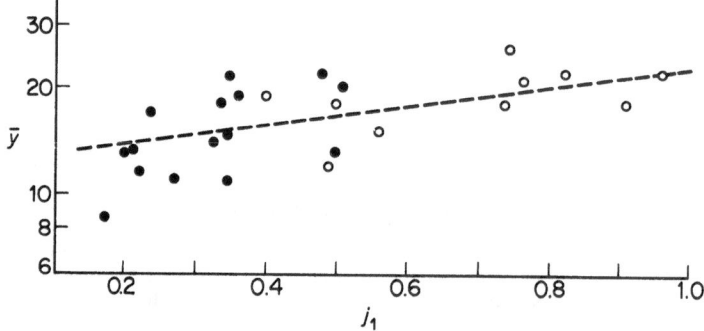

Figure 22.3. Correlation between fractional transcapillary transfer rate of plasma fibrinogen (j_1 per day) and interstitial fibrinogen ($\bar{y}$ mg per kg body weight). The open dots are the data in healthy female subjects and the solid dots are those obtained in the patients. Analysis showed a significant correlation between j_1 and $\bar{y}$ with a correlation coefficient of 0.604 ($p < .001$). The regression equation was $\bar{y} = 12.6 \exp(0.624 j_1)$.

Another interesting disease is disseminated intravascular coagulation (DIC). This is a relatively new concept and may be defined as a syndrome brought about by acute, subacute or chronic thrombosis which occurs in association with various diseases. Some of the characteristics of DIC are capillary thrombi, hypotension and hemorrhagic diathesis[11-13], and the following diseases tend to be accompanied by DIC: gram-negative bacterial infections, trauma, eclampsia, premature separation.of placenta, malignant tumour, haemolysis due to mismatched blood transfusion, malaria and others. Even DIC is no different from other thrombosis in that injury to vascular walls, retardation of blood flow or alteration of blood constituents are responsible. We studied 3 cases of DIC with respect to fibrinogen metabolism, using ^{125}I-fibrinogen. Two cases were studied in remission. Under this condition, the plasma behavior of intravenously injected ^{125}I-fibrinogen was identical with that under normal conditions. The only difference was that the plasma fibrinogen ($\bar{x}$) was greatly increased so that $j_1\bar{x}$ and $j_3\bar{x}$ were greatly increased. In the third case, in the middle of study, the patient developed a thrombotic episode which was immediately detected by a sudden shortening of the plasma half-life of ^{125}I-fibrinogen. This shows the intermittent nature of thrombus formation in DIC. Generally, in thrombotic conditions the amount of plasma fibrinogen ($\bar{x}$) should be decreased because of its consumption in thrombus formation, but whether it is actually decreased or not depends on the time elapsed after thrombus formation and the degree of fibrinogen consumption. This may be due to the fact that the liver is capable of increasing the rate of fibrinogen synthesis very rapidly[14].

Prothrombin metabolism under normal conditions

Prothrombin has a molecular weight of $68\,900$[2]. It generates thrombin under appropriate conditions which is directly responsible for fibrin formation. Therefore, a knowledge of prothrombin metabolism is of great interest and importance for the further understanding of thrombosis. Unfortunately, not many human studies have been reported partly because of difficulty in the purification of prothrombin and partly because of the stringent NIH regulations in the use of human subjects. However, in early days a considerable number of human studies were reported[15-20]. In these studies, subjects were injected with large amounts of prothrombin concentrates, and the decline of plasma prothrombin activity was followed in an attempt to determine the plasma half-life of prothrombin. In other cases, large amounts of prothrombin-depressant agents were given to subjects to block prothrombin synthesis completely, and the declining levels of plasma prothrombin activity were measured to determine the half-life of plasma prothrombin. These studies provided only the plasma half-life of prothrombin at best. Recently, a few studies of prothrombin metabolism in humans and animals has appeared, using ^{125}I-prothrombin[21-24]. We studied prothrombin metabolism in 13 healthy men, using ^{125}I-prothrombin. We also carried out similar studies in 9 healthy calves. Heparin was found to have no appreciable effects on the metabolism of prothrombin in 7 healthy calves. One of the conclusions from all these studies was that a major portion of prothrombin if not all is catabolised directly without the formation of thrombin under normal conditions.

Prothrombin metabolism under pathological conditions

Very few studies in human subjects are available[21]. Shapiro and Martinez studied prothrombin metabolism in hypocoagulable states using ^{125}I-prothrombin, and reported no appreciable difference between normal and hypocoagulable states. We studied prothrombin metabolism in 5 calves to which a single injection of typhoid vaccine was given[23]. These calves were used as a model for a gram-negative bacterial infection in man. The studies showed that the plasma half-life of intravenously administered ^{125}I-prothrombin was markedly shortened and the plasma prothrombin concentration as measured by a radial immunodiffusion method was decreased. However, a large amount of heparin did not prevent these changes suggesting that the observed alteration of prothrombin metabolism was not due to accelerated coagulation processes.

Plasminogen metabolism under normal conditions

Plasminogen has a molecular weight of about 130000[25-29]. However, other investigators reported values for the molecular weight which are somewhat different from ours. Thus, Shulman, Alkjaersig and Sherry[30] reported 143000, Robbins and Summaria[31], 81000, and Davies and Englert[32], 84000. It appears that the various investigators are dealing with different kinds of plasminogen. This seems probable because multimolecular forms of plasminogen have been reported to occur in purified plasminogen preparations[33,34]. However, our preparations were always uni-molecular although we were able to produce multimolecular forms out of a single molecular form by dialysing the latter against 0.01 M phosphate buffer (pH 8.0) for a few days. Our method for plasminogen purification is very similar to that of Shulman and his associates and therefore our preparation appears very similar to theirs. At any rate, plasminogen generates plasmin when acted upon by plasminogen activators. The only human study of plasminogen metabolism we were able to find was that of Collen and his associates[35]. They reported the plasma half-life of injected ^{125}I-plasminogen to be about 2.21 ± 0.29(SD) days, using multimolecular forms of labelled plasminogen. Another conclusion was that under normal conditions appreciable amounts of plasmin are not generated. We concentrated our efforts on the study of plasminogen metabolism in dogs. During the purification of human and canine plasminogens, we accidentally found that plasmin can readily be separated from plasminogen by polyacrylamide electrophoresis[25-29]. Based on this finding, we developed a method for detecting and quantifying the plasmin generated. Then, we proceeded to study the metabolism of plasminogen in 8 healthy dogs. We found that no appreciable amounts of plasmin are generated under normal conditions.

Plasminogen metabolism under pathological conditions

Human studies of plasminogen metabolism under thrombotic conditions with the use of ^{125}I-plasminogen are not available. We have been prohibited by NIH to carry out human studies. Therefore, our studies were all carried out in dogs[25-29]. First, we studied the effects of a single injection of urokinase on ^{125}I-plasminogen metabolism. It was found that large amounts of ^{125}I-plasmin were generated, but that after 6 h the plasma levels of generated ^{125}I-plasmin declined exponentially with a half-life of about 0.58 ± 0.05(SD) days on average. Then, we produced thrombophlebitis in dogs by injecting 1 ml of 95% phenol into a leg vein occluded by a gauze tourniquet and removal of phenol after 1 min[25-29]. Under this condition, large amounts of ^{125}I-plasmin were also generated and remained elevated for at least 5 days. Then, to determine the mechanism of plasmin generation, a number of *in vivo* and *in vitro* experiments were carried out. First, the effects of heparinisation on plasmin generation was studied after a single injection of urokinase. Heparinisation was found to markedly reduce the amount of ^{125}I-plasmin generated. Then, the effects of ε-aminocaproic acid on plasmin generation in response to the urokinase injection were studied. ε-Aminocaproic acid was found to completely inhibit the generation of plasmin. Similar effects of heparin and ε-aminocaproic acid were also found on plasmin generation in response to thrombophlebitis. These results strongly suggested that plasmin generation in thrombophlebitis was due to plasminogen activators released from injured veins. One of the findings in these studies was that only one kind of plasmin was generated *in vivo*, which had an electrophoretic mobility much faster than plasminogen on polyacrylamide electrophoresis and a plasma half-life of about 0.58 ± 0.05(SD) days. One kind of plasmin was also generated *in vitro* when purified plasminogen was incubated with urokinase for 5 min at 38 °C. However, when the incubation time was prolonged beyond 20 min, 3 kinds of plasmin were found to be generated. Therefore, we undertook extensive studies of purification of the 3 plasmins and of their physicochemical and biological properties. These plasmins were separated by the use of Sephadex G-200 or G-75 gel filtration, and their molecular weights were determined by the method of Andrews[36] and by SDS polyacrylamide electrophoresis[37]. The results were 125000 ± 5000(SD), 63000 ± 2000(SD) and 31500 ± 1000(SD), respectively. During these studies we also found that the smallest plasmin (M.W. 31500) can be further split into 2 pieces of about 16000 molecular weight and that 2 polymers of 95000 and 48000 were formed when the smallest plasmin was incubated for 30 min in SDS and β-mercaptoethanol. The electrophoretic behavior of the smallest plasmin was found to be identical with that of plasmin generated *in vivo*. We also needed undenatured ^{125}I-plasmins for studies of their biological properties, but we could not obtain satisfactory preparations of ^{125}I-plasmins by labelling purified plasmins directly with [125]I. ^{125}I-plasmins thus prepared did not even penetrate 7 g% polyacrylamide gel in polyacrylamide electrophoresis. Therefore, we had to first prepare ^{125}I-plasminogen, which was then activated in urokinase and 50% glycerol. The generated ^{125}I-plasmins were separated by the use of Sephadex G-200 or G-75 filtration. ^{125}I-plasmins thus obtained were entirely satisfactory by several criteria as discussed in detail[29]. Since the smallest ^{125}I-plasmin had an identical electrophoretic behavior with that generated *in vivo* we next determined its metabolic behavior in 5 healthy dogs. These results indicated that the smallest plasmin (M.W. 31500) had an identical half-life with those generated *in vivo*. All these results seemed to support that *in vivo* only one kind of plasmin is generated whose molecular weight is about 31500 and that others found during prolonged incubation are polymers of the smallest plasmin. Studies were then made of the affinity of the smallest ^{125}I-plasmin with albumin, γ-globulin, α_2-macroglobulin, α_1-antitrypsin, fibrinogen and fibrin. The methodology is described in detail elsewhere[29]. The results meant that albumin or gamma-globulin did not have any affinity, but the others did and that fibrinogen or fibrin had a greater affinity than α_2-macroglobulin or α_1-antitrypsin. These findings appeared to answer the question: 'How is it possible for plasmin to dissolve thrombus in the presence of so much plasmin inhibitors in plasma?' Another experiment was performed to study the nature of the binding between ^{125}I-plasmin and its inhibitors. 1 ml of normal dog plasma in the standard ACD solution was mixed with 0.01 mg of ^{125}I-plasmin, incubated at 38°C for 30 min and electrophoresed in polyacrylamide gel[38]. It might be expected that the electrophoretic mobility of ^{125}I-plasmin–inhibitor complexes will be different from that of pure ^{125}I-plasmin, if the binding is irreversible. The results showed that the mobility of ^{125}I-plasmin after its incubation with plasma was unchanged. This must mean that the ^{125}I-plasmin bound to inhibitors during its incubation was freed from inhibitors during electrophoretic analysis, indicating the reversible nature of the binding between the two. The proof that ^{125}I-plasmin does bind with inhibitors during incubation is given elsewhere[29]. This result is also in agreement with the *in vivo* finding that ^{125}I-plasmin generated during thrombophlebitis had the same electrophoretic mobility as that of pure ^{125}I-plasmin.

Tissue thromboplastins (TTPs)

TTP is a tissue factor which initiates blood coagulation via the extrinsic system. Considerable effort has been put into the purification and physiochemical characterisation of TTPs[39–49]. However, disagreements still exist even about the molecular weight of TTP. Earlier investigators mainly used saline extracts of homogenised tissues and differential centrifugation for removal of coarse materials[39–43], whereas more recent workers have utilised the extract of TTP by sodium deoxycholate (DOC) after its further purification by Sephadex gel filtration[44–49]. Thus, earlier preparations are generally less pure and larger in molecular size than more recent preparations. This is due to the use of DOC which enables TTP to be freed from microsomal fractions and to be purified further[44,45]. We also have been working on the purification of TTPs from canine brain, lungs, arteries and veins. Our method for preparation of TTPs consisted of mincing the tissues, removing lipids by acetone and heptane-butanol, solubilising TTPs by DOC, removing coarse materials by high-speed centrifugation, and further purifying by filtration through columns of Sephadex G-200, G-100 and G-75 in sequence, and final purification by the use of preparative polyacrylamide electrophoresis. DOC was present in veronal buffer (Barbital 2.2 g per 4 l and Sodium barbital 14 g per 4 l, pH 8.6) at a concentration of 0.625% during the Sephadex filtration and at a concentration of 0.25 g% in the tris-glycine buffer in the upper chamber and in the standard buffer for polyacryllamide gel during polyacryllamide electrophoresis. The final TTP products were kept in the veronal buffer containing 0.625 g% DOC. By this method we were able to prepare active TTPs from brain, lungs, arteries and veins which are homogeneous by immunoelectrophoresis as well as by polyacrylamide electrophoresis. In the immunoelectrophoretic analysis, we used antibodies against crude preparations of TTPs from brain, lungs, arteries and veins, which were produced in rabbits. It was found that crude TTP preparations from brain were the poorest antibody producers compared with the other preparations which produced very potent antibodies, and also that each tissue contained at least 3 different antibody producers. Then, antibody against the final product of arterial TTP was produced in rabbits and the reactivity of each TTP against the anti-arterial TTP antibody was tested by immunoelectrophoresis and by a double diffusion analysis. It was found that all the purified TTPs reacted against the anti-arterial TTP antibody, although brain TTP reacted most weakly. Then, each purified TTP was labelled with ^{125}I in a ratio of about 0.5 atom iodine per molecule of TTP by the use of iodine monochloride method[50], and its molecular weight was determined by the method of Andrews[36]. The gel of Sephadex G-200 was made in 0.625 g% DOC in the veronal buffer described above. The average results were as follows: $80\,000 \pm 4000$ (brain TTP), $113\,000 \pm 5000$ (lung TTP), $62\,000 \pm 3000$ (artery TTP), and $47\,000 \pm 2000$ (vein TTP). Then, each TTP was dialysed against veronal buffer for a few days to remove DOC, and the molecular weight was again determined by the same method. It was found that multimolecular forms were formed from each TTP, indicating the importance of DOC in the preservation of the original molecular form of each purified TTP. Next, each of the ^{125}I-TTP was screened in a puppy for 10 min and their metabolic behaviour was studied in 4 dogs. The average plasma half-lives of TTPs were 8.1 ± 0.24(SD) h (brain TTP), 14.6 ± 0.5(SD) h (lung) 7.38 ± 0.48(SD) h (artery) and 24.3 ± 0.9(SD) h (vein). These results indicated that every TTP is different from every other, although there are similarities.

Discussion

As described above, a number of studies were carried out with the objective of finding a simple and effective means for prevention, diagnosis and treatment of thrombosis. We believe that a promise exists for the early recognition of hypercoagulability, which is a condition of blood characterised by an increased tendency to coagulate. A number of factors are responsible for this, particularly the generation of thrombin. Our studies of prothrombin metabolism as described earlier showed that appreciable amounts of thrombin are not generated under normal conditions. Therefore, the detection of thrombin is sufficient for the diagnosis of hypercoagulability. Unfortunately, specific methods available for thrombin detection are time-consuming and not simple enough. Simpler methods are available but are not specific enough. In addition, the extreme lability of the prothrombin molecule makes it difficult to decide whether any thrombin detected is a product formed *in vivo* or was formed during *in vitro* analysis of plasma. Therefore, other means had to be looked for. A possibility lies in the detection of factors responsible for thrombin generation. Among them are injuries to vascular walls and other tissues, which contain variable amounts of plasminogen activators[52-54] and tissue thromboplastin. It might be expected that these substances would be released into the circulation when vessels and other tissues were injured. Our studies of plasminogen metabolism in thrombophlebitis strongly suggested that this is the case. Thus, the detection of plasminogen activators or tissue thromboplastin could be used as a means of diagnosing vascular and tissue injury. Also, plasminogen activators, when released into the circulation, will act on plasminogen to generate plasmin, and tissue thromboplastin will form activated factor X which in turn will generate thrombin from prothrombin. Thus, the vascular and tissue injury could also be diagnosed by detection of plasmin in the circulation. Our studies showed that large amounts of plasmin were generated in thrombophlebitis. However, the detection of activated factor X, if possible, would not serve as a means for the diagnosis of vascular and tissue injury, since it can be generated by both extrinsic and intrinsic pathways. Thus, the most promising method for the diagnosis of vascular and tissue injury and consequently of hypercoagulability seems to be the detection of plasminogen activators, plasmin, or tissue thromboplastin in the circulation. Our studies showed that plasmin can readily be separated from plasminogen by polyacrylamide gel electrophoresis because of the great difference in the mobility of the two. Namely, plasminogen travels about 2 mm distance through 7 g% gel at 5 mA per gel in 1 h, whereas plasmin migrates about 56 mm. Based on these findings, we were able to develop a sensitive method for the detection of plasmin using ^{125}I-plasminogen. This method consists of a single intravenous injection of about 30 μCi of ^{125}I-plasminogen into a suspected patient and analysis of the plasma by polyacrylamide electrophoresis 3 h after the injection. The electrophoretic analysis itself requires about 1–2 h, so that a total of about 5 h is required before a diagnosis can be made. The only drawback of this method is that it requires undenatured ^{125}I-plasminogen which is uncontaminated by plasmin. This may be too much of a requirement for a routine laboratory. Therefore, the method for plasmin detection may be regarded as sensitive and specific, but not simple enough. In an attempt to develop a simpler and specific method, we then turned our attention to the purification and metabolic studies of tissue thromboplastins. So far, we have purified tissue thromboplastins from canine brain, lungs, arteries and veins, and we have found that they are immunologically similar to each other, while being dissimilar in some other aspects. The next step is to develop a sensitive radioimmunoassay for nanogram or picogram quantities of tissue thromboplastins in the circulation. First, it has to be determined whether or not appreciable amounts of tissue thromboplastins are present in the circulation under normal conditions, and whether they enter the circulation when vessels are injured. However, our finding that tissue thromboplastins are immunologically similar means that their detection by a radioimmunoassay method in the circulation can only be used for the diagnosis of tissue injury and not vascular injury specifically, even if they were not present in normal blood. On the other hand, if some tissue thromboplastins were present in normal blood, it has to be determined whether or not the detected amount of tissue thromboplastins is abnormally high. A word of caution is necessary. The presence of abnormal amounts of tissue thromboplastin does not necessarily imply the immediate onset of thrombosis because of the presence of several inhibitor systems in plasma[55–59]. Another possibility is the detection of vascular plasminogen activators, but very little is known of their metabolism, although some studies are available on their partial purification[60,61]. Thus, this is a problem for the future.

References

1. Laki, K. (1968) *Fibrinogen*, Marcel Dekker, New York, p. 63
2. Seegers, W. H. (1962) *Prothrombin*, Harvard University Press, Cambridge, Massachusetts, p. 32, 449.
3. Reeve, E. B., and Roberts, J. E. The kinetics of the distribution and breakdown of ^{131}I-albumin in the rabbit. *J. Gen. Physiol.*, 43 (1959), 415
4. Takeda, Y. Studies of the metabolism and distribution of fibrinogen in healthy men with autologous ^{125}I-labelled fibrinogen. *J. Clin. Invest.*, 45 (1966), 103
5. Takeda, Y. Unpublished observation.
6. Takeda, Y., Chen, A. Y. Studies of the metabolism and distribution of fibrinogen in patients with hemophilia A. *J. Clin. Invest.*, 46 (1967), 1979
7. Takeda, Y. and Chen, A. Y. Fibrinogen metabolism and distribution in patients with the nephrotic syndrome. *J. Lab. Clin. Med.*, 70 (1967), 678
8. Takeda, Y. Studies of the metabolism and distribution of fibrinogen in patients with rheumatoid arthritis. *J. Lab. Clin. Med.*, 69 (1967), 624
9. Takeda, Y. Fibrinogen metabolism in patients with essential hypertension. *Fed. Proc.*, 25 (1966), 45
10. Takeda, Y. Unpublished observation
11. Penick, G. D. (1969). Blood states that predispose to thrombosis. In *Thrombosis*, Sol Sherry, K. M. Brinkhous, E. Gertor, J. M. Stengel, eds., National Academy of Sciences, Washington, D.C. p. 553
12. Hardaway, R. M. III (1966). *Syndromes of disseminated intravascular coagulation*, C. C. Thomas Publisher, Springfield, Illinois

13. McKay, D. G. (1965). *Disseminated intravascular coagulation*, Hoeber Medical Division, Harper and Row, New York

14. Reeve, E. B., Takeda, Y. and Atencio, A. C. Some observations on the mammalian fibrinogen system in non-steady and steady states. *Protides of the biological fluids*, **14** (1966), 293

15. Didisheim, P., Loeb, J., Blatrix, C. and Soulier, J. P. Preparation of a human plasma fraction rich in prothrombin, proconvertin, Stuart factor, and PTC and a study of its activity and toxicity in rabbits and man. *J. Lab. Clin. Med.*, **53** (1959), 322

16. Biggs, R. and Denson, K. W. E. The fate of prothrombin and Factors VIII, IX and X transfused to patients deficient in these factors. *Brit. J. Haemat.*, **9** (1963), 532

17. Frick, P. G. Studies on the turnover rate of stable prothrombin conversion in man. *Acta haemat.*, (Basel), **19** (1958), 20

18. Hasselback, R. and Hjort, P. F. Effect of heparin on *in vivo* turnover of clotting factors. *J. Appl. Physiol.*, **15** (1960), 945

19. Hjort, P. F., Egeberg, O. and Mikkelsen, S. Turnover of prothrombin, Factor VII and Factor IX in a patient with haemophilia A. *Scand. J. Clin. Lab. Invest.*, **13** (1961), 688

20. Loeliger, E. A., van der Esch, B., Mattern, M. J. and Hemker, H. C. The biological disappearance rate of prothrombin, Factors VII, IX and X from plasma in hypothyroidism, hyperthyroidism, and during fever. *Thrombos. Diathes, haemorrh.* (Stuttgart), **19** (1964), 267

21. Shapiro, S. S. and Martinez, J. Human prothrombin metabolism in normal man and in hypocoagulable subjects. *J. Clin. Invest.*, **48** (1969), 1292

22. Takeda, Y. Studies of the effects of heparin, coumadin, and vitamin K on prothrombin metabolism and distribution in calves with the use of [125]iodine-prothrombin. Characterisation of the prothrombin system. *J. Lab. Clin. Med.*, **75** (1970), 355

23. Takeda, Y. (1970). Effects of typhoid endotoxin on fibrinogen and pro-thrombin metabolism in calves. In *Plasma Protein Metabolism*, M. R. Rothschild and T. Waldmann, eds., Academic Press, New York, p. 443

24. Takeda, Y. Studies of the metabolism and distribution of prothrombin in healthy men with homologous ^{125}I-prothrombin. *Thrombos. Diathes. haemorrh.* (Stuttgart), **27** (1972), 472

25. Takeda, Y. ^{125}I-plasminogen responses in dogs to a single injection of urokinase and typhoid vaccine and to vascular injury. *J. Clin. Invest.*, **51** (1972), 1363

26. Takeda, Y., Parkhill, T. R. and Nakabayashi, M. Effects of heparin and ε-aminocaproic acid in dogs on ^{125}I-plasmin-generation in response to urokinase injections and venous injury. *J. Clin. Invest.*, **51** (1972), 2678

27. Takeda, Y., Parkhill, T. R. and Nakabayashi, M. (1973). Generation of ^{125}I-labelled plasmin in dogs in response to venous injury. In *Protein*

Turnover. Ciba Symposium 9 (New Series), Elsevier, Amsterdam, p. 203

28. Takeda, Y. Fibrinogen and plasminogen metabolism in relation to thrombosis. *Metabolism* (Japan), **10** (1973), 124

29. Takeda, Y. and Nakabayashi, M. Physicochemical and biological properties of human and canine plasmins. *J. Clin. Invest.*, **53** (1974), 154

30. Shulman, S., Alkjaersig, N. and Sherry, S. Physicochemical studies on human plasminogen and plasmin. *J. Biol. Chem.*, **233** (1958), 91

31. Robbins, C. K. and Summaria, L. (1970). *Methods in Enzymology*, G. E. Pearlman and L. Lorand, eds., Academic Press, New York, p. 184

32. Davies, M. C. and Englert, M. E. Physical properties of highly purified human plasminogen. *J. Biol. Chem.*, **235** (1960), 1011

33. Deutsch, D. G. and Mertz, E. T. Plasminogen: purification from human plasma by affinity chromatography. *Science*, **170** (1970), 1095

34. Heberlein, P. L. and Barnhart, M. I. Canine plasminogen: purification and a demonstration of multimolecular forms. *Biochim. Biophys. Acta.*, **168** (1968), 195

35. Collen, D., Tytgat, G., Clayes, H., Verstraete, M. and Wallen, P. Metabolism of plasminogen in healthy subjects: Effect of tranexamic acid. *J. Clin. Invest.*, **51** (1972), 1310

36. Andrews, P. The gel-filtration behaviour of proteins related to their molecular weights over a wide range. *Biochem. J.*, **96** (1965), 595

37. Weber, K. and Osborn, M. The reliability of molecular weight determination by dodecyl sulphate–polyacrylamide gel electrophoresis. *J. Biol. Chem.*, **244** (1969), 4406

38. Davis, B. J. Disc electrophoresis. II. Methods and application to human serum proteins. *Ann. N.Y. Acad. Sci.*, **121** (1964), 404

39. Chargaff, E., Moore, D. H. and Benedich, A. Ultracentrifugal isolation from lung tissue of a macromolecular protein component with thromboplastic properties. *J. Biol. Chem.*, **145** (1942), 593

40. Astrup, T., Albrechtsen, O. K., Claassen, B. A. and Rasmussen, J. Thromboplastic and fibrinolytic activities in vessels of animals. *Circ. Res.*, **7** (1959), 969

41. Astrup, T. and Buluk, K. Thromboplastic and fibrinolytic activities in vessels of animals. *Circ. Res.*, **13** (1963), 253

42. Williams, W. J. The activity of human placenta microsomes and brain particles in blood coagulation. *J. Biol. Chem.*, **241** (1966), 1840

43. Deutsch, E., Irsigler, K. and Lomoschlitz, H. Studien über Gewebethromboplastin. I. Reinigung, chemische Charakterisierung und Trennung in Eiweiss- und Lipoiden-teil. *Thrombos. Diathes. haemorrh.*, **12** (1964), 23

44. Hvatum, M. and Prydz, H. Studies on tissue thromboplastin. Its splitting into two separate parts. *Thrombos. Diathes. haemorrh.*, **21** (1969), 217

45. Hvatum, M. and Prydz, H. Studies on tissue thromboplastin. Electron-microscopy. *Thrombos. Diathes. haemorrh.*, **21** (1969), 223

46. Nemerson, Y. The phospholipid requirement of tissue factor in blood

coagulation. *J. Clin. Invest.*, **47** (1968), 72

47. Nemerson, Y. Characteristics and lipid requirements of coagulant proteins extracted from lung and brain: The specificity of the protein component of tissue factor. *J. Clin. Invest.*, **48** (1969), 322

48. Nemerson, Y. and Pitlick, F. A. Purification and characterisation of the protein component of tissue factor. *Biochem.*, **9** (1970), 5100

49. Zeldis, S. M., Nemerson, Y. and Pitlick, F. A. Tissue factor (Thromboplastin): Localisation to plasma membranes by peroxidase-conjugated antibodies. *Science*, **175** (1972), 766

50. McFarlane, A. S. Efficient tracer labelling of proteins with iodine. *Nature*, **182** (1958), 53

51. Gonmori, H., Takeda, Y. and Parkhill, T. R. Unpublished observation

52. Lack, C. W. Proteolytic activity and connective tissue. *Brit. Med. Bull.*, **20** (1964), 217

53. Todd, A. S. Localisation of fibrinolytic activity in tissues. *Brit. Med. Bull.*, **20** (1964), 210

54. Warren, B. A. Fibrinolytic activity of vascular endothelium. *Brit. Med. Bull.*, **20** (1964), 213

55. Tocantins, L. M. Demonstration of antithromboplastic activity in normal and hemophilic plasmas. *Am. J. Physiol.*, **139** (1943), 265

56. Fiala, S. A thermolabile inhibitor of plasma coagulation. *Nature*, **167** (1951), 279

57. Schneider, C. L. The active principle of placental toxin: Thromboplastin; its inactivator in blood: Antithromboplastin. *Am. J. Physiol.*, **149** (1947), 123

58. Lanchantin, G. F. and Ware, R. G. Identification of a thromboplastin inhibitor in serum and in plasma. *J. Clin. Invest.*, **32** (1953), 381

59. Mammen, E. F. (1967). Plasma anticoagulants or inhibitors. In *Blood Clotting Enzymology*, W. H. Seegers, ed., Academic Press, New York, p. 345

60. Aoki, N. and von Kaulla, K. N. The extraction of vascular plasminogen activator from human cadavers and a description of some of its properties. *Am. J. Clin. Path.*, **55** (1971), 171

61. Aoki, N. and von Kaulla, K. N. Dissimilarity of human vascular plasminogen activator and human urokinase. *J. Lab. Clin. Med.*, **78** (1971), 354

Discussion

REGOECZI

The first question is: Did you use the same phospholipids which were originally bound to the protein in the relipidation of your purified TTP proteins?

TAKEDA

Yes. During the purification procedures, when the phospholipids were being removed by the use of heptane–butanol, we always saved the filtrates containing phospholipids. We dried and used them.

REGOECZI

The next question is: Would this purified protein still be reactive with your antibodies when recombined? What I am trying to find out is whether or not the protein is completely covered with phospholipids in its native state.

TAKEDA

We have not studied that problem so far, but we intend to study it in the very near future. At the present we are trying to determine whether or not purified TTP proteins from brain, lung, artery and veins are identical.

REGOECZI

The next question is: Where do you precisely localise the tissue thromboplastin in the body? Is it on a membrane or more likely, intracellular?

TAKEDA

I am not a molecular chemist, but according to experts, it is supposed to be attached to microsomes and the reason for the use of deoxycholate is to free TTP from microsomes.

REGOECZI

The last question is: Are the clearance studies you have shown those of complete tissue thromboplastin or of the purified protein?

TAKEDA

The only reason for the clearance studies of purified proteins is to obtain additional evidence for dissimilarity among them. The evidence indeed indicated that they are different with respect to their plasma behaviour. The plasma behaviour of complete tissue thromboplastins may be quite different.

MILLER

You must have carried out some simple Ouchterlony type diffusion study. Did you get reactions of identity, or partial identity?

TAKEDA

I am not very good at interpreting these things, but they showed reaction of partial identity as far as I can remember. We were more interested in the quantitative difference of the affinity of each purified protein with the same antibody. I have shown that there was obvious difference in the affinity of each antigen with the same antibody. Also, immunoelectrophoretically they all reacted, but there was some visible difference.

REEVE

How active is your preparation? I mean, are thromboplastins dangerous? Are they likely to be involved in thrombosis? What is the lethal dose of your TTPs?

TAKEDA

We have undertaken this study with the idea that in a number of clinical conditions thromboplastins must enter the circulation to cause thrombosis. However, there is no absolute proof really that they do enter the circulation. As far as I know, no one has proven it yet.

We intend to develop a sensitive radioimmunoassay for TTP to prove or disprove our idea. If our idea was found correct, then the method of detection for TTP in the circulation would be a useful diagnostic method for hyper-coagulability. The specific activity of our purified TTP proteins after their relipidation was as follows: 1395 units per mg BTTP, 1130 units per mg LTTP, 630 units per mg VTTP and 435 units per mg ATTP. During the *in vivo* studies, we always injected 30 to 50 μg of each TTP protein, because when we injected 200 μg or more the dogs went into a state of shock and they crumbled down, although in a half a minute or less they stood up again and thereafter they looked alright. We never measured the lethal dose of our materials in dogs, because they are so precious.

PART 5

Plasma Proteins and Hormonal Transport

PART 3

Plasma Proteins and
Hormonal Transport

23

Preparation of biospecific supports for affinity chromatography and immunoadsorption

S. COMOGLIO, A. MASSAGLIA, E. ROLLERI and U. ROSA

Introduction

In recent years much attention has been devoted to the techniques of affinity chromatography for studying complex biological systems[1]. The basic principle is to immobilise one of the components of the interacting system by binding to an insoluble support, preferably through covalent bonds[2,3]. The biologically active support can be used to separate selectively from the reaction mixture the component with which it interacts. A successive elution yields pure components[4-6]. This technique consequently looks very promising both as an analytical method[7] and for studying the mode of action of biological substances[8,9].

The preparation of the biospecific supports involves different problems, depending on the kind of molecules to be insolubilised. The properties of the solid phase are strictly related to the choice of the matrix and of the coupling reaction, which must be determined by the nature of the biological molecule to be insolubilised. With most proteins a variety of carriers and coupling reactions can be used, since the presence of several interacting sites in the molecule may still lead to a biologically active preparation, provided that the ratio of the number of covalent bonds to the matrix is kept low. A more rigorous selection of a suitable carrier can be required for particular classes of substances: a. small-sized molecules, for which the whole structure can be visualised as being involved in the interaction; in this case the abolition of a single functional group and/or slight stereochemical modifications can completely inhibit the biological activity; b. hydrophobic structures, for which matching with a carrier of hydrophobic nature can hinder the interaction with macromolecules.

As an example of these two situations, this paper attempts to describe some relevant points in the preparation of the solid phase for immuno-γ-globulins and steroid molecules. These substances have been used as models in the present studies in order to outline the different kinds of problems encountered. In fact, the insolubilisation of IgG is representative of the class of proteins which require their immunological properties to be conserved, by using reactions which yields as few covalent bonds as possible. Conversely, the coupling of steroids on a matrix sets problems connected with insolubilisation of small-sized hydrophobic molecules. For both systems Sepharose was chosen as a matrix, as being a highly suitable material both for its hydrophilic nature and the wide possibilities of chemical coupling which it allows.

The efficacy of some techniques of preparation and the influence of some reaction parameters have been quantitated by means of the evaluation of the residual affinity (equilibrium constant) with respect to the original affinity. In the case of solid-phase IgG the recovered number of sites has also been measured.

Insolubilisation of IgG

To understand better the factors influencing the properties of insolubilised IgG, the immunoadsorbents were prepared under various experimental conditions, including coupling to CNBr-activated Sepharose[10] and to a long-armed Sepharose derivative[11]; different pH of the reaction medium and different mass ratio IgG/matrix. The conjugates were characterised through the measurement of binding parameters and results were correlated with those obtained with the same immuno-γ-globulins in solution.

The system antiestradiol–estradiol was chosen for these studies for the following reasons: the small size of the hapten, whose diffusibility through the agarose matrix should facilitate the interpretation of the results; the univalence of the antibody site; the absence of non-specific interactions of estradiol with the matrix; the substantial identity between tracer and native hormone. All these conditions imply a simplified treatment of data.

Coupling to the active ester of Sepharose

In their experiments with amino acids, Cuatrecasas and Parikh[11] have shown that the rate of hydrolysis of the active ester increases as the pH of the medium is raised. The occurrence of this competitive reaction strongly affects the coupling yields.

Taking this into account, a pH of 6.4 was chosen because of the greater stability of the ester at this pH. Since pH 8.6 is frequently used for the direct coupling of proteins to CNBr-activated Sepharose[10], experiments have been carried out also at this pH for comparison. Figure 23.1 shows the time course of the coupling reaction at the two different pHs, while in figure 23.2 the resulting coupling yields for different initial ratios of IgG/Sepharose are

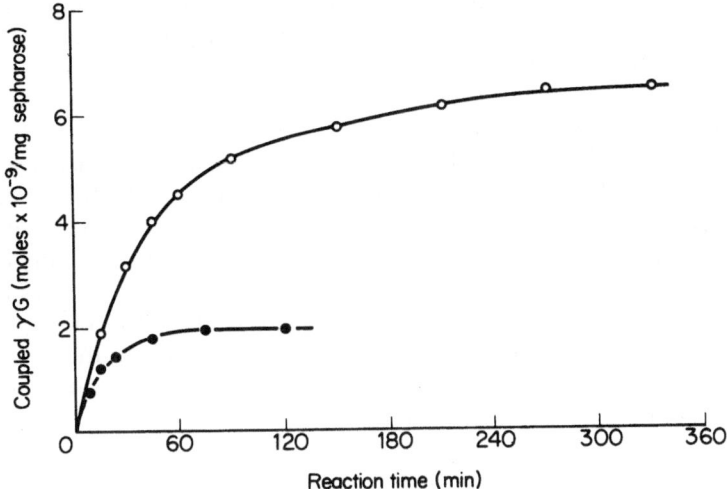

Figure 23.1. Time course of reaction of N-hydroxysuccinimide ester of agarose with IgG at 4°C in: (●) 0.1 M NaHCO₃ buffer pH 8.6 and (○) 0.1 M phosphate–citrate buffer pH 6.4. The data refer to an initial mass ratio IgG/Sepharose of 18×10^{-9} moles IgG/mg Sepharose assuming an average molecular weight of 150000 for IgG.

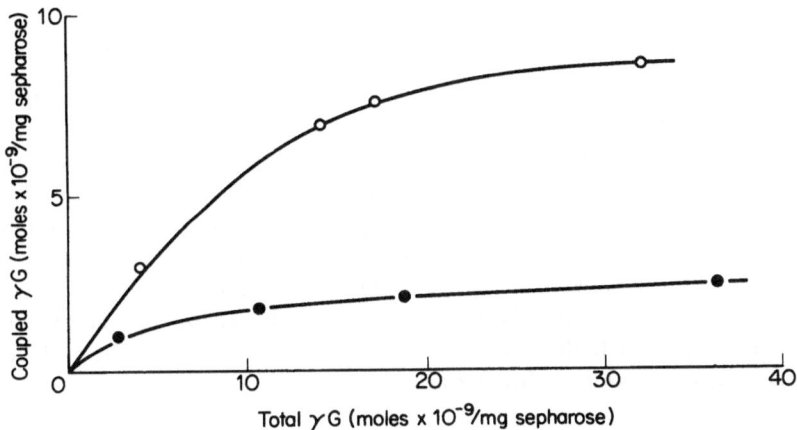

Figure 23.2. Effect of increasing the initial concentration of IgG on the coupling to N-hydroxysuccinimide ester of agarose at 4°C in: (●) 0.1 M NaHCO₃ buffer pH 8.6 and (○) 0.1 M phosphate–citrate buffer pH 6.4. The equilibrium values for a 300 min incubation time are reported.

reported. In agreement with the data reported by Cuatrecasas and Parikh for amino acids[11], the coupling reaction was found to be complete within 70–100 min at pH 8.6, while about 3 h were required at pH 6.4, and the amount of substitution was lower at pH 8.6 than at pH 6.4. For both pH conditions the maximum degree of substitution, expressed as nanomoles of IgG coupled to 1 mg of activated Sepharose, was obtained for an initial mass ratio of 20–25 nanomoles of IgG per mg of activated Sepharose.

Properties of insolubilised IgG

Table 23.1 shows the binding data, affinity constant K_0 and concentration of binding sites Ab_0, for the conjugates prepared by coupling the IgG directly to CNBr-activated Sepharose and to the active ester derivative of Sepharose, keeping the pH constant at 8.6. Within the limits set by the difficulty of controlling the coupling yield, the comparison was made for conjugates having similar IgG/Sepharose ratios, in order to separate the effect of the kind of reaction from those possibly arising from the degree of substitution. A higher K_0 value is apparent for the long-arm derivatives, whereas the site concentration recovered is substantially the same.

Table 23.1. Effect of the coupling reaction (pH 8.6, IgG/matrix ratio $\simeq 1.5$ nmole mg^{-1})

Matrix	$K_0 \times 10^9$ (M^{-1})	$Ab_0 \times 10^{-11}$* (M)
Sepharose–CNBr	1.3	0.91
Sepharose active ester	4.3	0.69

*Normalised to 1 nM of IgG.

Table 23.2 shows, in the case of the active ester of Sepharose, the effect of changing the pH of coupling on the binding data, under comparable conditions for the ratio of IgG to Sepharose. A net increase in the concentration of binding sites resulted in the case of coupling at pH 6.4, while the K_0 values essentially coincided. The superior quality of the immunoadsorbent obtained at pH 6.4 is probably attributable to the presence at this pH of a lower concentration of unprotonated amino groups which are required for the coupling reaction. Moreover, at this pH the reaction rate is lower than at pH 8.6, owing to the lower rate of hydrolysis of the active ester. Neither condition favours the occurrence of multiple linkages.

Table 23.2. Effect of the pH of coupling (long-armed Sepharose derivative, IgG/matrix ratio $\simeq 1.5$ nmole mg^{-1})

Coupling pH	$K_0 \times 10^9$ (M^{-1})	$Ab_0 \times 10^{-11}$* (M)
8.6	4.3	0.69
6.4	4.1	4.0

*Normalised to 1 nM of IgG.

Table 23.3 shows the effect of changing the IgG/matrix ratio for a coupling pH of 6.4 and using the long-armed derivative of Sepharose. It is apparent that both K_0 and Ab_0 values of the insolubilised IgG are strictly related to the mass ratio chosen. In fact, when one compares the binding parameters at the lowest mass ratio tested, with the data for IgG in solution (table 23.4) practically no loss of immunoreactivity can be found, while the data obtained for a 10-fold higher mass ratio closely resemble the data shown in table 23.1 for the direct coupling. These results suggest that an increase in the IgG/Sepharose mass ratio affects the accessibility of antibody sites to the hapten. This consideration is supported by the results given in figure 23.3 which show that the time taken to reach the equilibrium of the reaction with estradiol increases with increasing IgG/Sepharose mass ratios.

Table 23.3. Effect of the IgG/matrix ratio (long-armed Sepharose derivative, coupling pH 6.4).

IgG/Sepharose	$K_0 \times 10^9$ (M^{-1})	$Ab_0 \times 10^{-11}$* (M)
1.0	4.3	4.3
1.4	4.1	4.0
2.3	3.2	3.9
4.7	2.5	3.2
7.4	2.0	2.9
10.2	1.7	2.2

*Normalised to 1 nM of IgG.

Table 23.4. Comparison of binding data of antiestradiol in solid-phase and in solution.

	$K_0 \times 10^9$ (M^{-1})	$Ab_0 \times 10^{-11}$* (M)
Sepharose–antiestradiol	4.3	4.3
Antiestradiol in solution	4.8	5.0

*Normalised to 1 nM of IgG.

In conclusion, the results reported show that the nature of the coupling reaction, the reaction conditions, the interposition of a chemical arm between the protein and the matrix and the protein/matrix mass ratio are all factors which play a role in determining the loss of immunoreactivity associated with the insolubilisation of antibodies. An unequivocal assignment of an order of importance to these factors is difficult since, depending on the experimental conditions, the loss of immunoreactivity can be due either to a loss of active sites without modifications of the affinity of the residual sites, or to a combination of both effects. It is apparent, however, that in absolute terms the IgG/

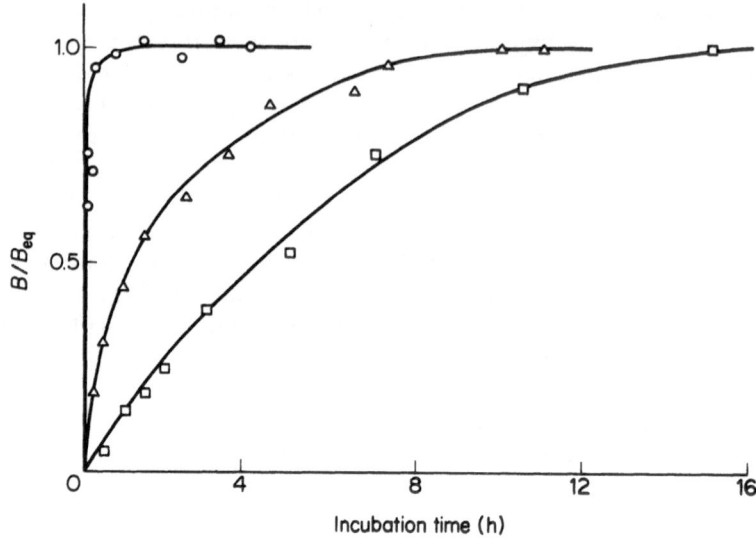

Figure 23.3. Binding of 17β-[2,4,6,7-H³] estradiol to IgG/Sepharose for different degrees of substitution: (△) 1.7×10^{-9} IgG moles/mg Sepharose; (■) 4.7×10^{-9} IgG moles/mg Sepharose as a function of incubation time. Open circles (○) refer to the IgG in solution. 0.05 nM H³-estradiol and 5 nM IgG solutions were used for all the experiments. The binding values are normalised to the fraction of bound radioactivity resulting for 20 h incubation (B_{eq}).

Sepharose ratio is by far the most critical factor in determining the behaviour of insolubilised antibodies and that by keeping the ratio low the antibody can be transferred from solution onto the solid matrix with a negligible loss in immunoreactivity.

Insolubilisation of testosterone

The application of affinity chromatography in the particular context of the isolation and fractionation of the antibodies contained in an antiserum to testosterone reflected well the different kinds of problem associated with the insolubilisation of small-sized hydrophobic molecules.

In addition to the immunoreactivity of the solid-phase, the parameters to be optimised in this case are the amount of ligand released in working conditions and the choice of the elution procedure. According to Ludens et al.[12], the design of the experiment was intended to fulfil three fundamental criteria: the complete desorption of proteins retained by the immunoadsorbent; minimal denaturation of proteins during the treatment; and maximal compatibility with the stability of the carrier.

Preparation and properties of the adsorbent

The position of attack on the steroid must be carefully chosen since the coupling must preserve, as far as possible, the full properties of the original molecule. In this particular case, the choice of the steroid derivative is however guided by the kind of antiserum to fractionate. Testosterone-3-oxime derivative was coupled to the matrix, since the antiserum had been raised against testosterone-3-BSA.

As for the matrix, the hydrophilic nature of Sepharose matches the requirement of minimising the interactions with hydrophobic structures, like steroids. In order to reduce the matrix effects, which are critical in this case[1], a chemical arm was interposed between ligand and matrix. This was done by using the 3,3-diamino-dipropylamine derivative of Sepharose.

The removal of the ligand loosely bound to the matrix is a crucial step in the preparation of the immunoadsorbent. The testosterone–Sepharose derivative was therefore submitted to exhaustive washing with solvents. To improve the desorption efficacy, further washing with albumin solution was carried out[13,14]. The characterisation of the immunoadsorbent was performed by testing its reactivity, i.e. its residual ability to interact with the specific antiserum when passing from the solution to the solid phase, and by measuring the fraction of ligand released in working conditions. Table 23.5 shows that immunoreactivity of testosterone was almost entirely preserved (70%) and that the stability of the adsorbent in the elution conditions chosen (see below) was acceptable. In any case, the fraction of released ligand was accounted for in the computation of binding parameters.

Table 23.5. Properties of testosterone–Sepharose.

Immunoreactive testosterone fraction (reactive μmoles/coupled μmole)	0.7
Ligand released on elution conditions* (released μmoles/coupled μmole)	
Phosphate–citrate buffer pH 7.4	0.05
pH 5	0.065
pH 3	0.09
Phosphate–citrate buffer pH 3, 10% dioxane	0.11

*Data referring to 10 mg of adsorbent suspended in 10 ml solution for 24 h at room temperature.

Design of the fractionation experiment

As for the conditions of elution of the antibody classes from the immunoadsorbent, they were assessed in preliminary experiments on the effects of various parameters (pH, solvent composition, etc.) upon the stability of the steroid–antibody complexes. The results of such experiments for testosterone are shown in figure 23.4. It is apparent that the binding ability of the antiserum is practically the same between pH 5 and 7.9, and that a sharp decline occurs between pH 4 and 3. However, over 50% of the maximal binding

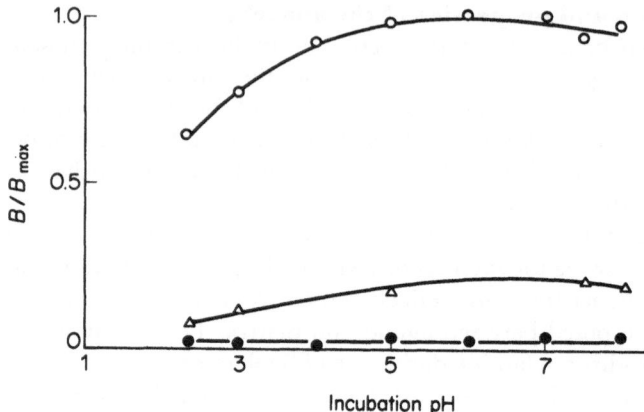

Figure 23.4. Effect of incubation pH on the reaction testosterone–antitestoster-one at 4°C: (○) phosphate–citrate buffer $\mu = 0.05$; (●) 10% (v/v) dioxane phosphate–citrate buffer; (△) 6 M guanidine in phosphate–citrate buffer. The binding values are normalised to the fraction of radioactivity bound in phosphate–citrate buffer at pH 7 (B_{max}).

ability is still found at pH 2.5. This indicates that for a complete desorption of the antibodies from the immunoadsorbent, the elution at gradient of pH must be followed by a final treatment with a powerful binding inhibitor. A complete inhibition occurs, for testosterone, in the presence of 10% dioxane, over the entire range of pH. 6 M guanidine, on the contrary, is a less effective inhibitor. Moreover, dioxane is to be preferred also for its lower denaturing effects. Table 23.6 lists the values of binding capacity obtained when the initial conditions are restored (phosphate–citrate buffer pH 7.4) after different times in the presence of the inhibitors. These values are expressed as % of the initial binding capacity. It is shown that over 80% of binding capacity is recovered in the case of dioxane after 300 min of contact with the inhibitor, while only 50% is recovered after treatment with guanidine for the same period of contact with the inhibitor.

Table 23.6. Binding ability of antiserum to testosterone recovered after treatment with inhibitors.

Inhibitor*	Contact time (min)	Binding ability recovered (%)
Guanidine, 6 M	30	83
	300	51
Dioxane, 10%	30	93
	300	82

*In phosphate–citrate buffer, pH 3.

Fractionation of antibodies to testosterone

The antiserum to testosterone used in fractionation experiments showed a nonlinear Scatchard plot. Linearisation using the Sips plot was obtained by assuming α to be equal to 0.88, which indicates a considerable degree of heterogeneity (*see* figure 23.5). The antiserum was incubated with a testosterone–Sepharose adsorbent in the appropriate quantity at pH 7.4. After washing, 5 fractions were eluted at pH 6, 5, 4, 3, and with a 10% dioxane buffer mixture at pH 3. Immediately after collection, each fraction was neutralised with Na_2HPO_4, ultrafiltered using hollow fibres, and freeze-dried. The first fraction was found to contain almost pure albumin, the 2nd β-globulins and traces of α-globulins, the 3rd mainly comprised α-globulins with some γ-globulins. Within the sensitivity limit of the electrophoretic method used,

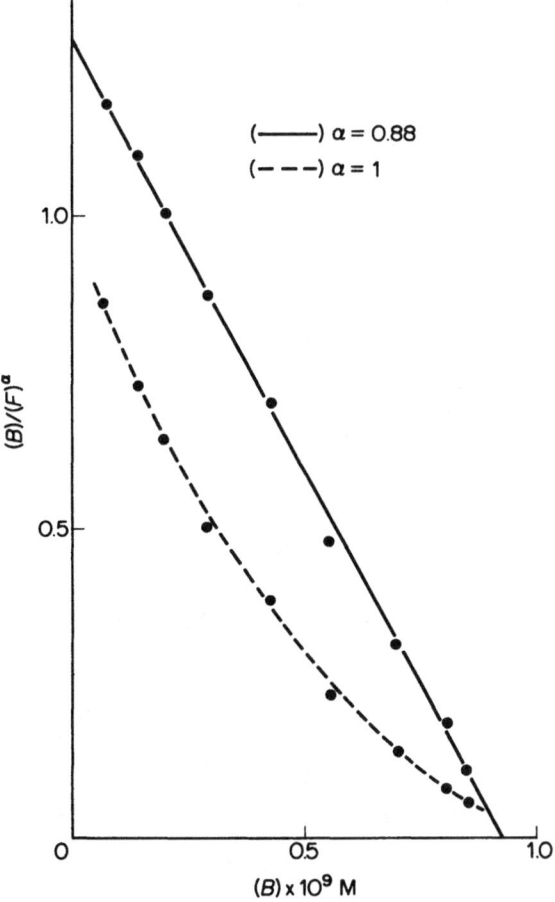

Figure 23.5. Scatchard plot (dotted line) and Sips plot (continuous line) for the antiserum to testosterone.

γ-globulins were the only components in the 4th and the 5th fractions, i.e. those eluted at pH 3 and with 10% dioxane, respectively.

As shown in figure 23.6, a linear B/F vs. B plot indicating homogeneity was obtained with each of these last fractions. The equilibrium constant and the concentration of binding sites of these fractions are reported in table 23.7. To evaluate the recovery of applied activity, proportional amounts of each fraction were mixed. The reconstituted antiserum showed a B/F vs. B plot very close to that of the original antiserum, and the same heterogeneity coefficient. The overall recovery in terms of immunoreactivity was 67%. The composition of the antiserum which can be derived from these data is that of two main families, one of which is about 10 times more concentrated than the other, accompanied by one or more families of low energy sites, which were eluted at pH 4, with an affinity constant of the order of 10^6–10^7 l mole^{-1}.

The result reported above show that homogeneous families of antibody sites

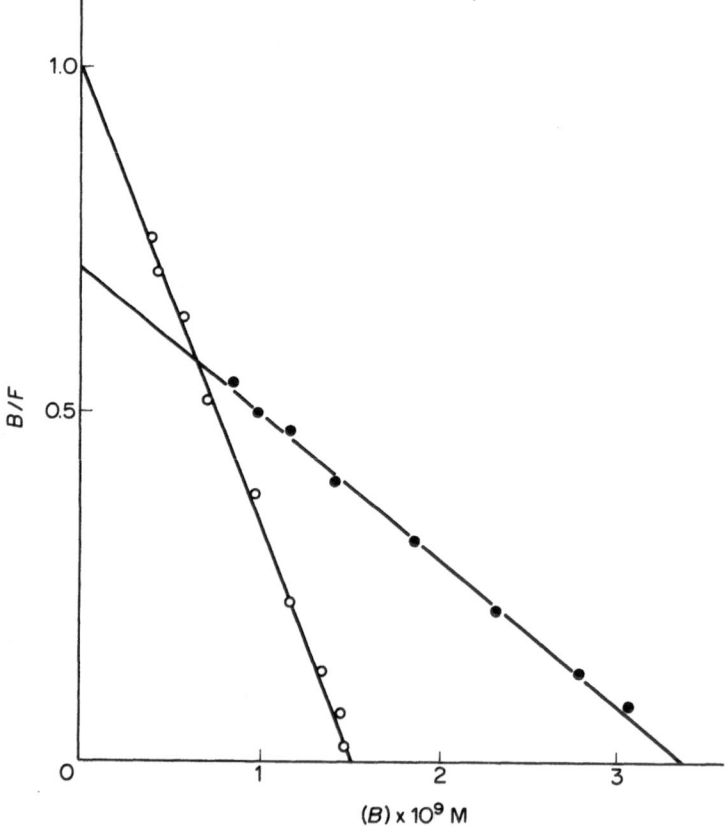

Figure 23.6. Scatchard plot for γ-globulins: (●) fraction IV and (○) fraction V. The amounts of released ligand were accounted for in the evaluation of the binding data.

can be isolated by affinity chromatography from antisera to testosterone but some denaturation occurs, probably due to the elution conditions required for the steroid–antibody complexes. Thus, the practical potentiality of affinity chromatography to enhance the quality of antisera to steroids must be regarded with caution.

Table 23.7. Fractionation of antiserum to testosterone

	$K_0 \times 10^8$ (M^{-1})	$Ab_0 \times 10^6$ (M)	α
Whole antiserum	7.5	3.5	0.88
Fraction IV*	2.1	0.84	1.0
Fraction V†	6.5	1.5	1.0
Reconstituted antiserum	5.8	3.1	0.88

*Elution with phosphate–citrate buffer, $\mu = 0.05$, pH 3.
†Elution with dioxane (0.022 molar fraction) in the same buffer.

However, the present results must be judged in the light of the reports published so far on the use of steroid–Sepharose adsorbents for the purification of steroid-binding proteins. As pointed out by Ludens *et al.*[12], the effective removal of steroid binding proteins from affinity columns has been reported in detail only with serum-binding proteins for corticosteroids (58% yield for CBG)[15] and for testosterone (0.4% yield for TBG)[16], while many attempts to recover active steroid-binding molecules of tissue origin with estradiol–Sepharose and androgen–Sepharose complexes have failed[12,17,18]. Ludens *et al.*[12] have provided strong evidence that it is the release of the steroid from the ligand-gel linkage which severely compromises the utility of affinity chromatography when minute amounts of high affinity proteins must be handled. On the other hand, Sica *et al.*[14] have recently shown that estradiol receptors can be isolated by affinity chromatography, using organic solvents such as methanol or dioxane and/or proteins to purify the adsorbent. Viewed against this background, the results of the present study indicate that the use of steroid–Sepharose adsorbents requires a careful design of the experiment and the adoption of the following precautions:

a. The affinity of albumin for the steroids can be exploited to remove with a single washing step the ligand which is not firmly bound to the Sepharose matrix.

b. The elution with extremely acid solution or with concentrated guanidine should be avoided in favour of milder eluting agents like dioxane.

c. The amount of Sepharose matrix used for the separation must be carefully proportionated.

d. A batchwise procedure should be preferred to column chromatography since it allows the use of smaller amounts of adsorbent and the shortening of the contact time with the various eluting and washing agents. In addition, the determination of free steroid in the eluates is critically important for the interpretation of the results.

References

1. Cuatrecasas, P. and Anfinsen, C. B. Affinity chromatography. *Ann. Rev. Biochem.*, **40** (1971), 259
2. Silman, I. H. and Katchalski, E. Water-insoluble derivatives of enzymes, antigens, and antibodies. *Ann. Rev. Biochem.*, **35** (1966), 873
3. Cuatrecasas, P. Protein purification by affinity chromatography. *J. Biol. Chem.*, **245** (1970), 3059
4. Sato, N. and Carcille, M. Separation of specific from nonspecific anti-FSH antibody by affinity chromatography on Sepharose-HCG. *Endocrinology*, **90** (1972), 302
5. Pensky, J. and Marshall, J. S. Studies on thyroxine-binding globulin (TBG). *Arch. Biochem. Biophys.*, **135** (1969), 304
6. Steers, E. Jr., Cuatrecasas, P. and Pollard, H. B. The purification of β-galactosidase from *Escherichia Coli* by affinity chromatography. *J. Biol. Chem.*, **246** (1971), 196
7. Weintraub, B. D. and Kadesky, Y. M. Fractionation of antibodies to human chorionic somatomammotropin: application to radioimmuno-assay. *J. Clin. Endocr.*, **33** (1971), 432
8. Rosenberry, T. L., Chang, H. W. and Chen, Y. T. Purification of acetyl-cholinesterase by affinity chromatography and determination of active site stoichiometry. *J. Biol. Chem.*, **247** (1972), 1555
9. Olesen, H., Molin, J., Hippe, E., Rye, M., Thomsen, J., Lee, L. and Haber, E. (1972). Studies on human intrinsic factor purified by affinity chromatography. 8th International Congress on Clinical Chemistry. Copenhagen, 18–23 June, abs. in *Scand. J. Clin. Lab. Invest.*, **29**, sup. 126 (1972).
10. Wide, L. Radioimmunoassay employing immunosorbents. *Acta Endocrin.*, Copenhagen, **63**, suppl. **142** (1969), 207
11. Cuatrecasas, P. and Parikh, I. Adsorbents for affinity chromatography. Use of N-hydroxysuccinimide esters of agarose..*Biochem.*, **11** (1972), 2291
12. Ludens, J. H., De Vries, J. R. and Fanestil, D. D. Criteria for affinity chromatography of steroid-binding macromolecules. *J. Biol. Chem.*, **247** (1972), 7533
13. Massaglia, A., Rolleri, E., Barbieri, U. and Rosa, U. Fractionation of antibodies to testosterone. *J. Clin. Endocrinol. Metab.*, **38** (1974), 820
14. Sica, V., Parikh, I., Nola, E., Puca, G. A. and Cuatrecasas, P. Affinity chromatography and the purification of oestrogen receptors. *J. Biol. Chem.*, **248** (1973), 6543
15. Rosner, W. and Bradlow, H. L. Purification of corticosteroid-binding globulin from human plasma by affinity chromatography. *J. Clin. Endocrinol. Metab.*, **33** (1971), 193

16. Burstein, S. H. The removal of testosterone binding globulin from plasma by affinity chromatography. *Steroids.*, **14** (1969), 263
17. Ludens, J. H., De Vries, J. R. and Fanestil, D. D. Studies on affinity chromatography of aldosterone-binding macromolecules. *Steroid Biochem.*, **3** (1972), 192
18. Vonderhaar, B. and Muller, G. C. Binding of oestrogen receptor to oestradiol immobilised on insoluble resins. *Biochem. Biophys. Acta*, **176** (1969), 626

Discussion

MILHAUD

Which was the best protein/sepharose ratio?

ROLLERI

One nanomole of IgG per mg of sepharose.

ROSA

I would like to stress that the major factor governing the behavior of the gammaglobulin on the solid phase is the protein/matrix ratio. As you know, there have been a lot of sophisticated versions of the method for insolubilising proteins especially gammaglobulin on solid phase, using chemical arms and specially adapted conjugation procedures. Finally, it has been recently confirmed by several laboratories, including that of Dr Hunter in England, that the only factor which plays an effective role in governing the properties of the insolubilised antibodies is the amount of immunogammaglobulin on the solid phase. Dr Hunter has shown that complete antisera can be transferred on solid phase practically without a loss of immunoreactivity[1]. This result is probably due to the dilution of IgG by albumin. As you know, our experiments have been carried out with gammaglobulin to avoid effects arising from competition with other proteins in the study of the binding parameters.

MILLER

Have you found any influence on the optimal ratio of IgG/sepharose if the ligand is a large protein molecule?

ROLLERI

We have dealt only with small hydrophobic ligands.

ROBBINS

We have been coupling thyroglobulin for the isolation of antithyroglobulin antibodies. The binding is very strong and the difficulty is in getting the antibody off.

REEVE

Is this a good method in preparing albumin?

ROLLERI

I think that it could be a promising method.

ROSA

We are trying to apply this procedure to the rapid purification of albumin. It should be possible by subfractioning the fraction we showed.

MILLER

Travis[2] has reported that he can obtain albumin better than 95% pure by the use of blue dextran coupled to sepharose. The elution is effected with sodium caproate.

WALDMANN

I think it is important to differentiate between the issue of whether a protein is pure, that is uncontaminated by other proteins, and whether it is undamaged and therefore, of value for metabolic studies. If you have to drop the pH very much or perform certain manipulations, you may get some pure but denatured material which would preclude the use of the preparation for turnover studies.

It is better to have a weak binding solid phase system with an antibody to bind your contaminant rather than an antibody to the material under study.

ROSA

I certainly agree with you. On the other hand one of the methods we have to characterise an antibody is that of measuring the binding data with antibody sites. It does not mean that the molecule has not been partially denatured or modified of course, but it means that the topology of the molecule near the binding side has been unmodified. This is one of the possible criteria of course for purity, that is to judge if a protein has been modified or not.

References

1. Bolton, A. E. and Hunter, W. M. The use of antisera covalently coupled to agarose, cellulose and Sephadex in radioimmunoassay systems for proteins and haptens. *Biochim. Biophys. Acta*, **329** (1973), 318
2. Travis, J. and Pannel, R. Selective removal of albumin from plasma by affinity chromatography. *Clinica Chimica Acta*, **49** (1973), 49

24
Transport of steroid hormones

A. VERMEULEN

In 1913 Oppenheimer[1] observed that cardiac glucosides are less toxic to the isolated frog heart when serum is added to the drug solution. This phenomenon he explained by adsorption of the digitalis steroids by the colloidal components of the serum. This was probably the first study indicating a steroid–protein interaction.

In 1931 Bennhold[2] reported on the binding of cholesterin to plasma globulins, and in 1938 on the binding of bile acids to albumin[3] but it was Brunelli[4] who in 1934 was probably the first to study the interaction of steroid hormones, namely estrone, with plasma proteins in his dialysis experiments using collodion membranes.

Steroid–protein interactions involve non-covalent bonds of relatively low energy; the association between steroid and protein is always reversible and the binding equilibrium is governed by the law of mass action. The interaction between a protein P and a steroid S can be represented by the equation

$$S + P \rightleftharpoons SP$$

and

$$k = \frac{[SP]}{[S][P]}$$

where [S] is concentration of free steroid at equilibrium, [SP] the concentration of bound steroid, [P] the concentration of free binding sites and k the equilibrium constant. When the binding protein has not been obtained in the pure state, the molar concentration is used instead of the number of binding sites and the association constant becomes then nK, n being the number of independent and equivalent binding sites per molecule.

Binding of steroid hormone to albumin

1. *The binding capacity* of plasma albumin for steroid hormones is far in excess of the physiological steroid hormone concentration. As a consequence, at all

A. Vermeulen

physiological concentrations, the ratio of albumin bound steroid to free steroid will be independent of total steroid concentration, and will be determined by the protein concentration of which it is a linear function. However, it should be mentioned here that Brunkhorst and Hess[5] obtained decreasing association constants at increasing albumin concentration for the interaction of cortisol with albumin, whereas at constant protein concentration nK decreased with increasing cortisol concentration. Whereas the former phenomenon might indicate heterogeneity of the albumin, the latter might be related to increased protein interaction in the more concentrated albumin solutions. Variations within the physiological concentration range of both albumin and the steroid do not appear to affect the K value appreciably. Nevertheless these studies indicate that when extrapolation of *in vitro* to *in vivo* conditions is considered, one should be careful to select the *in vitro* conditions as close as possible to physiological conditions.

2. Another consequence of the high binding capacity of albumin is, that, again at physiological concentrations, the binding of different steroid hormones to albumin may be considered to be *independent of each other*. However at supraphysiological concentrations this is no longer true[6,7]. Low affinity, high capacity, non-competitive binding of steroids to albumin is generally called non-specific binding, as opposed to the binding to high affinity, low capacity binding to specific protein.

3. *The association constant* of the interaction of a steroid hormone with albumin is rather low ($\sim 10^4$ M^{-1}) and as a rule decreases with increasing polarity of the steroid (polarity rule[8]), a consequence of the hydrophobic nature of the bond. Probably this low affinity albumin binding is of little biological significance.

The association constant varies however with the degree of purification of the albumin, the higher purity resulting in higher K value. Delipidation of HSA with organic solvents increases also the K value[9].

The interaction of the steroid with albumin appears to involve the lower or alpha side of the steroid which has a relatively plane surface[10]. As the large molecule binds at most a few steroid molecules and as one can assume that the binding sites will have approximately the size of the steroid molecule, the portion of the protein molecule involved in binding should be small. Some studies suggest that tryptophan residues and/or tyrosine residues may be involved[11,12].

4. Albumin seems to be the main protein responsible for the binding of steroid conjugates to plasma proteins; its binding capacity again exceeds greatly the steroid conjugates concentrations under physiological conditions. It is interesting that Puche and Nes[13] reported displacement of steroid sulphates, from albumin binding sites by increasing sulphate ion concentration, whereas Wang and Bulbrook[14] reported displacement by non-steroidal organic sulphates. Puche and Nes assume an interaction of the cationic sites of arginine with the anionic sulphate group.

Binding of steroids to the corticosteroid binding globulin or or transcortin

The presence in human plasma of a specific, high affinity but low capacity protein, binding corticosteroids, was shown independently by Daughaday[15] by Bush[16] and by Sandberg et al.[5,6] in 1956–1957.

This protein was called corticosteroid binding protein by Daughaday and transcortin by Sandberg and Slaunwhite. Studies of Daughaday showed that up to a concentration of 20 μg 100 ml^{-1} and at 4°C, cortisol is almost completely bound, the high affinity binding protein becoming subsequently saturated, whereas at still higher concentrations, cortisol becomes essentially bound to albumin. It was upon inspection of a similar curve published by Sandberg et al. that Murphy came upon the idea of measuring cortisol concentrations by competitive protein binding[17].

General properties of transcortin

Transcortin is a glycoprotein which on paper migrates with the α_1-globulin. Human CBG has only one binding site per mole[18]. The binding affinity for steroids is maximal at pH 8; below pH 5 irreversible denaturation takes place. In human plasma it is stable at 45°C[19]; pure CBG however subjected to gel-filtration at 45°C becomes inactivated[18]. Transcortin is destroyed by heating at 60°C for 10 minutes[20]; steroid binding however appears to protect CPB[19] against denaturation.

The association constant decreases with increasing temperature[21]; association is complete within minutes at 4°C, virtually instantaneous at 37°C. Half-dissociation times at 4°C were 25 min, 1 min at 27°C and about 10 s at 37°C[23]. The affinity[21] for cortisol at 37°C is about 3×10^7 M^{-1}, for progesterone 8×10^7 M^{-1}; for corticosterone 3×10^7 M^{-1}, for testosterone 1.5×10^6 M^{-1}, for estradiol 2×10^7 M^{-1}, for testosterone 1.5×10^6 M^{-1}, for estradiol 2×10^4 M^{-1}. It is interesting to note that corticosterone sulphate is bound to CBG with half the affinity of that of corticosterone[22] whereas the cortisol conjugates are not bound to transcortin[31]. Corticosteroid binding globulins, with however different specificity for the steroids, have been observed in virtually all vertebrate animal species[24–26].

Binding capacity

The approximate cortisol binding capacity of transcortin[27] in normal human serum is 20–25 μg 100 ml^{-1}, corresponding to a concentration of binding sites of $5–7 \times 10^{-7}$ M. As the molecular weight is about 52000[18,26] the concentration of CBG is about 36 mg l plasma^{-1}. There is no difference between males and females and the values do not change between 1 and 60 years; low values are found in cord plasma. The concentration in lymph is only slightly lower than in blood[28].

A decreased CBG concentration was observed in liver cirrhosis, in the nephrotic syndrome and in some forms of Cushing's syndrome. An hereditary familial decrease of CBG level has been described by Doe *et al.*[29] and by De Moor[30]. *Increased CBG* levels are observed in pregnancy, and after estrogen administration[31]. Recently Hammerstein *et al.*[32] observed a close correlation between estrogenic activity and increase of CBG levels and suggested the increase of transcortin levels as a parameter of estrogenicity. Plasma levels start to increase after about 3 days of estrogen treatment, and return to pretreatment values 7–10 days after cessation of therapy[28]. The increase of CBG under the influence of estrogen is a consequence of an increased synthesis or at least of an activation of binding sites, and not of decreased turnover as shown by plasma clearance studies with ^{125}I-transcortin[33]. The increase of cortisol levels during pregnancy, up to about four times the normal level during the last trimester of pregnancy, without the development of Cushing syndrome, leads to the hypothesis that the transcortin bound cortisol is biologically inactive. This was confirmed by studies of the MCR[34].

Notwithstanding the increase of transcortin, the free cortisol concentration is moderately increased during pregnancy[35,36]. These results were confirmed by us (figure 24.1). As plasma cortisol is only bound to albumin and transcortin, knowing the free cortisol concentration and the albumin concentration, distribution of cortisol over different plasma proteins may be calculated. It should however be realised that the association constant of albumin depends on its degree of purification. Moreover the competition of other steroids should be taken into account.

Nevertheless, in this view of the low association constant of albumin, even an error in this value of 100% would not effect greatly the distribution. At normal cortisol concentration between 10 and 20 μg 100 ml^{-1}, we observed a free cortisol fraction between 4 and 8%, yielding, at normal plasma albumin concentration, an albumin bound fraction of 5–11%, the rest being bound to transcortin. This is in contradiction with the results published by Westphal[9], who gives an unexplicably high albumin bound fraction of about 30%, which

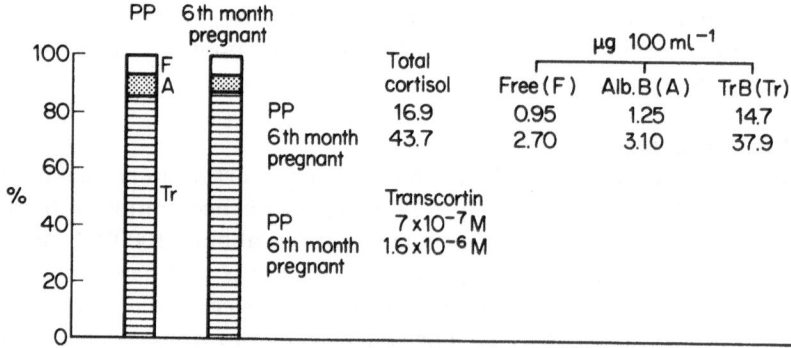

Figure 24.1. Distribution of cortisol and transcortin in plasma, post-partum (pp) and in the 6th month of pregnancy.

supposes a 3–6 times higher association constant. Our values are more in agreement with those of Rosenthal *et al.*[36,37] and of Tait and Burstein[34].

Sandberg showed that the increase in CBG *precedes* the rise in peripheral corticosteroids[33]. The site of CBG production is not yet definitely known, but experimental evidence from partial hepatectomy suggests it to be synthesised in the liver. Finally it should be mentioned that in the pregnant guinea pig there exists a specific protesterone binding globulin which is different from CBG and which does not bind either cortisol or corticosterone[38].

Binding of testosterone and estradiol to TeBG

Until 1958 it was generally assumed that testosterone in plasma was only bound by albumin. In 1958 Daughaday[39] showed that with increased concentration, plasma binding of testosterone decreased; this observation was incompatible with an exclusive binding by albumin. The same year, this author reported that Cohn's fraction IV (mainly α- and β-globulin) bound testosterone more firmly than albumin. Subsequently several publications confirmed these findings and in 1965 Mercier *et al.*[40] reported the presence of a specific testosterone binding globulin in plasma. In numerous subsequent publications, the binding of steroids to this plasma protein was extensively studied[41–44]. It was shown that the presence of a 17-β-hydroxyl group was required for binding[41]. A 5α-saturated ring A strengthened the binding, whereas in steroids with a 5β-configuration binding was strongly decreased (table 24.1)[41,44]. The protein has a β-globulin nature[45]; it is stable up to 45°C but is irreversibly destroyed at 60–65°C[41]. The binding affinity is higher at 4°C than at 37°C: at 37°C we found a binding affinity for testosterone of 8×10^8 M^{-1} against 1.2×10^9 M^{-1} at 4°C. The binding affinity for estradiol at 37°C is somewhat lower, 6×10^8 M^{-1}. Testosterone and estradiol appear to compete for the same binding sites[46]. Binding remains constant at pH values between

Table 24.1. Influence of structure on binding of natural steroids to TeBG relative binding in comparison to testosterone

	17β-OH ⟷ 17α-OH		
Testosterone	100	17α-OH(epi)testosterone	4
17β-hydroxypregn-4-ene-3,20 dione	33	17α-hydroxypregn-4-ene-3,20 dione	2.5
	17β-OH ⟷ 17 OXO		
Testosterone	100	Δ^4-Androstenedione	5
Oestradiol	62	Oestrone	4
5α-androstane-3α-17β-diol	160	Androsterone	3
5β-androstane-3α-17β-diol	24	Etiocholanolone	0
Δ^5-androstene-3β-17β-diol	50	DHEA	4.5
	Influence of substitution in C_{16}		
Oestradiol	62	Oestriol	32
Δ^5-androstene-3β-17β-diol	50	Δ^5-androstene-3β-17β-diol-16-one	1
	Influence of 5α/5β configuration		
5α-androstane 3α,17β-diol	160	5β-androstane 3α,17β-diol	24
17β-hydroxy-5α-androstane-3 one	300	17β-hydroxy-5β-androstane-3 one	10

6 and 8; it decreases sharply[41] at pH below 5. A M.W. of this protein of 52000[47] respectively 98 and 102000[48,49] has been reported. As far as binding capacity is concerned, this is higher in adult females than in adult males[50]. In pre-pubertal children of both sexes it is as high or higher than in females. In males it increases in old age, towards values as found in normal females. Similarly, in male hypogonadism, binding capacity is increased[52]. In the female, increased binding capacity, to levels 4 to 10 times basal levels is observed during pregnancy[50] and in women taking estrogen-containing contraceptives, whereas in hyperthyroidism, high levels are observed both in males and females[51].

Murphy[53] reported increased testosterone binding capacity in male patients with liver cirrhosis and Rosenbaum *et al.*[54,55] reported augmented estradiol binding in these subjects; this might be related to the high estrogen levels which have been reported in cirrhotic patients; in an own group of cirrhotic patients however, binding capacity was rather variable and seems to correlate inversely with testosterone levels. Decreased binding capacity is observed in virilised females[54]; also in newborn infants TeBG capacity is low. Among the pharmacological factors that influence this binding capacity, it is known that estrogen as well as thyroid hormone administration[51] increases binding capacity. Androgen administration as well as high doses of corticosteroids on the other hand, decrease the binding capacity; a similar decrease was also observed after administration of medroxyprogesterone acetate[50].

As far as other animal species are concerned, high affinity binding for testosterone has been observed with human, bovine, frog[53] and guinea pig[38] plasma, as well as in the atlantic salmon and in thorny skate[57]. No activity was found in rat, rabbit, dog and duck plasma[38]. It should be mentioned however that in the rat foetus, but not in the adult rat, a specific estradiol and estrone binding plasma protein (EBP) has been described[58]. Recently, an androgen binding protein has been isolated from rabbit plasma[61].

Apart from binding to albumin and TeBG, for all practical purposes, at 37°C and at physiological concentrations, binding of testosterone and estradiol to other plasma proteins such as transcortin or α_1-acid glycoprotein (orosomucoid or AAG) is negligible: indeed the nK value[59] of AAG for testosterone is $\pm 2 \times 10^5$ M^{-1} and for estradiol $\pm 5 \times 10^4$ M^{-1} and its concentration is $\pm 1.8 \times 10^5$ M^{-1}, whereas the binding affinity[52] at 37°C of transcortin for testosterone is 1.5×10^6 M^{-1} and its concentration is $\pm 5 \times 10^{-7}$ M. As only TeBG and albumin bind testosterone and estradiol to any significant degree, it is possible, having determined the free hormone concentration by dialysis and knowing the albumin content, to calculate the distribution of these hormones among plasma proteins. The values obtained are approximate, however, as the association constant of albumin for testosterone and estradiol, (3.5 and $6 \times 10^4 \, M^{-1}$ respectively) as it occurs in plasma, is only approximately known. Nevertheless for comparative purposes the values may be valid. Figures 24.2 and 24.3 give some examples of the distribution of testos-

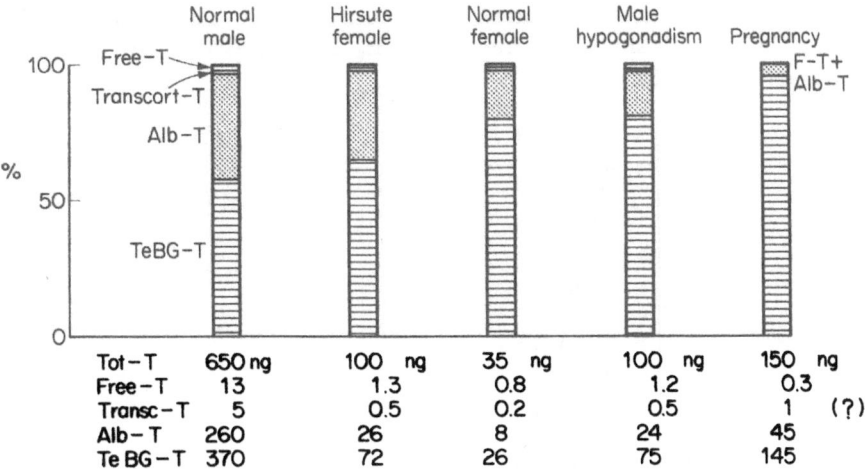

Figure 24.2. Distribution of T over different plasma proteins.

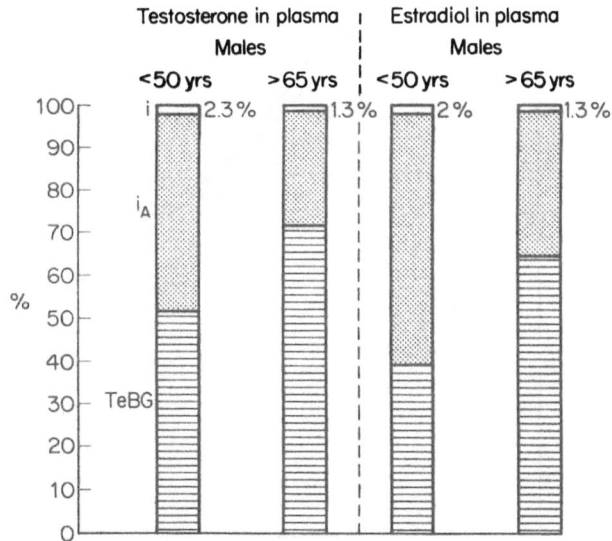

Figure 24.3. Distribution of T and E$_2$ in plasma protein in young and elderly males.

terone and estradiol in different physiopathological conditions. An interesting corollary of the differences in affinity of TeBG and albumin for estradiol and testosterone, is the shift in free fraction of both steroids which occurs with increasing binding capacity of TeBG. Indeed, whereas at low binding capacity (young males for example) the free testosterone fraction (in % of total T) is higher than the free E$_2$ fraction, at high TeBG levels (elderly males and in women) the E_2 fraction becomes the more important. This is illustrated in figure 24.4 which shows the free T and E_2 fraction as found in 100 males between 15 and 95 years old.

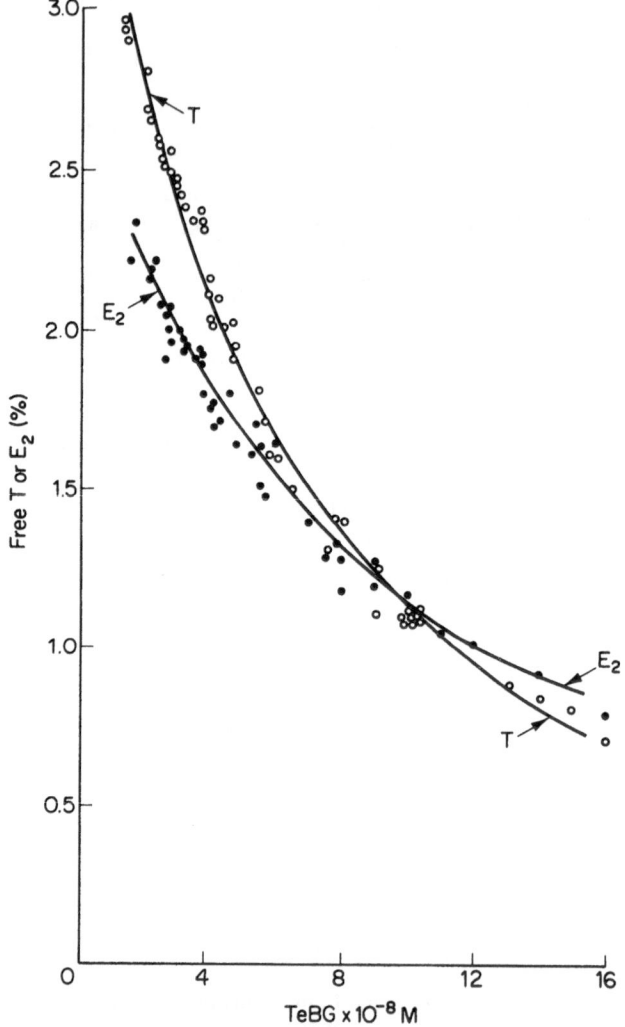

Figure 24.4. Free T over free E_2 in function of TeBG capacity in males age 20–95.

Finally a few words should be said about the possible role of the binding proteins. The original idea of a steroid solubilising role is untenable in the light of the low concentration of these steroid hormones in plasma. On the other hand there is ample evidence that the specifically bound hormone is not available to the tissues: for example women taking oral contraceptives do not develop a Cushing syndrome, notwithstanding high total cortisol levels, and hyperthyroid women do not become virilised, notwithstanding high testosterone levels. Moreover there is an inverse correlation between the MCR and the degree of binding of a steroid; finally *in vitro* studies have confirmed the biological unavailability of the bound hormone[59]. What then might be the role of the protein bound steroid hormone? It appears to us, that the most meaningful role of these proteins would be a protection of the organism against the high hormonal levels, that are necessary for specific target tissues. The latter would be able to dissociate the steroid plasma protein complex by virtue of the competing effects of specific receptor proteins with high affinity.

References

1. Oppenheimer, E. Zur Frage der Fixation der Digitaliskörper im tierischen Organismus und besonders deren Verhalten zum Blut. *Biochem. Z.*, **55** (1913), 134

2. Bennhold, H. Uber die Bindung des Cholesterins an die Globuline: zugleich ein weiterer Beitrag zur Frage der Funktion der Serumeiweisskörper. *Verh. Deutsch. Ges. inn. Med.*, **43** (1932), 213

3. Bennhold, H. (1938). Die Vehikelfunktion der Bluteiweisskörper. In *Die Eiweisskörper des Blutplasmas Hrsg.*, H. Bennhold, E. Kylin and S. Rusznyak, eds., Steinkopf, Dresden, p. 220

4. Brunelli, B. Sulla funzione veicolanto delle proteine plasmatiche per l'ormone folliculare. *Arch. Intern. pharmacodyn.*, **49** (1934), 262

5. Brunkhorst, W. H. and Hess, E. I. Interaction of cortisol with serum albumin. *Arch. Biochem.*, **111** (1965), 54

6. Sandberg, A., Slaunwhite, W. R. and Antoniades, H. M. The binding of steroids and steroid conjugates to human plasma proteins. *Rec. Progr. Horm. Res.*, **13** (1957), 209

7. Westphal, U. (1971). *Steroid protein interactions*. Springer-Verlag, Berlin, p. 129

8. Westphal, U. (1961). Interactions between steroids and proteins. In *Mechanism of action of steroid hormones*. Villee C. A. and Engel, U., eds., Pergamon Press, N.Y., p. 33

9 Westphal, U. Steroid protein interactions. XII—Distribution of progesterone and corticosteroid hormones among serum proteins. *Hoppe—Seylers Z. Physiol. Chem.*, **346** (1966), 243

10. Westphal, U. and Ashley, B. D. Steroid protein interactions. VI—Stereochemical aspects of interaction between Δ^4-3 ketosteroids and human serum albumin. *J. Biol. Chem.*, **234** (1959), 284

11. Ryan, M. T. and Gibbs, G. Analysis of ultraviolet spectral perturbations arising in the interaction of steroids and human serum albumin. *Arch. Biochem.*, **136** (1970), 65

12. Oyakawa, E. K. and Levedahl, B. H. Testosterone binding to bovine and human serum albumin: the role of tyrosine groups. *Arch. Biochem.*, **74** (1958), 17

13. Puche, R. C. and Nes, W. R. Binding of dehydroepiandrosterone sulphate to serum albumin. *Endocrinology*, **70** (1962), 857

14. Wang, D. Y. and Bulbrook, R. D. Binding of sulphate esters of dehydroepiandrosterone, testosterone, 17-acetoxypregnenolone and pregnenolone in the plasma of man, rabbit and rat. *J. Endocr.*, **39** (1967), 405

15. Daughaday, W. H. Evidence for two corticosteroid binding systems in human plasma. *J. Lab. Clin. Med.*, **48** (1956), 799

16. Bush, I. (1957) The physicochemical state of cortisol in blood. In *Hormones in blood*. Ciba Foundation Colloquium on Endocrinology, Vol. 11, 263

17. Murphy, B. E. P. Application of the property of protein binding to the assay of minute quantities of hormones and other substances. *Nature*, **201** (1964), 679

18. Muldoon, T. G. and Westphal, U. Steroid protein interactions. XV. Isolation and characterisation of corticosteroid binding globulin from human plasma. *J. Biol. Chem.*, **242** (1967), 5636

19. Doe, R. P., Fernandez, R. N. and Seal, U. S. Measurement of corticosteroid binding globulin in man. *J. Clin. Endocr.*. **24** (1964), 1029

20. Daughaday, W. H. Steroid protein interaction. *Physiol. Rev.*, **39** (1959), 885

21. Westphal, U. Steroid protein interactions. XIII. Concentration and binding affinity of corticosteroid binding globulins in sera of man, monkey, rabbit and guinea pig. *Arch. Biochem.*, **118** (1967), 556

22. Lebeau, M. C. and Baulieu, E. E. Binding of steroid conjugates to human corticosteroid binding globulin. *J. Clin. Endocr.*, **30** (1970), 166

23. Dixon, D. F. The kinetics of the exchange between transcortin bound and unbound cortisol in plasma. *J. Endocr.*, **40** (1968), 457

24. Seal, U. S. and Doe, R. P. Vertebrate distribution of corticosteroid binding globulin: species distribution and small scale purification. *Endocrinology*, **73** (1963), 371

25. Seal, U. S. and Doe, R. P. Vertebrate distribution of corticosteroid binding globulin and some endocrine effects on concentration. *Steroids*, **5** (1965), 827

26. Seal, U. S. and Doe, R. P. (1966). Corticosteroid binding globulin: biochemistry physiology and phylogeny. In *Steroid dynamics*. Pincus, G., Nakao, T. and Tait, J. F. eds., Academic Press, N.Y., p. 63

27. De Moor, P., Heirwegh, K., Heremans, J. and De Clercq-Raskin, M. Protein binding of corticoids studied by gel filtration. *J. Clin. Invest.*, **41** (1962), 816

28. Sandberg, A. A. and Carter, A. C. Transcortin: a corticosteroid binding globulin III. The effects of various steroids. *J. Clin. Invest.*, **39** (1960) 1914

29. Doe, R. H., Lohrenz, F. M. and Seal, U. S. Familial decrease in cortico-steroid binding globulin. *Metabolism*, **14** (1965), 940

30. De Moor, P. and Hendrickx, A. Clinical features observed in patients with an unexplained low cortisol binding capacity. *Acta Endocr.*, **48** (1965), 26

31. Sandberg, A. A. and Slaunwhite, W. R. Transcortin: a corticosteroid binding globulin of plasma. II. Levels in various conditions and the effects of estrogens. *J. Clin. Invest.*, **38** (1959), 1290

32. Schwarz, U. and Hammerstein, J. The oestrogenic potency of various contraceptive steroids as determined by their effects in transcortin bind-ing capacity. *Acta Endocr.* **76** (1974), 159

33. Sandberg, A. A., Woodruff, M., Rosenthal, H., Nienhouse, S. and Slaunwhite, W. R. Transcortin: a corticosteroid binding protein of plasma VII. Half life in normal and estrogen treated subjects. *J. Clin. Invest.*, **43** (1964), 461

34. Tait, J. F. and Burstein, S. (1969). *In vivo* studies of steroid dynamics in man. In *The hormones*, Vol 5, Pincus, G., Thimann, K. V. and Astwood, E. G. eds., Academic Press, N.Y., p. 441

35. Plager, J. E., Schmidt, K. G. and Staubitz, W. J. Increased unbound cortisol in the plasma of estrogen treated subjects. *J. Clin. Invest.*, **43** (1964), 1066

36. Sandberg, A. A., Rosenthal, H. and Slaunwhite, W. R. Transcortin: a corticosteroid binding protein of plasma. VIII. Parameters affecting cortisol metabolism. *Proc. Sec. Congr. Horm. Steroids Exc. Med. I.C.S.*, **132** (1967), 707

37. Rosenthal, H. E., Slaunwhite, W. R. and Sandberg, A. A. Transcortin: a corticosteroid binding globulin of plasma. Cortisol and progesterone interplay and unbound levels of these steroids in pregnancy. *J. Clin. Endocr.*, **29** (1969), 352

38. Diamond, M., Rust, M. and Westphal, U. High affinity binding of progesterone, testosterone and cortisol in normal and androgen treated guinea pig during various reproductive stages; relationship to masculinisa-tion. *Endocrinology*, **84** (1969), 1143

39. Daughaday, W. H. Binding of corticosteroids by plasma proteins. II. The binding of corticosteroid and related hormones by human plasma and plasma protein fractions as measured by equilibrium dialysis. *J. Clin. Invest.*, **37** (1958), 511

40. Mercier, C., Alfsen, A. and Baulieu, E. E. A testosterone binding globulin. *2nd Symp. on Steroid Hormones.* Ghent—Belgium—Exc. Med. FCS, **101** (1966), 212

41. Vermeulen, A. and Verdonck, L. Studies on the binding of testosterone to human plasma. *Steroids*, **11** (1968), 609

42. Steeno, O., Heyns, W., Van Baelen, H. and De Moor, P. Testosterone binding in human plasma. *Ann. Endocr.*, **29** (1968), 141

43. Horton, R. and Kato, T. Studies of testosterone binding globulin. *3rd Int. Congr. Endocr. Exc. Med. ICS*, **757** (1968), 160

44. Vermeulen, A. and Verdonck, L. Studies on the binding of testosterone to plasma. *Ann. Endocr.*, suppl. **29** (1968), 149

45. Guérigan, J. and Pearlman, W. H. Separation of testosterone binding protein of human pregnancy serum from CBG. *Fed. Proc.*, **26** (1967), 757

46. De Moor, P., Steeno, O., Heyns, W. and Van Baelen, H. The steroid binding β-globulin in plasma: pathophysiological data. *Ann. Endocr.*, **30** (1969), 233

47. Mercier-Bodard, C., Alfsen, A. and Baulieu, E. E. Sex steroid binding plasma protein (SBP). *Trans. 2nd Karol. Symp. on Res. method. in Endocr. Steroid assay by protein binding*, (1970), 204

48. Corvol, P. L., Chrambach, A., Rodbard, D. and Bardin, C. W. Physical properties and binding capacity of testosterone–estradiol-binding globulin in human plasma, determined by polyacrylamide gel electrophoresis. *J. Biol. Chem.*, **246** (1971), 3435

49. Hansson, V., Larsen, J. and Reusch, L. Physiochemical properties of the 5α-dihydrotestosterone binding protein in human male serum. *Steroids*, **20** (1972), 555

50. Vermeulen, A., Verdonck, L., Van der Straeten, M. and Orie, M. Capacity of the testosterone binding globulin in human plasma and influence of specific binding of testosterone on its metabolic clearance rate. *J. Clin. Endocr.*, **29** (1969), 1470

51. Dray, F., Mowszowicz, I., Ledru, M., Crepy, O., Delzant, G. and Sebaoun, J. Anomalies de l'affinité de liaison de testosterone dans le serum des sujets thyreotoxiques et dans le virilisme pilaire idiopathique. Augmentation de l'affinité de liaison après l'administration d'hormones thyroidiennes chez le sujet dont l'affinité est normale ou basse. *Ann. Endocr.*, **30** (1969), 323

52. Vermeulen, A., Rubens, R. and Verdonck, L. Testosterone secretion and metabolism in male senescence. *J. Clin. Endocr.*, **34** (1972), 730

53. Murphy, B. E. P. Binding of testosterone and oestradiol in plasma. *Canad. J. Biochem.*, **46** (1968), 299

54. Rosenbaum, W., Christy, M. P. and Kelly, W. G. Electrophoretic evidence for the presence of an oestrogen binding β-globulin in human plasma. *J. Clin. Endocrin.*, **26** (1966), 1399

55. Rosenbaum, W., Christy, M. P., Kelly, W. G., Roginsky, M. S. and Tavernetti, R. R. Augmented binding of estradiol in the plasma of male patients with cirrhosis of the liver. *2nd Int. Congr. Horm. Ster. Exc. Med. I.C.S.* no. 111, (1966), 270

56. Southren, A. L., Gordon, G. G., Tochimoto, S., Olivo, Y., Sherman, D. H. and Pinzon, G. Testosterone and androstenedione metabolism in the polycystic ovary syndrome: studies on the percentage binding of testosterone in plasma. *J. Clin. Endocr.*, **29** (1969), 1356

57. Idler, D. R. and Freeman, H. C. Binding of testosterone, 1α-hydroxy-corticosterone and cortisol by plasma proteins of fish. *Gen. Comp. Endocr.*, **11** (1968), 366

58. Raynaud, J. P., Mercier-Bodard, C. and Baulieu, E. E. Rat estradiol binding plasma protein (EBP). *Steroids*, **18** (1971), 767

59. Kerkay, J. and Westphal, U. Steroid protein interactions. XIX. Complex formation between α₁-acid glycoprotein and steroid hormones. *Bioch. Biophys. Acta*, **170** (1968), 329

60. Billiar, R. B., Tanaka, Y., Knappenberger, M., Hernandez, R. and Little, B. Influence of transcortin and albumin on the rate of reduction of progesterone by human placental 20α-hydroxysteroid dehydrogenase. *Endocrinology*, **84** (1969), 1152

61. Hansson, V., Ritsen, E. M., Weddington, S. C., McLean, W. S., Tindall, D. J., Mayfeh, S. N. and French, F. S. Preliminary characterisation of a binding protein for androgen in rabbit serum. Comparison with the testosterone-binding globulin (TeBG) in human serum. *Endocrinology*, **95** (1974), 690

Discussion

GORDON

What happens to the steroid hormones when, either tightly or loosely bound, they reach the plasma membrane of the liver cell? Can one envisage them all switching over both types on to a receptor-protein and going on from there bound to this other protein; or is it possible that some of the tightly bound material remains bound to the same protein and is carried into the liver cell with its plasma constituent?

VERMEULEN

There is good evidence that at least at the level of the liver cell there is no dissociation of the tightly bound steroid and that only the free or albumin-bound steroid is metabolised by the liver cell. It has been shown by Tait[1] a few years ago that the metabolic clearance rate of cortisol for example corresponds very well to the hepatic extraction of only the free and albumin-bound cortisol. This applies probably also for testosterone and estradiol.

However, what happens at the more specific receptor sites for testosterone and estradiol I do not know, but it is my theory that the role of these binding proteins, these transport proteins, might be to protect the non-target cells from the high levels of hormones that are circulating and that these high levels would only be available to the specific target organs having a receptor protein with an affinity constant which is higher than the affinity constant of the transport protein. This is just an hypothesis, I have no experimental data for this. But at the level of the liver cell it appears that only the free steroid is metabolised.

ROBBINS

I would like to make some comments on this point. I do not know if there are any data on this with respect to the steroid hormones, but in the case of thyroxine Hillier has measured the rate of dissociation and association for both bovine and human thyroxine binding globulin and prealbumin[2]. Despite the very high affinity, the rate constants are extremely fast. One can calculate that the free thyroxine in the capillaries would turn over at the rate of about 50 times per second. And even though the transport of thyroxine into the liver is very fast, this dissociation rate is rapid enough to allow for it.

There is no evidence that the complex itself goes into the cells.

VERMEULEN

The dissociation of the transcortine–cortisol complex has been studied by Dixon. He found half-dissociation time at 4° of 25 min, at 27° of 1 min, and of 10 s at 37°[3].

GORDON

The main evidence concerns the bound forms going into the lysosomes of the liver cells, but the general picture is that every plasma protein must go in. Unless the material is specifically split off on the way into the liver cell, it will reach the lysosomes.

ROBBINS

Yes, I am sure that is true, but the proteins with their transported small molecules must be going at a much slower rate than the free hormones.

References

1. Tait, J. F. and Burstein, Sh. (1964). *In vivo* studies of steroid dynamics in man. In *The hormones, physiology, chemistry and applications.* Vol. V. Pincus, G. K. V. Thimann and E. B. Astwood, eds., Academic Press, New York
2. Hillier, A. P. and Balfour, W. E. Human thyroxine-binding globulin and thyroxine-binding pre-albumin: dissociation rates. *J. Physiol.*, **217** (1971), 625
3. Dixon, P. F. The kinetics of the exchange between transcortin-bound and nubound cortisol in plasma. *J. Endocr.*, **40** (1968), 457

25
Genetic control of thyroxine transport

J. ROBBINS

Almost all of the thyroid hormone in the plasma of primates is associated with transport proteins, with an extremely small proportion ($\sim 0.03\%$) in the unbound state[1]. Thyroxine-binding globulin (TBG) is quantitatively the most important, carrying approximately 75% of circulating thyroxine (T_4) and triiodothyronine (T_3). Thyroxine-binding prealbumin (PA) is another relatively specific transport protein, but is involved mainly in T_4 transport. Serum albumin also binds T_4 and T_3, as it does many other small molecules in blood. Recently, Hoch and Lewallen[2] have confirmed earlier suggestions that the plasma lipoproteins transport a minor proportion of the circulating hormones, and in the foetus[3] the so-called foetal post-albumin also shows an affinity for T_4. Up to the present time, however, no specific role for any of these proteins has been demonstrated, and the major ones appear to function solely as a buffering system for the extrathyroidal hormones. Hillier[4] has shown that the rates of dissociation of the hormones from protein are extremely rapid, and indeed rapid enough to permit the expected flux of free hormone in the microcirculation. Although it is possible that in particular tissues some of the transport proteins may function more specifically, the subject I will discuss cannot be related in any obvious way to the control of thyroid hormone action. Nevertheless, the genetic alterations affecting the transport proteins are significant in clinical testing of thyroid function, and also represent an interesting system for the study of genetic influence on plasma protein levels.

I will limit my remarks to observations concerning PA and TBG. Prealbumin is affected by a genetic polymorphism which has so far been seen only in subhuman primates. TBG is affected by genetically determined variations in its plasma concentration, but no obvious polymorphism has been detected. A more fully referenced review of this subject has been published recently[5].

PA polymorphism was first observed in rhesus monkeys (*Macacus mulatta*) some years ago[6] and has been the subject of further study in our own and other laboratories during the past few years[7-12]. Three polymorphic forms are clearly distinguished by polyacrylamide gel electrophoresis (figure 25.1). The homozygous types (Pt1-1 and Pt2-2) reveal a single PA band, the first of which is

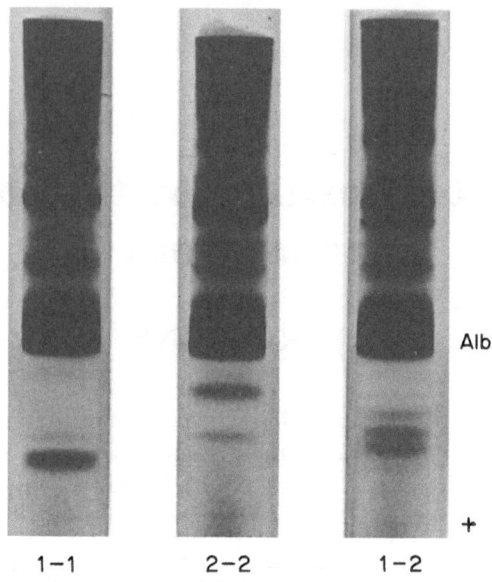

Alb

+

1–1   2–2   1–2

Figure 25.1. Polyacrylamide gel electrophoresis of *M. mulatta* serum, demonstrating the genetic polymorphism in prealbumin. Phenotypes Pt1-1 and Pt2-2 are homozygous and Pt1-2 is heterozygous for the two alleles. (From Bernstein *et al.*[8]).

very similar to human PA. The heterozygote pattern (Pt1-2) is more complicated, consisting of five components, all of which bind thyroxine. This finding was the first indication that PA might be a tetrameric protein, and this was later verified by chemical and structural studies of human PA[13–15]. Direct evidence on this point in the case of rhesus PA was obtained from hybridisation studies using either whole monkey serum[7,8] or purified rhesus PA[11]. Mixtures of Pt1-1 and Pt2-2 slowly gave rise to an electrophoretic pattern indistinguishable from that of Pt1-2, and the rate of hybridisation was greatly accelerated by 6M guanidine HCl.

By analogy with the studies on human PA it seems evident that the homozygous proteins consist of tetramers of identical subunits, and this is consistent with the findings from partial amino acid sequence done in collaboration with Morgan *et al.*[12]. Peptide mapping and amino acid analysis indicated a single amino acid difference per subunit in Pt1-1 compared with Pt2-2. The NH_2-terminal amino acid sequence revealed this to be a valine–isoleucine interchange in residue 5 (table 25.1). Since this would not explain the difference in charge between Pt1-1 and Pt2-2, it is evident that another difference (e.g., amide content) must also exist. Nevertheless, the polymorphism can be explained by a point mutation in the structural gene for prealbumin.

Studies with several *M. mulatta* colonies[8,9] gave unequivocal evidence for an autosomal dominant mode of inheritance for the PA polymorphism, involving

Table 25.1. NH$_2$-terminal amino acids of monkey and human prealbumin*

	1	5	10	15	20
Pt1-1	Gly–Pro–Thr–Gly–Val–Asp–Glu–Ser–Lys–Cys–Pro–Leu–Met–Val–Lys–Val–Leu–Asp–Ala–Val–Arg–				
Pt2-2	Gly–Pro–Thr–Gly–Ile–Asp–Glu–Ser–Lys–Cys–Pro–Leu–Met–Val–Lsy–Val–Leu–Asp–Ala–Val–Arg–				
Human	Gly–Pro–Thr–Gly–Thr–Gly–Glu–Ser–Lys–Cys–Pro–Leu–Met–Val–Lys–Val–Leu–Asp–Ala–Val–Arg–				

*From van Jaarsveld et al.[12]

two alleles. The gene frequency of the allele which directs synthesis of the slow variant (Pt²) was low, ranging from 0.05 to 0.3, making Pt1-1 the most common PA phenotype (usually 80–90% of the population)[7–9]. In other species of macaque monkeys, however, the gene frequency of Pt² is highly variable; in the Japanese species, *M. fuscata*, the slow variant was present in 100% of individuals[9]. Since criteria for Hardy–Weinberg equilibrium were met in most colonies, there appeared to be no genetic advantage for any of the three types of PA. The wide variation in gene frequency raises the possibility that physiological factors play a role; however, this variation may simply reflect random genetic drift or so-called founder effects[9,10]. Although this subject is controversial[16], it has been thought that most genetic polymorphisms result from random events[17].

Aside from factors which might affect gene frequency, it is of interest to determine whether the polymorphism is associated with variations in serum concentration of the protein. We have investigated this question in the rhesus monkey by measuring PA by radial immunodiffusion and indirectly by thyroxine-binding capacity[12]. In monkeys homozygous for the slow variant, the PA level was about half that in Pt1-1, whereas the level in heterozygotes was intermediate (table 25.2). Presumably these variations are caused either by differences in the rate of PA synthesis or in the rate of degradation, but no data are available to choose between these two alternatives.

The observed differences in protein level also raise a question concerning possible effects on the transport function of PA, but first we must know whether binding to PA is affected by the polymorphism. Direct experiments with purified Pt1-1 and Pt2-2 revealed no detectable difference in thyroxine-

Table 25.2. Plasma prealbumin concentration in *M. mulatta*

Phenotype	Prealbumin Concentration
	(mg dl^{-1})
Pt1-1	22.4 ± 2.9
Pt1-2	17.5 ± 3.0
Pt2-2	11.9 ± 3.4

Values are mean ±s.d. calculated from the thyroxine-binding capacity ($n = 1$, M.W. = 54 000). Adapted from Van Jaarsveld et al.[12].

binding as determined by equilibrium dialysis[11]. A second transport function of this protein was also examined. Human PA has been shown to interact with retinol-binding protein (RBP), which functions as a specific transport protein for vitamin A[18-20]. We measured the interaction of human RBP with each of the homozygous types of monkey PA by polarisation of fluorescence, and found no differences in this binding reaction[11]. It is thus apparent that the genetic polymorphism does not affect either of these interactions, and any physiological differences in the transport functions, therefore, should be attributable solely to altered PA levels. No specific function of PA is known in the case of thyroxine, but in vitamin A transport the PA–RBP interaction retards renal loss of the smaller RBP (M.W. 21 000). Whether the changes in PA concentration are sufficient to have an effect on vitamin A metabolism, however, is uncertain since serum normally contains a considerable molar excess of PA.

Inherited variations which affect thyroxine-binding globulin differ from those involving PA in three respects: they have so far been observed only in man; they are detected by measurements of the serum level of TBG rather than by any evident polymorphism; and they are linked to the X-chromosome. There is a large literature on this subject, which has been thoroughly reviewed[5,21]; therefore, this discussion will be an attempt to integrate the findings and speculate upon them rather than to document them.

Both decreased and increased TBG plasma levels occur. Although not all family studies give sufficient information to allow an unequivocal judgment as to the mode of inheritance, all are consistent with involvement of the X-chromosome[5,21] and several large pedigrees clearly show X-linkage (cf. reference 21). Thus, the female offspring of affected males all carry the trait, the male offspring of female heterozygotes are abnormal in about half the cases, and affected males show the trait in its full expression and do not transmit it to male offspring. Three types of affected males are seen: those with totally absent TBG; those with diminished levels of TBG and those with increased TBG. In families of the third type, it is of special interest that the average TBG level in the homozygote is increased to 3–4 times normal[21]. As expected from X-linked inheritance, affected females have levels intermediate between those in affected males and in unaffected family members.

Some efforts have been made to identify linkages of the TBG variants with other traits related to the X-chromosome, such as colour blindness, Xg blood types and glucose-6-phosphate dehydrogenase alterations. Although families have been found with more than one abnormality[5,21-23], the frequency of their recombination indicates that their loci on the X-chromosome are not very close. Thus, the TBG anomalies can provide information on mapping the X-chromosome, which no doubt will be exploited in future studies.

One family is of special interest since elevated TBG capacity was found in association with goitre[24]. Although both traits were carried on the X-chromosome, the authors concluded that they were probably inherited independently.

As I remarked earlier, there is no evidence that these variations in TBG have any effect on thyroid hormone action. Indeed, the normality of individuals with totally absent TBG, or TBG elevated to 4 times normal, constitutes the strongest evidence available that the T_4–TBG complex is not the hormonally active species. Furthermore, the tendency for the total hormone concentration to adjust to a level at which free hormone concentration is normal is a strong indication that the free hormone level is the significant determinant in thyroid hormone availability to tissue sites.

In order to understand the pathogenesis of the genetically determined variation in TBG level, it is first necessary to know whether the protein is normal. Until the protein is isolated from these subjects and subjected to appropriate chemical and physical studies, we must depend on indirect answers to this question. From such indirect studies, mainly by Refetoff *et al.*[21], we know the following:

1. The TBG level measured by radioimmunoassay is proportional to the TBG level measured by thyroxine-binding capacity[21,25-27].

2. TBG from patients with the genetic alteration is indistinguishable from normal in electrophoretic mobility, heat stability, immunodiffusion and affinity for thyroxine[21].

3. Labelled TBG injected into such patients[28] disappears from blood at a normal rate ($t_{1/2} \simeq 5$ days). Since no abnormality has been found, it is attractive to consider the possibility that the inherited variations are due to mutation of a gene which controls expression of the structural gene coding for TBG. However, no precedent for such control genes in eukaryotes is known, and this postulate must be made with caution.

In normal man, the plasma concentration of TBG is increased by oestrogens and decreased by androgens, probably through effects on the rate of protein synthesis. If these hormones act on a control gene which is defective in the subjects with genetically altered TBG level, one might expect no steroid effects in affected males, but proportionately normal responses in females, assuming that the Lyon hypothesis of random X-chromosome inactivation is valid. The experimental observations[5,21] show that males with altered TBG levels respond very little or not at all. Heterozygous females, however, also show blunted responses in most cases. In many such subjects, cortisol-binding globulin, a plasma protein with strong chemical similarity to TBG, as well as ceruloplasmin and testosterone-binding globulin, have been tested and show normal responses to oestrogen. Thus, there is no convincing evidence that the abnormality involves a control site at which the steroid hormones exert their effect.

On the other hand, there is ample evidence that structural gene variations can affect the rate of protein synthesis, and the findings with prealbumin already described are a case in point. Most commonly, the synthetic rate of the affected protein is decreased compared to normal. There are several theoretical mechanisms for such an effect: for example, the base change in DNA itself or in its mRNA product may alter the rate of release of mRNA, the stability of mRNA or the rate at which mRNA is translated[29]. Furthermore, it is apparent that total absence of TBG synthesis could result from gene deletion or, perhaps more likely, a less extensive change such as a frame shift mutation. The inherited variation in which TBG is elevated is a much more unusual type of genetic defect. A structural gene mutation could increase protein synthesis if the normal gene product was a regulator of gene expression, or by the converse of any of the mechanisms mentioned previously. Recently an abnormality of erythrocyte glucose-6-phosphate dehydrogenase has been described which resembles that seen with TBG in several ways[30–32]. G-6-PD Hektoen is associated with an activity level which is four times normal, and its gene is located on the X-chromosome. Although the variant enzyme is indistinguishable from normal in almost all of its physical and enzymatic properties, it shows abnormal electrophoretic mobility at pH 6.5. Peptide and amino acid analysis have identified an exchange of one tyrosine residue for histidine in the normal enzyme, indicating that the genetic variant is the result of a point mutation in a structural gene. With such a striking precedent, it seems quite possible that the increased-TBG variant also results from a structural gene mutation with secondary effects on the rate of TBG synthesis. To establish this as a fact, it will be necessary to identify an altered amino acid composition in TBG.

Whichever mechanism proves to be the case, study of these TBG variants can be important for understanding the control of protein synthesis in man.

References

1. Robbins, J. and Rall, J. E. (1966). In *Hormones in Blood*, 2nd ed., Gray, C. H. and Bacharach, A. L., eds., Academic Press, London, p. 383
2. Hoch, G. H. and Lewallen, C. *J. Clin. Endocrinol. Metab.*, **38** (1974), 663
3. Andreoli, M. and Robbins, J. *J. Clin. Investig.*, **41** (1964), 1070
4. Hillier, A. P. *J. Physiol.*, **217** (1971), 625
5. Robbins, J. *Mt. Sinai J. Med.*, **40** (1973), 511
6. Blumberg, B. S. and Robbins, J. *Endocrinol.*, **67** (1960), 368
7. Alper, C. A., Robin, N. I. and Refetoff, S. *Proc. Nat. Acad. Sci.*, **63** (1969), 775
8. Bernstein, R. S., Robbins, J. and Rall, J. E. *Endocrinol.*, **86** (1970), 383
9. Weiss, M. L., Goodman, M., Prychodko, W. and Tanaka, T. *Primates*, **12** (1971), 75
10. Weiss, M. L. and Goodman, M. *J. Hum. Evolution*, **1** (1972), 41
11. VanJaarsveld, P., Branch, W. T., Robbins, J. and Edelhoch, H. *J. Biol. Chem.*, **248** (1973), 4706
12. VanJaarsveld, P., Branch, W. T., Robbins, J. Morgan, F. J., Kanda, Y. and Canfield, R. E. *J. Biol. Chem.*, **248** (1973), 7898

13. Branch, W. T., Robbins, J. and Edelhoch, H. *J. Biol. Chem.*, **246** (1971), 6011
14. Blake, C. C. F., Swan, I. D. A., Rerat, C., Berthou, J., Laurent, A. and Rerat, B. *J. Mol. Biol.*, **61** (1971), 217
15. Morgan, F. J., Canfield, R. F. and Goodman, DeW. S. *Biochem. Biophys. Acta*, **236** (1971), 798
16. Kolata, G. B. *Science*, **184** (1974), 452
17. Yamazaki, T. and Maruyama, T. *Science*, **178** (1972), 56
18. Kanai, M., Raz, A. and Goodman, DeW. S. *J. Clin. Investig.*, **47** (1968), 2025
19. Raz, A., Shiratori, T. and Goodman, DeW. S. *J. Biol. Chem.*, **245** (1970), 1903
20. Peterson, P. A. *J. Biol. Chem.*, **246** (1971), 34
21. Refetoff, S., Robin, N. I. and Alper, C. A. *J. Clin. Investig.*, **51** (1972), 848
22. Bode, H. H., Rothman, F. J. and Danon, M. *J. Clin. Endocrinol. Metab.*, **37** (1973), 25
23. Leiba, S., Landau, B., Ber, A., Adam, A. and Sterling, K. *J. Clin. Endocrinol. Metab.*, **38** (1974), 569
24. Shane, S. R., Seal, V. S. and Jones, J. E. *J. Clin. Endocrinol. Metab.*, **32** (1971), 587
25. Levy, R. P., Marshall, J. S. and Valayo, N. L. *J. Clin. Endocrinol. Metab.*, **32** (1971), 372
26. Hamada, S., Takemura, Y. and Sterling, K. *J. Clin. Endocrinol. Metab.*, **33** (1971), 326
27. Hansen, J. M. and Siersbaek-Nielsen, K. *J. Clin. Endocrinol. Metab.*, **35** (1972), 461
28. Refetoff, S., Fang, V. S., Robin, N. I. and Marshall, J. S. *Endocrinol.*, **92** (1973), T-8 (*Abstract, Amer. Thyroid Assoc.*)
29. Stanbury, J. B., Wyngaarden, J. B. and Fredrickson, D. S. (1972). *The Metabolic Basis of Inherited Disease*, 3rd ed., McGraw-Hill, New York, p. 10
30. Dern, R. J. *J. Lab. Clin. Med.*, **68** (1966), 560
31. Yoshida, A. *J. Mol. Biol.*, **52** (1970), 483
32. Dern, R. J., McCurdy, P. R. and Yoshida, A. *J. Lab. Clin. Med.*, **73** (1969), 283

Discussion

HOFFENBERG

Two questions to Dr Robbins: first, is there any explanation for the role of the acute reaction of TBPA in stress, where the concentration drops perhaps to 30% of normal? Do you think this has any part to play in thyroxine metabolism? It seems unlikely since TBPA binds about 15% of the circulating thyroxine, and I think it has been calculated that only about 1% of the molecules of

TBPA actually participate in binding. Therefore, it seems that the binding of thyroxine is minor. I wonder if you have views about it.

Secondly, as I recall, some patients with absent TBG have been shown to have high absolute concentrations of free thyroxine—so it has been claimed at any rate. Do you think it can be true, and if so, how do they remain euthyroid?

ROBBINS

These are very significant questions and I do not really have any good answers. There is one interesting observation though with respect to the prealbumin question. From the dissociation rate data of Hillier that I mentioned, and the known distribution of thyroxine between prealbumin and thyroxine binding globulin, one can calculate that, for any given interval of time, about half of the thyroxine that is going from the bound to the free form is coming from prealbumin. So even though it *seems* to be a very unimportant binder, in fact it is very important in the delivery of thyroxine. Now, whether this has any role in the response to acute disease or stress or chronic disease, is totally unknown as far as I am aware.

Just a secondary point on that topic: there has been evidence to show that the changes in prealbumin, which occur quite rapidly, are due to changes in prealbumin synthesis rather in degradation, as was shown by measuring degradation rates of prealbumin[1].

As regards the increased free thyroxine level in the patients with absent TBG, this has been reported and I do not know the significance. I think I could say that the variations from the normal are very small and, if my recollection serves me, much smaller than one would expect for free thyroxine level in, for example, hyperthyroidism. So, I think they are above the normal range, but without producing hyperthyroidism.

Reference

1. Socolow, E. L., Woeber, K. A., Purdy, R. H., Holloway, M. T. and Ingbar, S. H. Preparation of ^{131}I-labelled human serum prealbumin and its metabolism in normal and sick patients. *J. Clin. Invest.*, **44** (1965), 1600

26
The impact of proinsulin on insulin research

N. R. LAZARUS

Introduction

Proinsulin was discovered as the precursor of insulin by Steiner *et al*. in 1967[1]. A selective review of the literature eight years after the discovery of proinsulin shows that this hormone has had considerable influence on insulin research.

Insulin synthesis

The synthesis of insulin was first accomplished by the Aachen group[2]. With present technology the commercial synthesis of insulin is not feasible. With the ever increasing prices of materials that go to make up animal feedstuffs, coupled with their scarcity, the pressure to synthesise insulin on a commercial scale is growing. This pressure is further compounded by the poor ability of domestic animals to convert food to protein. Cattle, for example, need 3 kg of food to make 0.5 kg of protein, pigs 1.25 kg of food for 0.5 kg of protein. It may well be that in the future it will become too expensive to rear cattle. Under these circumstances alternative sources of insulin will become mandatory. The discovery of proinsulin has hastened the possible synthesis of insulin. Insulin has previously been synthesised by making the two chains, A and B, and then coupling these chains under appropriate conditions. Yields of biologically active insulin were very low. Under the conditions employed, the A and B chains tended to form oligomers rather than couple one with the other. In the β-cell insulin is synthesised via proinsulin. Proinsulin has an additional polypeptide (C-peptide) consisting of approximately 33 amino acid residues (porcine). In proinsulin this peptide links the glycine residue at position A1 to the B chain's terminal amino acid residue at position 30. Apparently this connecting peptide aligns and holds the A and B chains in a spatial configuration about 5 to 8 Å apart so as to favour proper disulphide bridging between the two chains. Three groups, those at Hoechst pharmaceuticals, Sussex University and the German Wool Research Institute at Aachen approached the synthesis of insulin by trying to mimic what happens in nature. At Sussex, Lindsay, using bishydroxysuccinamide succinate on a commercial sample of bovine insulin managed to link residues A1 and B29[3]. Brandenburg at Aachen has used a variety of bifunctional reagents varying in length from 2 to 13 carbon atoms in order to produce crosslinking. The most suitable linker was found to be suberic acid[4]. One drawback in these reagents is that they cannot be removed without damaging insulin. The problem was solved independently by Hoechst and the Aachen group. Insulin was cross-linked with $\alpha\alpha^1$-diamino-suberic acid, between residues A1 and B29. This link can be removed by standard techniques. When this 'mimi-proinsulin' was reduced and then oxidised and the bridge removed yields of 20–40% of recovered insulin were obtained. Geiger and Obermeier at Hoechst then went a step further. They first isolated A and B chains from insulin and then connected the separated A and B chains with diaminosuberic acid. The crosslinked product was then oxidised and the bridge finally removed. Biologically active insulin was obtained in yields of up to 40%. However exciting this research may be, it is well to remember that 40% yields of product, although very much better than obtained before the proinsulin era, are not yet of the order to entice a venture into commercial synthesis. Undoubtedly, however, the technology of linking A and B chains will improve and one more step on the pathway to complete synthesis will have been solved.

C-Peptide

Immunoassay

It has already been stated that in the β-cell single chain proinsulin facilitates the correct orientation of the A and B chains relative to one another. Proinsulin is then proteolytically cleaved to yield insulin and C-peptide. C-peptide is secreted with insulin in a mole : mole ratio and enters the circulation. The C-peptides of various species, including the guinea-pig, duck, man, pig and ox have different primary structures. Immunological comparisons of the three insulins that are of practical importance, i.e. human, bovine and porcine, show very little cross-reaction. Thus if an immunoassay could be developed to the C-peptide antigenic determinant alone it should be possible to measure human C-peptide in both normal people and in diabetics who have been receiving injections of porcine and/or bovine insulin. In the latter patients, antibodies to the injected insulin are invariably formed, making the measurement of endogenously secreted insulin very difficult. However, patients would secrete C-peptide along with their own insulin and if the C-peptide could be measured, then the quantity of insulin secreted could be deduced by extrapolation.

Such an immunoassay has been developed by Melani et al.[5]. Using this assay in insulin treated diabetics, who have developed circulating antibodies that interfere with the insulin immunoassay, Block *et al.*[6] have shown that serum C-peptide immunoreactivity increases during the remission of diabetes suggesting that the improvement is due to renewed β-cell activity. Eventual relapse was associated with decreasing C-peptide immunoreactivity. They were thus able to confirm the long held view that the temporary remission phase of diabetes is due to resumption of β-cell activity.

Another use of the assay could be in the diagnosis of patients with insulinomas. Insulinomas usually show autonomous endocrine function. Such function is generally better diagnosed by suppression tests rather than by the more often used stimulatory tests. Probably the best suppressor of insulin secretion is hypoglycaemia. Hypoglycaemia is induced easily by the injection of insulin. However, since most insulins cross-react with the antibodies used in the immunoassay it becomes impossible to differentiate between exogenous and endogenous insulin. The C-peptide immunoassay provides a tool to differentiate injected and secreted insulin. Since human C-peptide antibodies will only react with human C-peptide it is possible to measure β-cell output without simultaneously measuring injected insulin.

Some metabolic and biological studies have been performed with C-peptide[7,8]. However, the metabolism of this peptide in the human is in its infancy. It remains to be seen whether the extrapolation of C-peptide concentration to insulin secretion is as facile as it presently seems. One suspects that some conditions will come to light in which C-peptide metabolism may be influenced separately from that of insulin. Furthermore, the assay is neither cheap nor easy to set up and while useful as a research tool its diagnostic value is unproven. In the diagnosis of insulinoma, for example, a method that is just as effective as the C-peptide immunoassay is to use fish insulin rather than bovine or porcine insulin to induce hypoglycaemia. Fish insulin does not cross-react with the usual insulin antibodies thus allowing the responsiveness of endogenous insulin secretion to be measured by the usual immunoassay techniques[9].

Evolution

A comparison of the amino acid sequences in the C-peptides of various mammalian species shows much greater variations in primary structure than occur in the insulins of these same species. Since insulin has been used to construct a phylogenetic tree[10] it should also be possible to utilise C-peptide structures in the same way. Duck and guinea-pig C-peptide differ considerably from other mammalian C-peptides and at first glance would not appear to be evolutionary ancestors. However, Markussen and Volund[11] have developed a new method of calculating evolutionary rates of proteins and applied this theory to insulin and C-peptide. Using this theory they found that duck C-peptide had the expected differences from the mammalian C-peptides, and could be placed on the same evolutionary tree. However, guinea-pig

C-peptide did not fit in and this finding confirms the already unclassifiable structure of guinea-pig insulin. It will be interesting to see whether C-peptides from other hystricomorph rodents are similar to that of guinea-pig.

Insulin purification

Steiner *et al.*[12] showed that commercial crystalline insulins were contaminated by a number of proinsulin and insulin-like substances. These products arose from the pathway of proteolytic cleavage of proinsulin to insulin. The pathway in humans is not known but is probably similar to that delineated by Tager *et al.*[13] for beef insulin. The contaminants of crystalline insulin are proinsulin, desdiproinsulin, which arises from the cleavage at only one end of the connecting peptide, diarginyl insulin, which had two arginine groups still attached to it, monarginine insulin and ethyl insulin which arises from the extraction process itself. There is also present a dimer the nature of which is not completely known but appears to be a covalent linking of two insulin molecules. This linkage is not through the disulphide bridges because reduction of these chains does not release insulin[12].

The demonstration that crystalline insulins were 'impure' (i.e. contaminated with about 2% proinsulin breakdown products) prompted the idea that the development of antibodies by those diabetics receiving insulin might be due to these contaminants. The removal of these substances should yield a pure insulin that would perhaps be non-antigenic. To this end Novo developed a purification procedure for the manufacture of monocomponent insulins[14]. Since then several other companies have produced purified insulins.

The aims and rationale for insulin purification have been admirably set out by Schlichtkrull[15]. Results from extensive clinical trials appear to be contradictory but a balanced view has been published by Bruni *et al.*[16]. While reduction of antibody levels in diabetics is desirable there is as yet no recognisable correlations between antibody titre and some clinical conditions such as duration of insulin treatment, daily insulin requirement, types of insulin used, presence or not of microangiopathy and insulin lipoatrophy. High insulin requirements are not always accompanied by elevated levels of IgG binding. When patients are switched from 'regular' insulin to porcine monocomponent insulin there is a reduction in antibody levels. Just what the practical benefit of such a drop in antibody titre means is not known. It remains to be seen whether beef monocomponent insulin is as non-antigenic as porcine monocomponent insulin. If one is permitted to hazard a guess it would be that the greatest impact of proinsulin on insulin manufacture will be that it demonstrated that commercial insulins are contaminated with unwanted insulin-like substances and, from the patients' point of view, it would be better to be treated with a pure preparation rather than an impure one.

Prohormones

Since the discovery of proinsulin and its identification in the circulation, evidence has been accumulated that other peptide hormones are present both in their glands and in the circulation in more than one form. Techniques of investigation derive almost directly from those used in the now classic proinsulin studies. These involve the fractionation of serum and plasma on gel columns. The fractions are then immunoassayed and profiles obtained. One of the most surprising findings was reported by Yalow and Berson[17] who claimed that there was an insulin molecule both in pancreas and in the circulation of some tumour patients that had a M.W. of about 100000. It is rather difficult to fit such a molecule into the elegant pathway of insulin synthesis as postulated with a proinsulin precursor. This big molecule must then be a proinsulin precursor and the only possible locations of the link between proinsulin and big, big insulin would be either at the carboxyl or amino terminal end of the molecule. Presumably, just as the conversion of proinsulin to insulin is accompanied by intermediate species it would be expected that the conversion of 'big, big' insulin to proinsulin would result in the release of some intermediates. None of these has yet been located. A recent report on big, big insulin[18] suggests that big, big insulin is an aggregate of insulin and that this association is only amenable to dissociation under very special conditions.

The following hormones have now been shown or have been claimed to be derived from precursor molecules. Parathormone, ACTH, gastrin and growth hormone. It has also been claimed that glucagon is synthesised via a precursor[19,20]. However, critical evaluation of these data shows that the conclusion that a precursor of glucagon has been identified is premature. Besides other criticisms, the most telling is that pulse chase experiments have as yet been unable to demonstrate this precursor relationship.

Proinsulin in the circulation

Roth *et al.*[21] filtered human serum on G-50 Sephadex and reported the presence of a large molecule that had insulin-like properties. They called this molecule big insulin. Subsequently it was shown that big insulin and proinsulin are identical.

Fasting values of proinsulin-like material in normal subjects are of the order $0-0.4$ ng ml^{-1}. The level of proinsulin in the serum is higher than the percentage of proinsulin in the pancreas. This discrepancy is accounted for by the longer circulatory half-life of proinsulin as compared with insulin[22]. Since the conversion of proinsulin to insulin involves the formation of intermediates it would be interesting to know whether these intermediates are present in the serum. The only report so far is that of Gutman *et al.*[23] who were unable to show any intermediates in normal subjects but did find immunoreactive material in the serum of obese subjects that migrated in a position compatible with intermediate proinsulin species. This study as yet remains uncorroborated, but it would seem intermediates play no great role in insulin biology.

There is, however, a considerable body of evidence to show that patients with islet cell tumours have a large proportion of proinsulin-like material in the circulation[24-27]. This is true irrespective of whether total insulin is moderately or greatly raised. The monitoring of proinsulin levels in the blood of suspected insulinoma patients is thus of diagnostic value. Further the disappearance from the circulation of the proinsulin component after treatment can also be used as an indication of the efficacy of the treatment[26-28]. While high proinsulin levels have so far been associated almost exclusively with insulin tumour patients, high proinsulin levels can also be found in subjects with hypokalaemia and it is probable that several other conditions will eventually be found that also have high circulating levels thus diminishing the diagnostic value of a high proinsulin level.

Insulin secretion

Although proinsulin is not located in the insulin granules, it is secreted into the circulation. The mechanism of this secretion is not understood. Various authors have presented evidence that there may be a mechanism of insulin secretion other than that involving β-granule extrusion[29]. As far as can be judged, this mechanism operates almost exclusively in insulinomas that have dedifferentiated sufficiently so as not to form insulin granules. However, this 'alternate' pathway probably does not exist under normal conditions. It may be that a study of this tumour secretory process will throw some light on the mechanism of proinsulin secretion.

References

1. Steiner, D. F., Cunningham, D. D., Spigelman, L. and Aten, B. Insulin biosynthesis: evidence for a precursor. *Sciences*, **157** (1967), 697
2. Meienhoffr, J., Schnabel, B., Bremer, H., Brinkhoff, O., Zabel, R., Sroka, W., Klostermeyer, H., Brandenburg, D., Okuda, T. and Zahn, H. Synthese der Insulinketten und ihre Kombination zu insulinaktiven Praparaten. *Z. Naturforsch.*, **18b** (1963), 1120
3. Lindsay, D. G. Intramolecular cross-linked insulin. *FEBS Letters*, **21** (1972), 105
4. Brandenburg, D., Busse, W-D., Gattner, H-G., Zahn, H., Wollmer, A., Glieman, J. and Puls, W. (1973). Structure-function studies with chemically modified insulin. In *Peptides 1972*, H. Hanson and H. D. Zakubke, eds., North-Holland, p. 270

5. Melani, F., Rubenstein, A. H., Oyer, P. E. and Steiner, D. F. Identification of proinsulin and C-peptide in human serum by a specific immunoassay. *Proc. Nat. Acad. Sci.*, **67** (1970), 148

6. Block, M. B., Rosenfield, R. L., Mako, M. E., Steiner, D. F. and Rubenstein, A. H. Sequential changes in β-cell function in insulin-treated diabetic patients assessed by C-peptide immunoreactivity. *New Eng. J. Med.*, **288** (1973), 1144

7. Katz, A. I. and Rubenstein, A. H. Metabolism of proinsulin, insulin and C-peptide in the rat. *J. Clin. Invest.*, **52** (1973), 1113

8. Stoll, R. W., Touber, J. L., Menahan, L. A. and Williams, R. H. Clearance of porcine insulin, proinsulin and connecting peptide by the isolated rat liver. *Proc. Soc. Exp. Biol. Med.*, **133** (1970), 894

9. Turner, R. C. and Johnson, P. C. Suppression of insulin release by fish-insulin induced hypoglycaemia with reference to the diagnosis of insulinomas. *Lancet*, 30th June (1973), 1483

10. Dayhoff, O. M. Atlas of protein sequence and structure 1973. *Nat. Biomed. Res. Fdn.* (1972), D-176

11. Markussen, J. and Volund, A. A new method of calculating evolutionary rates of proteins applied to insulin and C-peptides. *Int. J. Peptide Protein Res.*, **6** (1974), 79

12. Steiner, D. F., Clark, J. L., Nolan, C., Rubenstein, A. H., Margoliash, E., Aten, B. and Oyer, P. E. Proinsulin and the biosynthesis of insulin. *Recent Progress in Hormone Research*, **25** (1969), 207

13. Tager, H. S., Emdin, S. O., Clark, J. L. and Steiner, D. F. Studies on the conversion of proinsulin to insulin. II. Evidence for a chymotrypsin like cleavage in the connecting peptide region of insulin precursors in the rat. *J. Biol Chem.*, **248** (1973), 3476

14. Schlichtkrull, J. and Jorgensen, K. H. (1969). South African Patent, 031256

15. Schlichtkrull, J. (1973). Monocomponent insulin and its clinical implications. 5th E.F. Copp Lecture. Novo Research Institute

16. Bruni, B., Castellazzi, R., Osenda, M. and Turco, G. L. Serum IgG ^{125}I-insulin-binding capacity in long-term insulin treated diabetics, as an indication for monocomponent insulin therapy. *Panminerva Medica*, **15** (1973), 231

17. Yalow, R. S. and Berson, S. A. 'Big, big insulin'. *Metabolism*, **22** (1973) 703

18. Beischer, W., Melani, F. and Keller, L. 'Big, big insulin', a reality? *Diabetologia*, **10** (1974), 358

19. Tung, A. K. and Zerega, F. Biosynthesis of glucagon in isolated pigeon islets. *Biochem. Biophys. Res. Commun.*, **45** (1971), 387

20. Noe, B. D. and Bauer, G. E. Evidence for glucagon biosynthesis involving a protein intermediate in islets of the Anglerfish (*Lophius Americanus*). *Endocrinology*, **89** (1972), 642

21. Roth, J., Gorden, P. and Pastan, I. 'Big insulin': A new component of

plasma insulin detected by immunoassay. *Proc. Nat. Acad. Sci.*, **61** (1968), 138

22. Katz, A. I. and Rubenstein, A. H. Metabolism of proinsulin, insulin and C-peptide in the rat. *J. Clin. Invest.*, **52** (1973), 1113
23. Gutman, R., Lazarus, N. R. and Recant, L. Electrophoretic characterisation of circulating human proinsulin and insulin. *Diabetologia*, **5** (1972), 136
24. Goldsmith, S. J., Yalow, R. S. and Berson, S. A. Significance of human plasma insulin Sephadex fractions. *Diabetes*, **18** (1969), 834
25. Gorden, P. and Roth, J. Plasma insulin fluctuations in the 'big' insulin component in man after glucose and other stimuli. *J. Clin. Invest.*, **48** (1969), 2225
26. Melani, F. Proinsulin secretion by a pancreatic β-cell adenoma: proinsulin and C-peptide secretion. *New Eng. J. Med.*, **283** (1970), 713
27. Gutman, R. A., Lazarus, N. R., Penhos, J. C., Fajans, S. and Recant, L. Circulating proinsulin-like material in patients with functioning insulinomas. *New Eng. J. Med.*, **284** (1971) 1003
28. Blackard, W. G., Garcia, A. R. and Brown, C. L., Jr. Effect of streptozotocin on qualitative aspects of plasma insulin in a patient with a malignant islet cell tumour. *J. Clin. Endocrinol. Metab.*, **31** (1970), 315
29. Creutzfeldt, W., Creutzfeldt, C. and Frerichs, H. (1970). Evidence for different modes of insulin secretion. In *The Structure and Metabolism of the Pancreatic Islets*, S. Falkmer, B. Hellman and I-B. Taljedal, eds., Pergamon Press, p. 181

Discussion

SHAFRIR

I have one question which is not directly connected with the subject of this lecture and this concerns immunoreactive insulin components in animals which have had their pancreas destroyed by treatment with streptozotocin. We find that despite the fact that the insulin content of their pancreas had been reduced to one hundredth of the normal level, there are still some immunoreactive components circulating in the plasma. What would they be? Are they part of the insulin molecule, or are these incomplete peptides of insulin? They react in the standard ways and run in immunoassay.

LAZARUS

I can't really answer that question because I don't know. How long afterwards do you find this immunoreactive insulin?

SHAFRIR

Between 3 days and a week.

LAZARUS

And then it disappears?

SHAFRIR

It goes down.

LAZARUS

It may be it is some residual leakage from cells which have been extremely damaged by the streptozotocin and that it's these intermediates which are coming out and are circulating. But I can't give you any better explanation than that.

JAMES

Has proinsulin similar properties to insulin?

LAZARUS

Proinsulin appears to have all the biological properties of insulin, except at a very much lower level, namely about 2–5% of the insulin action, i.e. 2–3 times the half-life of insulin.

MILLER

I just have a question about the possible source of confusion if one is attempting to prepare his own insulin for immunoassays. Do you feel from your experience that it is important to critically examine the heterogeneity of the alledgedly purified insulin that one uses as the material on which to base a radioimmunoassay.

LAZARUS

I don't think so. These contaminations are extremely small, but of course if you want a purer preparation, the more you purify, the better it is. But in general for 99.9% of cases, it's unnecessary to purify the insulin to that extent. I would say that you only need to begin to purify when you begin to do membrane binding studies where the contaminant might begin to influence some of your results.

Pathophysiology of Plasma Proteins

27
Lipoprotein synthesis in hypoproteinemia of experimental nephrotic syndrome and plasmapheresis

E. SHAFRIR and T. BRENNER

The nephrotic syndrome is characterised by three salient features: proteinuria, hypoproteinemia and hyperlipoproteinemia. The massive protein loss imposes an extreme compensatory burden on the liver, which has the task of filling a virtually bottomless barrel: most of the newly synthesised protein is promptly excreted in the urine. Epstein in 1917[1] used the term 'diabetes albuminuricus' for nephrosis, because of parallelism to the glucosuria: today we may say that the futile hepatic gluconeogenesis in diabetes, magnifying the spillover of glucose in the urine, seems to represent a more appropriate analogy.

We wish to focus attention to the least understood aspect of the nephrotic syndrome—the hyperlipoproteinemia. Why is the marked hypoproteinemia of nephrosis always associated with an elevation of plasma lipoproteins? Is this a retention lipidemia due to a delayed tissue disposal of the lipoprotein-borne lipids? The starvation-like pattern of metabolism in this situation, evidenced by the enhanced peripheral release of amino acids and other substrates required to support the increased hepatic protein synthesis, or a specific defect in the lipoprotein lipase-mediated lipid uptake may be responsible for this type of hyperlipidemia. Or are we faced with an increased hepatic synthesis of lipoproteins, along with other plasma proteins, so that the nephrotic hyperlipoproteinemia represents an actual elevation of apolipoproteins in the circulation? The lipoproteins being of large molecular size do not pass through the glomerular membrane, and accumulate in the circulation even when damaged enough to permit the seepage of most of the smaller plasma proteins. A support for the latter possibility has been obtained from previous studies which demonstrated an increased channeling of several substrates into both the lipid and protein moieties of the lipoproteins in experimental nephrotic syndrome of rats[2-5].

It is quite probable that the stimulus for the increased hepatic synthesis of albumin in nephrosis is the low oncotic pressure of the plasma and extra-cellular fluid proteins[6]. It is also probable that such stimulus is not limited to the specific replacement of albumin, or other proteins passing into the urine, but may well be a general one eliciting the synthesis of all proteins the liver is capable of contributing to the plasma. In fact, the production of fibrinogen, which is not excretable to the urine, is likewise enhanced in the nephrotic syndrome[5].

In our recent studies we have investigated the possibility that the loss of oncotic pressure, due to hypoproteinemia of nephrosis or direct plasma re-moval, results in the enhancement of hepatic synthesis of both the protein and lipid moieties of the lipoproteins. By following the incorporation of ap-propriate precursors we wished to obtain a clue as to the primary factor respon-sible for the hyperlipoproteinemia—enhanced synthesis of apoproteins or elaboration of lipids. We were also interested in the nature of the lipid pre-cursor(s) in hypoproteinemia.

We have used *ad libitum* fed rats, weighing about 250 g, 5 to 7 days after the nephrotic syndrome was produced by a series of 6 daily subcutaneous injec-tions of aminonucleoside of puromycin. As shown in table 27.1, our nephrotic animals had markedly reduced serum albumin level and increased triglyceride and cholesterol content. Table 27.2 shows that the total amount of circulating

Table 27.1. Serum constituents in nephrotic and plasmapheretic rats

Rats	Total Protein Albumin g 100 ml^{-1}		Cholesterol Triglycerides mg 100 ml^{-1}	
Normal	6.7	3.9	73	96
Nephrosis	4.5	1.3	378	695
Plasmapheresis	6.2	3.3	72	101

Values are means for groups of 12 to 20 animals.

Table 27.2. Lipoprotein composition in serum from aminonucleoside nephrotic rats*

Fraction	Normal			Nephrotic		
	Triglycerides Cholesterol mg 100 ml^{-1}		Protein	Triglycerides Cholesterol mg 100 ml^{-1}		Protein
VLDL ($d < 1.019$)	47	7	6	461	115	74
LDL ($1.019 > d > 1.063$)	8	25	18	50	258	170
HDL ($1.063 > d > 1.21$)	5	27	86	6	21	53
all lipoproteins†	60	59	110	517	394	297

*The lipoproteins were isolated by tube slicing after successive flotations in rotor Ti50 of Model L$_3$50 Spinco Ultracentrifuge from pooled serum of 4 rats.
†Representing a 75–90% recovery of total serum lipids.

lipoproteins was markedly increased in the nephrotic rats. On the basis of apoprotein content the very low density lipoproteins (VLDL) and the low density lipoproteins (LDL) increased approximately ten-fold in concentration, whereas the high density lipoproteins (HDL) were somewhat decreased. The total lipoprotein protein was increased three-fold in the serum of nephrotic rats. It may be pointed out that such an increase in total amount of lipoproteins seems unique for the nephrotic syndrome, since in Type IV hyperlipidemia, of genetic or nutritional origin, the additional serum lipids are mostly carried by the existing apoproteins as judged from lack of appreciable changes in the total apolipoprotein content[7].

To ascertain whether the observed changes are specific to the nephrotic syndrome or may be generally applicable to the condition of hypoproteinemia we have also experimented in rats after plasmapheresis. We have removed about a quarter of rat plasma, twice within 24 h, and usually performed the experiments 3 h after the second plasma removal. Table 27.1 contains the data on the serum constituents of these rats; the albumin level was decreased by about 20% and there was a slight increase in serum triglyceride content. The latter change was not evident in all experimental groups.

Incorporation of amino acids by the perfused liver

The liver was perfused in a modified Miller system, using a buffered 4% bovine albumin solution containing an amino acid mixture at physiological concentration. ^{14}C-amino acids from algal protein hydrodysate (Radiochemical Centre, Amersham, Great Britain) served as the label. We have determined the incorporation of radioactivity into the albumin and lipoproteins secreted into the perfusate after 3 h of perfusion. The lipoproteins were isolated by two successive ultracentrifugations at the density of 1.063 and 1.21. The VLDL and LDL in the $d < 1.063$ fraction were not separated since these two classes of lipoproteins in the rat overlap to a large extent in the composition of their apoprotein peptides[8], particularly when fractionated by density.

As may be seen in table 27.3 we obtained a two-fold increase in the incorporation of amino acid label into the apoproteins (VLDL + LDL) from the liver perfusate of nephrotic rats. There was also a tendency, albeit insignificant, toward increased incorporation into the HDL. The increase in apoprotein label paralleled the increase in albumin label and was of similar extent. In contrast, there was no increase in the radioactivity of the lipid portion of the lipoproteins and in the albumin-bound lipids. In fact, a decrease in amino acid incorporation into the lipids of liver perfusate from nephrotic rats was noted.

As in the experiments with nephrotic animals, the amino acids were incorporated at an increased rate into albumin and lipoproteins of the liver perfusate of rats after plasmapheresis (table 27.3). In agreement with the

Table 27.3. Incorporation of ^{14}C-amino acids into liver perfusate proteins and lipids

Rats	VLDL + LDL ($d < 1.063$)	HDL ($1.063 < d < 1.21$)	Albumin $d > 1.21$
($^0/_{00}$ of initial radioactivity per 3 h per 10 g liver)			
PROTEIN			
Normal	1.11	0.55	32.4
Nephrosis	2.42*	0.70	65.0*
Plasmapheresis	1.93*	0.79	70.6*
LIPID			
Normal	2.81	0.39	1.70
Nephrosis	1.10*	0.25	1.81
Plasmapheresis	2.27	0.32	1.84

Values are means for groups of 8 to 12 rats. Asterisk denotes statistically significant difference at $P < 0.05$ at least. The lipids were extracted from the lipoproteins isolated by ultracentrifugation into chloroform–methanol 2:1. Albumin-bound lipids in the $d > 1.21$ fraction were mostly free fatty acids. The initial perfusate radioactivity was 6.5×10^6 counts min^{-1}.

observation in nephrotic rats there was no increase in the lipid label of the various fractions.

Table 27.4 indicates that the increased label incorporation into the apoproteins was accompanied by an augmented secretion of the lipoproteins by the liver. The specific activity and the quantitative secretion of albumin could not be determined since bovine albumin was a constituent of the perfusion medium.

We have repeated our perfusion experiments using U-^{14}C-leucine as the amino acid label to ascertain that there really is no increase in lipid production from amino acids. Leucine is a potential precursor of fatty acids because of the

Table 27.4. Apolipoprotein production by liver perfused with ^{14}C-amino acids

Rats	VLDL + LDL ($d < 1.063$)	HDL $1.063 < d < 1.21$
mg protein per 3 h per 10 g liver		
Normal	8.2	16.9
Nephrosis	12.4*	24.1*
Plasmapheresis	10.9*	18.3
specific activity (% initial ^{14}C per mg protein)		
Normal	0.12	0.033
Nephrosis	0.18*	0.029
Plasmapheresis	0.16	0.043

Values are means for groups of 8 to 12 rats. Asterisk denotes statistically significant difference of $P < 0.05$ at least. Specific activity calculated on the basis of values from table 27.3.

acetyl CoA intermediate upon degradation. As with the labelled amino acid mixture, in the liver perfusate of nephrotic rats we have obtained a three-fold increase in incorporation of leucine into the apoproteins (VLDL + LDL) and into the albumin, but no significant increase in lipid radioactivity. These results led to the conclusion that amino acids are poor substrates for the elaboration of the lipid moiety of the lipoproteins. Since it is impossible to secrete lipoproteins without co-synthesis of the lipid complement, we decided to look for lipid precursors other than amino acids.

Incorporation of fatty acids by the perfused liver

When the liver of nephrotic rats was perfused with albumin-bound labelled oleate a two- to three-fold increase in the incorporation of the free fatty acid into the lipoproteins secreted in the perfusate was observed (table 27.5). This finding implies that, in the nephrotic syndrome, the liver prefers to utilise preformed fatty acids for the synthesis of the lipid moiety of the lipoproteins, rather than to synthesise fatty acids *de novo* from amino acids. It should be mentioned that, in nephrotic rats, liver lipid radioactivity did not increase during the perfusion with oleate over that in control rats (table 27.5). The fatty acid increment taken up was promptly esterified and released into the perfusate as lipoprotein-borne lipid, most probably due to the presence of increased amounts of the apoprotein carriers.

Table 27.5. Incorporation of 9,10-^{3}H oleate into lipoprotein-borne lipids secreted by perfused rat liver

Rats	VLDL + LDL ($d < 1.063$)	HDL ($1.063 < d < 1.21$)	Total liver lipid
	($^0/_{00}$ initial ^{3}H radioactivity per h per 10 g liver)		
Normal	23.9	5.6	163
Nephrotic	71.6*	14.0*	155

Values are means for groups of 6 rats. Asterisk denotes a statistically significant difference. The initial radioactivity was 5.7×10^7 counts min^{-1} for 200 μmole of potassium oleate per 100 ml of perfusate, containing 4% bovine albumin.

The fact that free fatty acids constitute an important precursor of the lipid moiety of the lipoproteins in nephrosis is also illustrated by the time course of the development of aminonucleoside-induced nephrosis. Figure 27.1 shows a gradual decrease in the protein level in the plasma, with an abrupt fall between 5th and 7th day of aminonucleoside injections. After cessation of the injections the protein level in the plasma stabilised at about 3 g 100 ml^{-1}. The proteinuria rapidly increased between the 5th and the 7th day, then reached its peak on the 10th or 14th day. The onset of hyperlipidemia coincided well

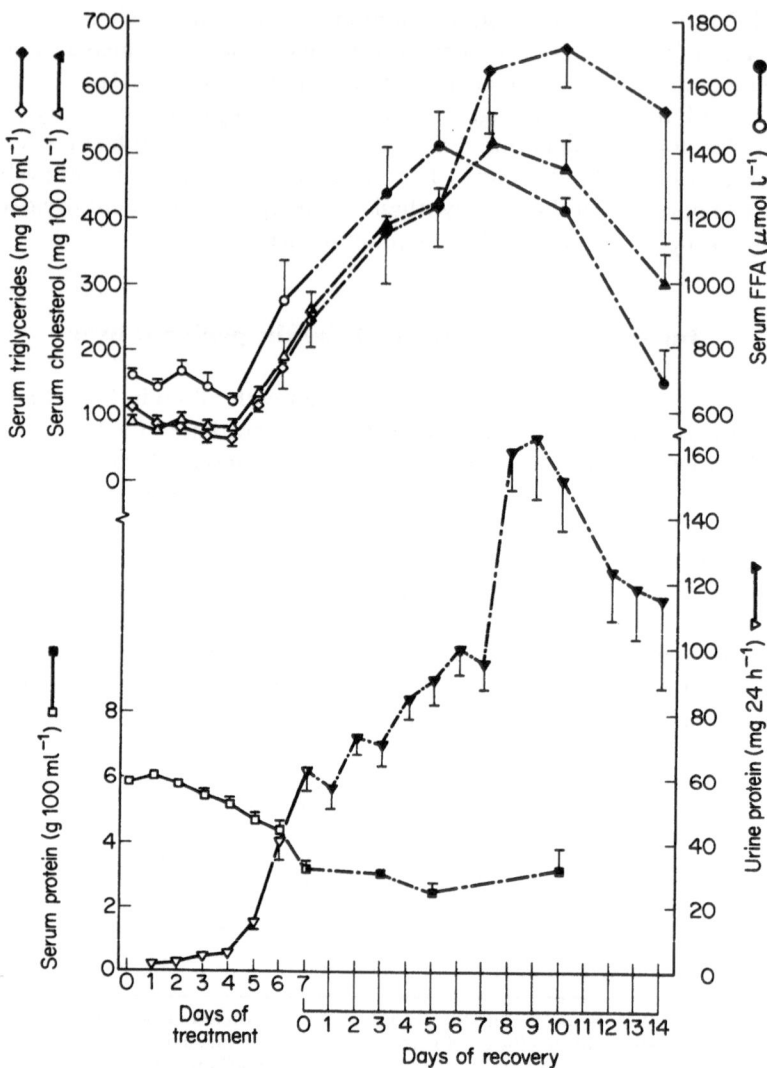

Figure 27.1. Time course of the hypoproteinemia, proteinuria and hyperlipidemia during and after aminonucleoside injections to rats[14]. *Left side* (O)
treatment with seven daily subcutaneous injections of 2 mg 100 g^{-1} aminonucleoside to induce the nephrotic syndrome. *Right side* (●) changes in protein
and lipid levels in nephrotic rats recovering from the side-effects of aminonucleoside injections. Mean ±S.E. values are given for 12 to 20 rats in each
group. FFA = free fatty acids.

with that of proteinuria and hypoproteinemia. This relationship has been
observed earlier[9] but the important finding here was a marked, and likewise
coincident, increase in circulating free fatty acids. Enhanced outflow of free
fatty acids from adipose tissue in nephrosis has also been demonstrated in *in*

vitro experiments[10,11] while the entry and metabolism of free fatty acids in the liver are well known to be determined by the rate of their external supply.

Thus, free fatty acids, provided by increased mobilisation from peripheral adipose tissue, appear to massively participate in the hepatic production of lipoprotein-borne lipids *in vivo*, at least at this stage of aminonucleoside nephrosis.

Fatty acid synthesis in nephrotic syndrome

The lack of significant rise in amino acid incorporation into lipids, as contrasted with the enhanced appearance of preformed fatty acids in the lipoproteins, raised the question as to the hepatic capacity of fatty acid synthesis in nephrosis. In previous experiments we have encountered, in aminonucleoside-nephrotic rats, a decrease in the activity of several liver enzymes regulatory for glycolysis, NADPH generation and fatty acid synthesis[12,13]. We have recently studied this aspect more carefully[14] and came to the conclusion that the decrease in the activity of enzymes of lipogenesis is produced by aminonucleoside independently of nephrosis, most probably due to its capacity to inhibit RNA synthesis in selected systems. When sufficient time elapses from the termination of aminonucleoside injections, the rate-limiting enzymes of lipogenesis in the liver of nephrotic rats regain their activity and often significantly exceed the activity of the corresponding enzymes in control rats (table 27.6). The rise in the activity of these enzymes may represent an adaptive response to the condition of hyperlipogenesis imposed by the availability of apolipoproteins.

Table 27.6. Activity of liver enzymes of lipogenesis pathway during the course of nephrosis induced by aminonucleoside injections

Enzyme	Preinjection	6 to 7 days of aminonucleoside injections	5 to 7 days after cessation of injections	10 to 14 days after cessation of injections
		(nmole min^{-1} per mg cytosol protein)		
NADP-malate dehydrogenase ('malate enzyme')	74	27	29	96
ATP-citrate lyase ('citrate cleavage enzyme')	36	17	18	66*
Acetyl-CoA carboxylase	19.2	8.5	11.5	28.3*
Fatty acid synthetase	6.4	3.8	7.1	12.0*

Nephrosis was induced by 7 daily subcutaneous injections of aminonucleoside of puromycin, 2 mg per 100 g rat weight. Full blown nephrotic syndrome was apparent on the 6th or 7th day of injections. For the time course of changes in protein and lipid levels see figure 27.1. Asterisk denotes a statistically significant increase in the activity of enzymes as compared with the preinjection activity. For details of experimental procedures and methods see ref. 14.

Incorporation of citrate and leucine into lipoproteins in the intact rat

To further explore the sources for the synthesis of both moieties of the lipoproteins, we have injected ³H-leucine together with ¹⁴C-citrate, as an immediate precursor of fatty acids, into the whole animal. One hour after the injection the plasma was taken for lipoprotein isolation. As in the experiments with
perfused liver we have found a three-fold increase in the incorporation of
leucine into the apoproteins (VLDL + LDL), when the nephrotic and normal
rats were compared (table 27.7). The HDL apoprotein also became excessively
labelled which seems important since this lipoprotein is usually decreased in
quantity in nephrosis. HDL appears to be small in size enough to be lost in
urine in the nephrotic animal; it may be decreased also as a result of participa-
tion in the formation of VLDL lipoproteins since VLDL contains a peptide
identical to HDL[8].

Table 27.7. *In vivo* incorporation of 4,5-³H-leucine and 1,5-¹⁴C-citrate
into lipoproteins

Rats	VLDL + LDL ($d < 1.063$) count min⁻¹ per 10 ml serum	HDL ($1.063 < d < 1.21$)	Liver count min⁻¹ per 10 g liver
1. Leucine to proteins			
Normal	2061	3314	11400
Nephrotic	7816*	6927*	12600
2. Citrate to lipids			
Normal	1349	495	22800
Nephrotic	3904*	654*	26400

Measured one hour after intravenous injection to rats of 10 μCi and
10 μmole of labelled leucine together with 15 μCi and 50 μmole of
labelled citrate. Asterisk denotes significant increase in protein or
lipid radioactivity compared to normal rats.

We wish to stress that the increase in the incorporation into the lipoproteins
in nephrosis was not accompanied by an increase in leucine label in liver
proteins. There was also no appreciable leucine-derived radioactivity in liver
or plasma lipids.

Table 27.7 shows that the incorporation of citrate into the lipid moieties of
the VLDL + LDL was also markedly increased in nephrosis. This means that
in vivo there is an increased channelling of substrates through the lipogenesis
pathway if they originate from citrate and not from amino acids. It is also
interesting that it occurs despite the possibly reduced activity of citrate cleavage enzyme and acetyl CoA carboxylase (table 27.6), at this time period after
aminonucleoside injection. Presumably, the availability of apolipoproteins as

final acceptors of the end product enhances the flow of substrates through the lipogenesis pathway, even if the maximal capacity of the enzymes is reduced. Preferential channelling of citrate into lipoprotein-borne lipids is further stressed by the fact that there was no increase in the incorporation into liver lipid of the nephrotic rats.

We have injected leucine together with citrate to rats after plasmapheresis at various time intervals after the plasma removal. In these animals there is the opportunity to follow the kinetics of the protein replacement response from the rapid onset of the oncotic stimulus and to see the temporal relation between the incorporation into the protein and lipid moieties of the lipoproteins and into albumin. Figure 27.2 shows that 3 h after plasmapheresis is the peak of

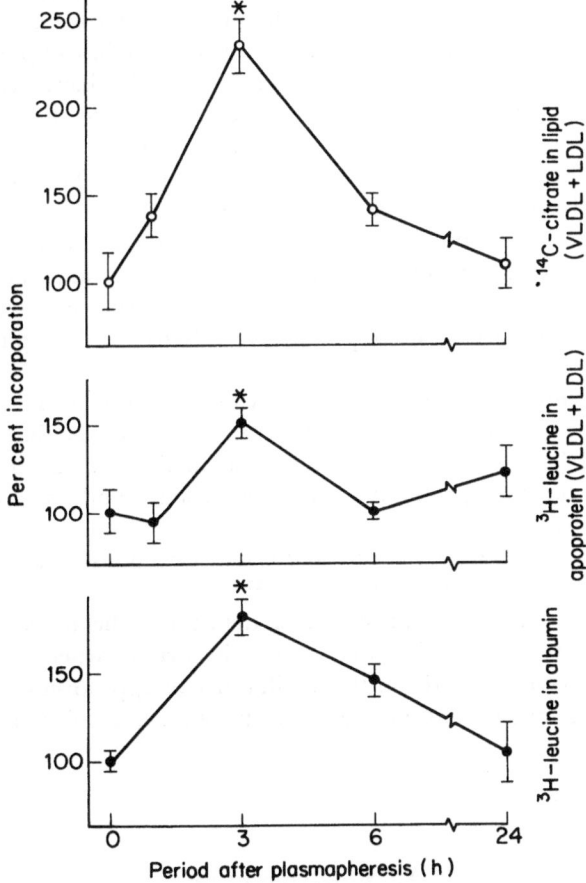

Figure 27.2. Time course of 4,5-³H-leucine and 1,5-¹⁴C-citrate incorporation into lipoproteins and albumin in rats after plasmapheresis. Labelled leucine and citrate were injected one hour prior to sacrifice at the indicated times after the second plasma removal, in amounts given in table 27.7. (Shafrir and Brenner, unpublished experiments.)

leucine incorporation into the apoproteins (VLDL + LDL), as well as into the albumin. The increase above normal is still significant at 6 h and returns to the baseline range 24 h after the removal of plasma. There is a striking coincidence between the incorporation of leucine into the proteins and the incorporation of citrate into the lipids, hinting at the interdependence of these processes elicited by the hypoproteinemia.

Conclusions

Our results demonstrate an increased incorporation of amino acids into apolipoproteins in rats with hypoproteinemia of nephrosis or after plasmapheresis in tight association with the increased channelling of preformed fatty acids and of citrate into the lipoprotein lipids. Thus, the increased lipoprotein synthesis is very likely a part of the general response to the loss of plasma oncotic pressure and a compensatory adjustment to the hypoproteinemia. It is not evoked by protein seepage specifically through the kidney.

The hyperlipogenesis in nephrosis or after plasmapheresis occurs to fulfil the immediate requirement of lipid complement for the apoliproproteins synthesised in excess. In other words, the excessive production of apolipoproteins is the causative factor for hyperlipogenesis from several sources:

1. Trapping and utilisation of preformed free fatty acids originating in the periphery.

2. *De novo* synthesis of fatty acids by drawing upon precursors such as citrate or glucose[5] but not amino acids (although there is evidence that amino acids are mobilised from muscle and metabolised in the liver at an augmented rate[13,15] they are most probably spared for protein synthesis and pathways other than lipogenesis).

3. In some cases there may be utilisation of endogenous hepatic cytoplasmic lipids with a decrease in liver lipid pool[16,17].

The overwhelming phenomenon which determines the demand for lipids is the increased rate of synthesis of the apoprotein carriers, whereas the nature of the lipid precursors depends on the availability of appropriate precursors in deference to the nutritional state and the efficiency of the fatty acid synthesis pathway.

Acknowledgement

The studies reported in this lecture have been supported in part by the United States–Israel Binational Science Foundation, Research Grant Agreement No. 205.

References

1. Epstein, A. A. *J. Amer. Med. Ass.*, **69** (1917), 444
2. Marsh, J. B. and Drabkin, D. L. *Metabolism*, **9** (1960), 948
3. Marsh, J. B. and Whereat, A. F. *J. Biol. Chem.*, **234** (1959), 3196
4. Radding, C. M. and Steinberg, D. *J. Clin. Invest.*, **39** (1960), 1560
5. Bar-on, H. and Shafrir, E. *Isr. J. Med. Sci.*, **1** (1965), 365
6. Rothschild, M. A. and Waldman, T. (1970). In *Plasma Protein Metabolism*, Academic Press, New York, p. 233
7. Ruderman, N. B., Jones, A. L., Krauss, R. M. and Shafrir, E. *J. Clin. Invest.*, **50** (1971), 1355
8. Bersot, T. P., Brown, W. V., Levy, R. I., Windmueller, H. G., Fredrickson, D. S. and LeGuire, V. S. *Biochemistry*, **9** (1970), 3427
9. Dubach, V., Recant, L., Hatch, F. and Koch, M. *Proc. Soc. Exp. Biol. Med.*, **105** (1960), 392
10. Gutman, A. and Shafrir, E. *Amer. J. Physiol.*, **205** (1963), 702
11. Tashimo, M. and Matsuda, I. *Clin. Chim. Acta*, **41** (1972), 67
12. Shafrir, E., Brenner, T., Gutman, A., Orevi, M., Diamant, S. and Mayer, M. *Isr. J. Med. Sci.*, **8** (1972), 271
13. Shafrir, E., Brenner, T., Gutman, A. and Orevi, M. *Amer. J. Physiol.*, **226** (1974), 162
14. Diamant, S. and Shafrir, E. *Biochim. Biophys. Acta.*, **360** (1974), 241
15. Mayer, M. and Shafrir, E. *Isr. J. Med. Sci.*, **8** (1972), 859
16. Heyman, W. and Hackel, W. B. *Metabolism*, **8** (1959), 169
17. Rosenman, R. H., Friedman, M. and Byers, S. O. *J. Clin. Invest.*, **34** (1955), 700

Discussion

MILLER

Dr Shafrir, one question that disturbs me relates to the use of uniformly labelled leucine as a label for protein synthesis, particularly when conditions are such as to encourage both lipogenesis and synthesis of the protein moiety of the lipoprotein. Similarly in the citrate experiment, the possibilities for incorporation into both amino acid and lipids are real and it seems difficult to me to quantitate the exact precursor–product relationships under those conditions.

In experiments we carried out a number of years ago, we came to the conclusion that massive plasmapheresis in rats was not an innocuous procedure even when carried out with skill. The removal of large volumes of blood and the simultaneous return of red cells was traumatic and was associated with an acute phase response, as revealed in the examination of 4 to 5 different acute phase proteins, fibrinogen, haptoglobin and so on. They all increased in response to simple plasmapheresis.

SHAFRIR

As to your second remark I can say that in our rats, which were bled twice by cardiac puncture, the serum electrophoretic pattern did not suggest an acute phase reaction. With regard to your first question, what is the specific objection to the use of leucine as a protein precursor?

MILLER

It is not the use of leucine as a protein precursor, but the use of uniformly labelled leucine, which may give rise to the appearance of fragments labelled in the 4 or 5 position which are potential lipid precursors. We have said many times that carbon one of leucine, or any amino acid labelled in carbon one, is a much more appropriate precursor for protein labelling studies than an amino acid with the label in other positions. The carbons of a uniformly labelled amino acid always run the possibility of getting mixed up in the metabolic pool.

SHAFRIR

I thank Dr Miller for bringing this to our attention. We have intentionally used uniformly labelled leucine because of our initial purpose to study the channelling of this amino acid into proteins and lipids as well. However, in experiments in which we have perfused the liver with leucine or intravenously injected leucine there was so little incorporation into the lipid components that we felt it was justified to make the combined injection of ^{3}H-leucine and ^{14}C-citrate. Similarly in the animals which we have injected with citrate only, there was very little incorporation of the label into plasma or liver protein during the one hour interval used to measure this incorporation.

As far as the immediate precursor-product relationships are concerned, we have also determined leucine and citrate levels in the liver. There was no significant change between the injected and control animals and we wish strongly to imply that the incorporation data presented here can be interpreted as an actual increase in the synthesis of both lipid and protein components of the lipoproteins.

JAMES

The fact that your leucine studies seem all right surely reflects the absence of significant catabolism of the leucine within the liver.

TAVILL

Is my understanding correct that LDL can be formed from VLDL? If that is correct, have you been able in your ^{14}C-leucine incorporation studies to show a temporal relationship between VLDL and LDL synthetic responses?

SHAFRIR

You seem correct in the assumption that there may be a LDL–VLDL interchange in certain instances. We have not investigated this aspect thoroughly. From several experiments we have evidence that both LDL and VLDL proteins and lipids are excessively synthesised in nephrosis and plasmapheresis.

However, we are reluctant to draw conclusions as to the synthetic sequence of these lipoproteins on the basis of label incorporation, since their separation by the conventional ultracentrifugal procedure is not sharp, particularly in the rat. Immunochemical procedures are needed to tackle this question.

TAVILL

In the time course of changes in free fatty acids, cholesterol and triglycerides in your nephrotic rats, you emphasised the early increase in free fatty acids but did not provide the evidence that free fatty acids might be the precursors of cholesterol and triglycerides. Is this what you implied?

SHAFRIR

In the *in vivo* study we can only suggest, but not prove, a precursor–product relationship between free fatty acids and triglyceride or cholesterol. The rise in free fatty acids does somewhat precede the hyperlipidemia but we have to take into account that there could be more than one type of precursor of lipids in the nephrotic rat and there is the additional factor of retarded peripheral removal of the lipids. However, in the isolated perfused liver we could definitely obtain this precursor–product relationship.

28

Albumin transcapillary escape rate as an approach to microvascular physiology in health and disease

N. ROSSING, H.-H. PARVING and
N. A. LASSEN

Kinetic studies with labelled proteins have been concerned mostly with measurements of metabolic rates and the distribution of plasma proteins between the intravascular and extravascular space. Such studies have contributed considerably to our realisation of the fact that from the synthesis to the disappearance from the body these proteins circulate at individual rates between plasma and interstitial fluid. Much time and effort have been devoted to kinetic models to see how many pools, or rather how many separate rates of exchange between intravascular and extravascular compartments can be distinguished for one individual protein. Since the animal experiments of the Wasserman and Mayerson group[1-5] there have been few investigations into the nature of the transcapillary escape of endogenous macromolecules and the pathophysiological changes hereof in vascular disease. Most information on the microvascular permeability has been extracted from studies employing small tracer molecules or dextrans.

We shall report studies in man on the overall transcapillary plasma protein escape in normal, experimental and pathological states. The studies were conducted to elucidate the transport mechanisms involved with special reference to hypertension and diabetes mellitus[6-10]. We wanted to evaluate the hypothesis that increased extravasation of plasma substances and subsequent deposition in the microvascular wall is the basis for hypertensive and diabetic microangiopathy. This is the concept of plasmatic vasculosis originally put forward by Lendrum[11].

Methods

All subjects were males. The subjects to be examined fasted 12 h and rested in the supine position at least $\frac{1}{2}$ h before and during the examination to minimise lymph flow and plasma volume fluctuations.

5–10 μCi of albumin and in some cases also IgG labelled electrolytically with ^{125}I or ^{131}I was injected intravenously and 8 plasma samples were collected over the next 60 min, the first one being drawn after 10 min. The radioactivity in each 3 ml sample was related to the total protein concentration as determined by refractometry to cancel out the influence of plasma volume fluctuation.

The transcapillary escape rate (TER) was calculated as the rate constant, k, of the assumed monoexponential decrease in plasma specific activity over the 60 min and expressed as that percentage of intravascular albumin that passes to extravascular compartments per hour.

Intravascular masses (IVM) of plasma proteins were calculated from the plasma volume and plasma protein concentrations determined immuno-chemically.

To assume that the transcapillary escape rate of the intravascular albumin equals the fractional disappearance rate of intravascular tracer during 60 min implies two approximations:

1. We ignore the return of tracer from extravascular compartments during the investigation period.

2. Disappearance due to catabolism is neglected. Concerning the problem of tracer return, prolonged studies have shown that only if we continue for $1\frac{1}{2}$ h, or more in the case of albumin, do we get anything distinguishable from a monoexponential decrease. Furthermore, we have collected thoracic duct lymph in dogs after i.v. albumin injection, and we were not able to collect any activity at all during the first 20 min. Finally, average calculations on the dilution of tracer in the interstitial space, and the rate of lymph flow warrant us to assume that the approximation is justified. As for catabolism, its influence can be neglected since TER is 10–12 times greater than the fractional catabolic rate.

The calculations were modified in experiments where changes in the intravascular mass or plasma volume were deliberately induced. In these cases changes in plasma volume were calculated from hematocrit values or the transcapillary escape rate was calculated from net changes in the intravascular mass.

Normal values

An average calculation of albumin TER values in all 28 male control subjects gave a value of 5.4% h^{-1} (s.d. 1.1% h^{-1}). This value is close to that found in dogs of 4.9% h^{-1} by Wasserman and Mayerson[2]. Repeated studies with a one year interval showed an intra-individual variation coefficient of 14.4%. Reproducibility experiments showed a variation coefficient of 8.5%.

Hypertension

Essential hypertension

Eighteen patients with untreated benign essential hypertension and a mean blood pressure of 136 mm Hg had a mean TER of albumin of 7.6% h^{-1} (figure 28.1). This was significantly above the values of 5.6% h^{-1} found in 10 matched control subjects. Figure 28.2 shows the significant correlation between TER and blood pressure in the hypertensive group.

Ten of the 18 patients were reinvestigated when on effective treatment with a β-blocking agent. Figure 28.3 shows that TER decreased to normal values, when blood pressure was normalised.

Figure 28.4 shows the correlation between the reduction in blood pressure and TER.

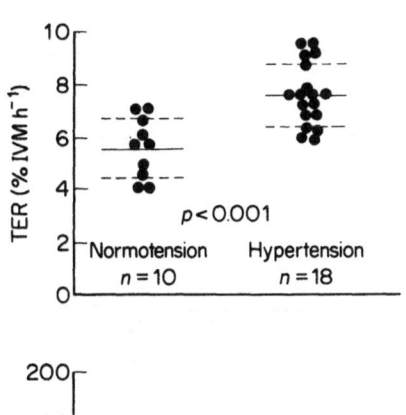

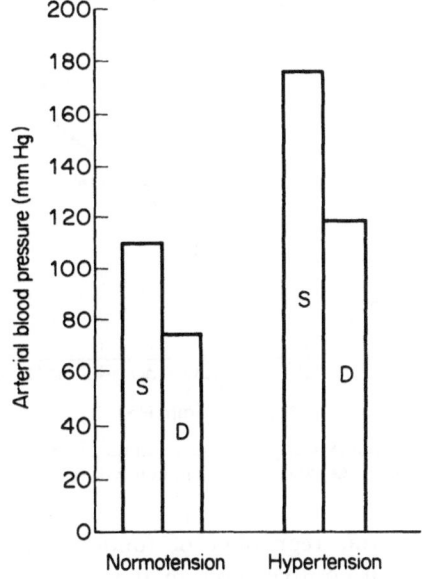

Figure 28.1. Transcapillary escape rate of albumin in essential hypertension.

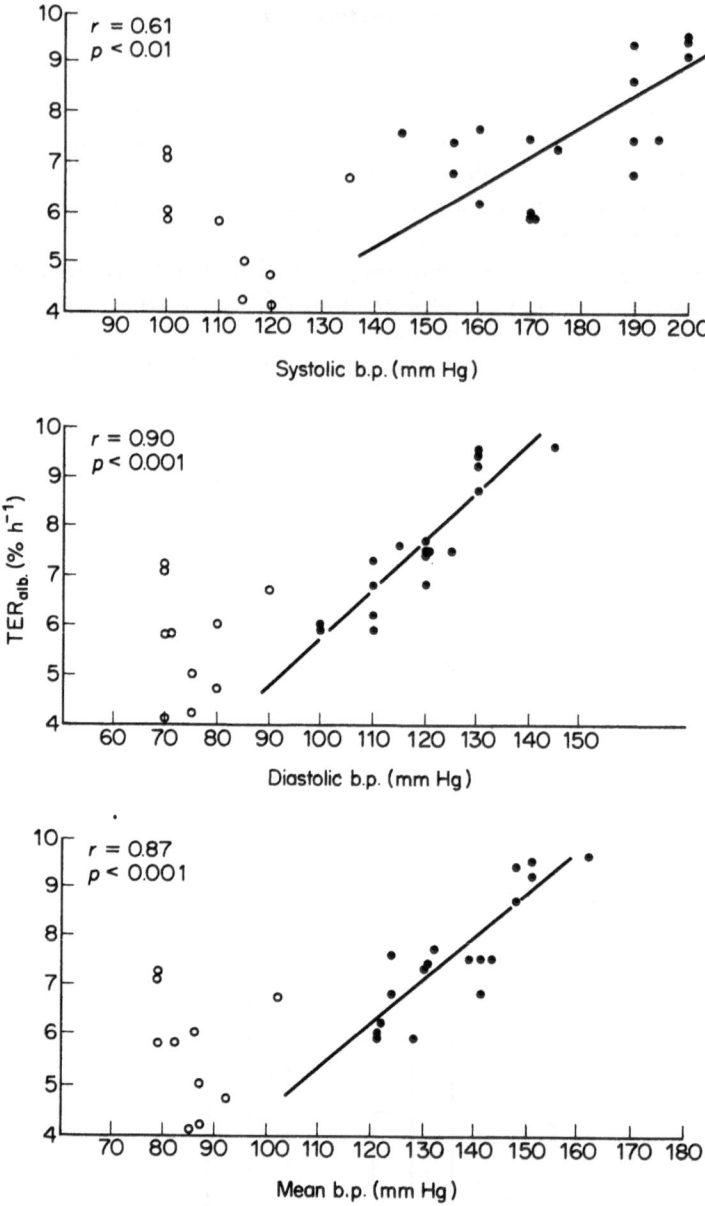

Figure 28.2. Transcapillary escape rate of albumin in 18 essential hyperten-
sives (●) and in 10 normotensives (○).

Decrease in proteinuria, regression of fundoscopic changes, and disappearance of fluorescein-positive spots in the retina during treatment of
pre-eclampsia and severe essential hypertension further demonstrate the

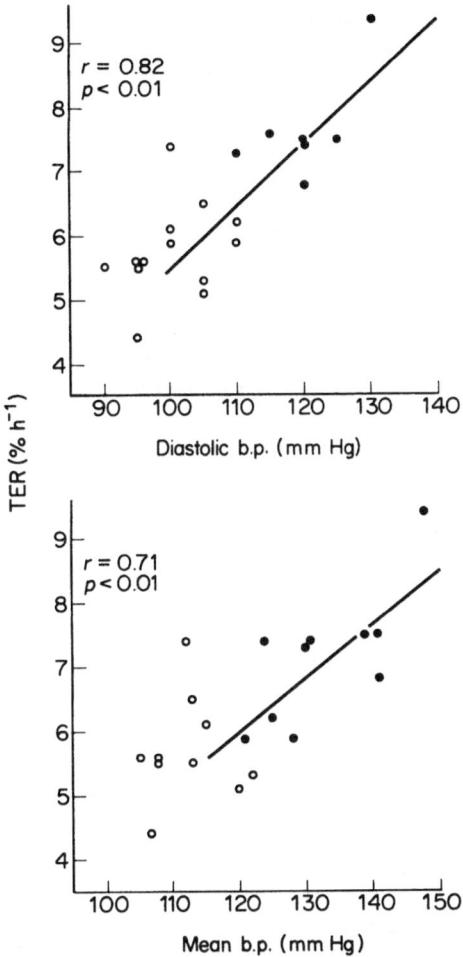

Figure 28.3. Transcapillary escape rate of albumin before (●) and after (○) treatment of essential hypertension.

reversible nature of the processes involved. So, if extravasation of plasma substances is the basis for hypertensive microangiopathy then our results support the case for an early and effective treatment of the disease.

The mechanism behind the increased TER in essential hypertension cannot be evaluated by these experiments. Accepting restricted diffusion[12] as the transport mechanism one may suggest enlargement of the interendothelial pores due to stretching to account for the abnormality. If filtration is involved the increased transmural hydrostatic pressure in hypertension will cause an increased filtration rate.

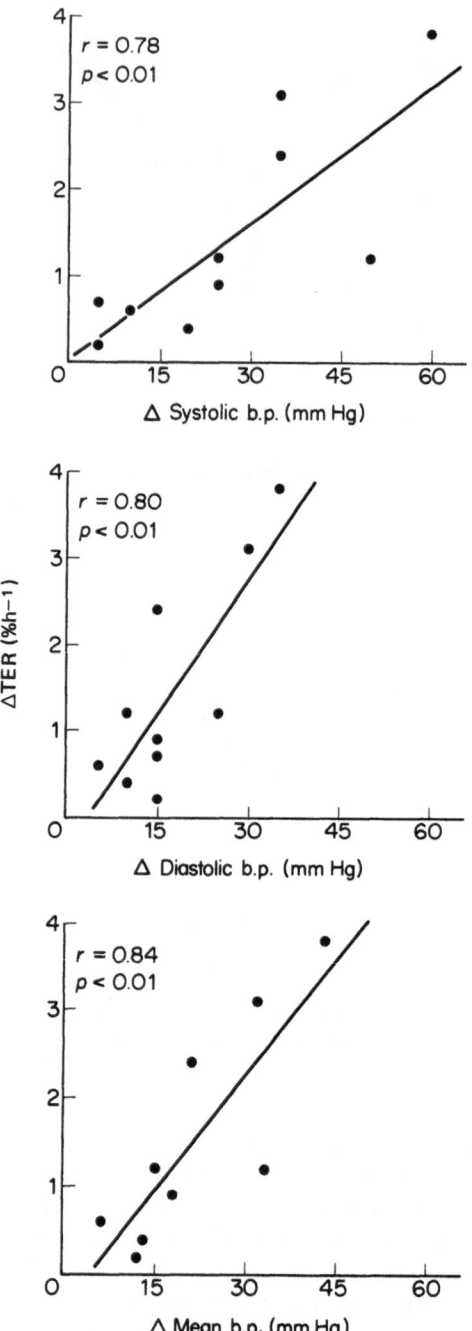

Figure 28.4. Comparison between the reduction in transcapillary escape rate of albumin and the decrease in arterial blood pressure induced by treatment of essential hypertension.

Experimental hypertension

To obtain further information on the pressure related plasma protein leakage we measured TER for albumin in 6 healthy subjects before and during a controlled increase in blood pressure induced by infusion of angiotensin II. In 5 out of 6 cases TER increased markedly (figure 28.5).

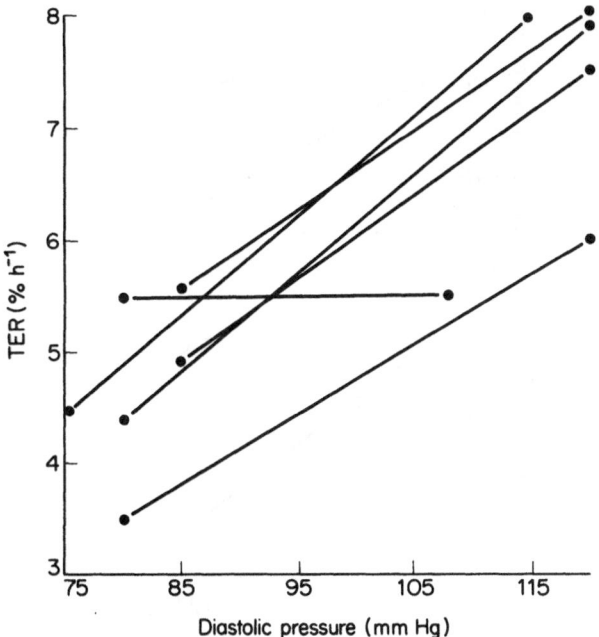

Figure 28.5. Diastolic blood pressure and transcapillary escape rate of albumin before and during angiotensin II infusion. The solid lines connect values in the same individual.

Figure 28.6 shows that 90 min of acute hypertension caused plasma volume and the intravascular masses of albumin IgG and IgM to decrease significantly. The almost parallel decrease corresponds to a bulk efflux of plasma. Considering that normal TER of IgG is 3% h^{-1} and that of IgM supposedly[13] only between 1 and 2% h^{-1} the permeation has been facilitated relatively more for IgM (M.W. 1 000 000) than for albumin (M.W. 70 000) with IgG (M.W. 175 000) in between.

That this can occur through 'stretched pores', as described by Wasserman *et al.*[3], has recently gained electron microscopic support by Pietra *et al.*[14] and Goldby *et al.*[15]. Both groups found that widening of the pores between the endothelial cells in the arterioles takes place during acute hypertension in animals. It has been a controversial subject whether the protein transport through the enlarged pores takes place by diffusion or by filtration. The fact

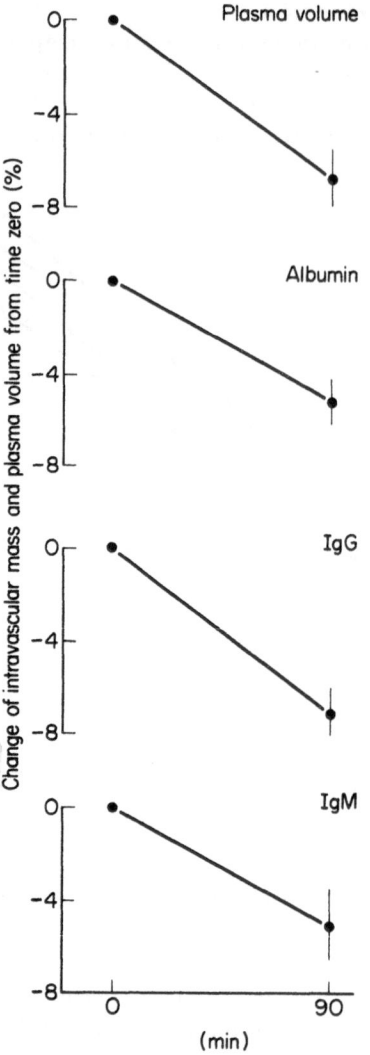

Figure 28.6. Plasma volume and intravascular mass of albumin, IgG and IgM
during acute hypertension in 6 subjects.

that we observed a bulk flow is, however, inconsistent with the concept of
restricted diffusion and warrants us to conclude that sieving i.e., restricted
filtration[3], is the mechanism responsible for the increased protein leakage in
hypertension. Of course pinocytosis could explain our findings if this process
were pressure dependent. Electron microscopic investigations have not been
able to substantiate this hypothesis[14,16].

Diabetes mellitus

We examined TER of albumin and IgG simultaneously in 10 long-term juvenile diabetics with microangiopathy, and 9 matched control subjects. Their data are given in table 28.1.

Table 28.1. Clinical data of non-diabetic and diabetic subjects

	Number of cases	Age (Years)	Weight (kg)	Duration of diabetes (Years)	Retino-pathy	Nephro-pathy	Neuro-pathy	Blood pressure (mm Hg)
Non-diabetics	9(M)	31 ±8	74 ±9					118/80 ±8/6
Long-term diabetics	10(M)	35 ±8	64 ±8	20 ±5	3(III) 7(II)	5	5	122/84 ±15/9

Figure 28.7 shows that both for albumin and for IgG (normal TER: 3.0% IVM h^{-1}) TER values are significantly increased in the diabetic subjects. Figure 28.8 shows that TER is increased proportionally for both proteins. For albumin TER was $7.4 \pm 1.1\%$ h^{-1} compared with $5.2 \pm 1.0\%$ h^{-1} in the control group. The values for IgG were 3.0 ± 0.7 and $4.4 \pm 1.0\%$ h^{-1}, respectively. There is no indication of an increased transmural filtration pressure, so

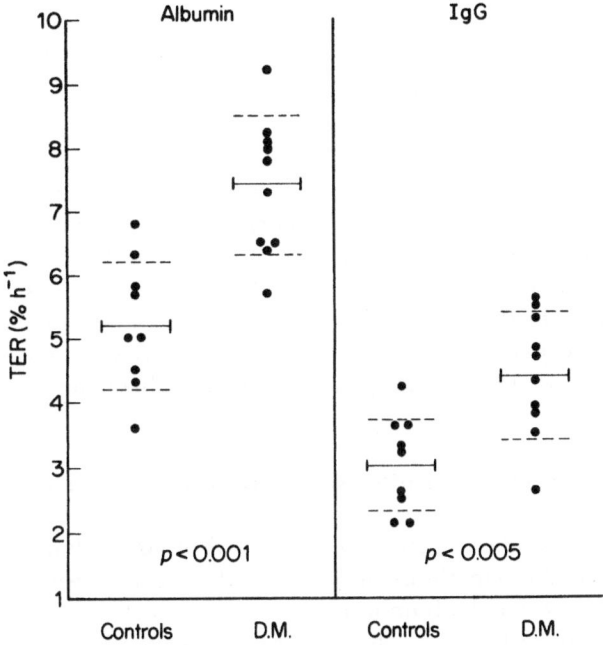

Figure 28.7. Studies in long-term juvenile diabetics (D.M.).

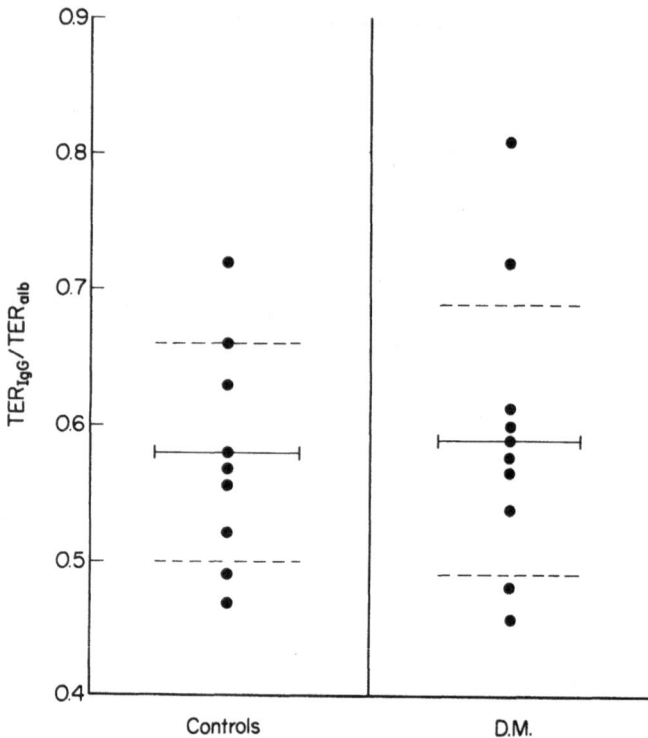

Figure 28.8. Ratio between transcapillary escape rates of albumin and IgG in controls and long-term juvenile diabetics (D.M.).

the increased TER should be ascribed either to an increased microvascular surface area or to an increased number of pores per unit surface area. Even in patients with proliferative retinopathy there is no indication that neovascularisation is a generalised feature[17]. Therefore, we believe that the equally increased TER of albumin and IgG is the consequence of an increased number of pores per unit surface area. We associate the increased leakage of plasma substances with the thickening of the basement membrane and the deposition of PAS-positive material in the arterioles, capillaries, and venules. These changes constitute the morphological features of diabetic microangiopathy.

Metabolic studies in essential hypertension

Being intrigued by the hypothesis put forward by McFarlane in 1967[18] that vascular endothelium might be the site of a diffuse catabolism of albumin, we performed conventional albumin turnover studies with ^{131}I-labelled albumin in 10 hypertensive subjects[19]. They all had increased TER-values. So, would these patients with an increased fraction of the intravascular albumin mass passing across the vessel wall per time unit also catabolise an increased fraction of the intravascular mass per time unit, as would be expected if albumin catabolism is spatially or temporally related to the permeation of the endothelial membrane? All subjects had normal plasma albumin concentrations, but a mean reduction in plasma volume and hence also in intravascular albumin mass of 10%. Figure 28.9 shows that this was the case, normal values for FCR being from 6–12% 24 h^{-1}. This finding may imply that whether the intravascular-to-extravascular albumin exchange rate is normal or increased, the 'average molecule' makes the passage from intravascular to extravascular compartments 10–12 times before it is degraded.

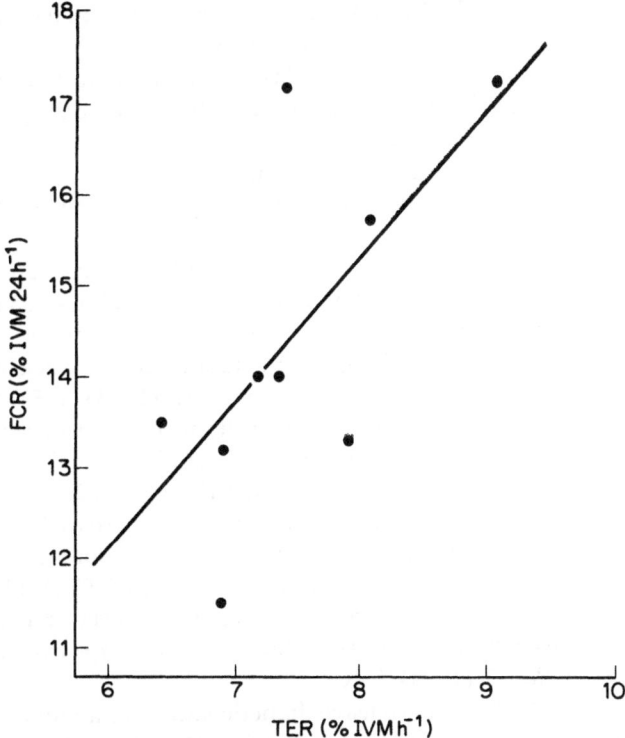

Figure 28.9. Connection between transcapillary escape rate and fractional catabolic rate of albumin in 10 hypertensive subjects ($r = 0.67$; $p < 0.05$)[19].

Conclusion

In hypertension and long-term diabetes mellitus the microcirculation is abnormally leaky to high molecular plasma substances. This is consistent with the hypothesis that increased extravasation of plasma proteins and subsequent deposition in the walls of the microcirculation is the basis for hypertensive and diabetic microangiopathy[11]. Metabolic albumin studies show that there is a significant correlation between fractional catabolic rate and transcapillary exchange rate in hypertension.

References

1. Wasserman, K. and Mayerson, H. S. Exchange of albumin between plasma and lymph. *Am. J. Physiol.*, **165** (1951), 15–26
2. Wasserman, K. and Mayerson, H. S. Dynamics of lymph and plasma protein exchange. *Cardiologia* (Basel) **21** (1952), 296
3. Wasserman, K., Loeb, L. and Mayerson, H. S. Capillary permeability to macromolecules. *Circulat. Res.*, **III** (1955), 594
4. Shirley, H. H., Wolfram, C. G., Wasserman, K. and Mayerson, H. S. Capillary permeability to macromolecules: Stretched pore phenomenon. *Am. J. Physiol.*, **190** (1957), 189–193
5. Mayerson, H. S., Wolfram, C. G., Shirley, H. H. and Wasserman, K. Regional differences in capillary permeability. *Am. J. Physiol.*, **198** (1960), 155–160
6. Parving, H.-H. and Gyntelberg, F. Transcapillary escape rate of albumin and plasma volume in essential hypertension. *Circulat. Res.*, **33** (1973), 643
7. Parving, H.-H. and Gyntelberg, F. Albumin transcapillary escape rate and plasma volume during long-term beta-adrenergic blockade in essential hypertension. *Scand. J. Clin. Lab. Invest.*, **32** (1973), 105–110
8. Parving, H.-H. and Rossing, N. Simultaneous determination of the transcapillary escape rate of albumin and IgG in normal and long-term juvenile diabetic subjects. *Scand. J. Clin. Lab. Invest.*, **32** (1973), 239–244
9. Parving, H.-H., Nielsen, S. L. and Lassen, N. A. Increased transcapillary escape rate of albumin, IgG and IgM during angiotensin II induced hypertension in man. *Scand. J. Clin. Lab. Invest.*, **34** (1974), 111–118
10. Lassen, N. A., Parving, H.-H. and Rossing, N. Editorial. Filtration as the main mechanism of overall transcapillary protein escape from the plasma. *Microvascular Res.*, **7** (1974), i–iv
11. Lendrum, A. C. The hypertensive diabetic kidney as a model of so called collagen diseases. *Canad. Med. Ass. J.*, **88** (1963), 442–452
12. Pappenheimer, J. R., Renkin, E. M. and Cornoero, L. M. Filtration, diffusion and molecular sieving through peripheral capillary membranes. *Am. J. Physiol.*, **167** (1951), 13–46
13. Vaerman, J.-P. and Heremans, J. F. Origin and molecular size of immunoglobulin-A in the mesenteric lymph of the dog. *Immunology*, **18** (1970), 27–38
14. Pietra, C. G., Szidan, J. P., Leventhal, M. M. and Fishman, A. P. Hemoglobin as a tracer in hemodynamic pulmonary edema. *Science*, **166** (1969), 1643–1646

15. Goldby, F. S. and Beilin, L. J. How an acute rise in arterial pressure damages arterioles. *Cardiovascular Res.* **6** (1972), 569–584

16. Pietra, C. G., D'Amodio, M. D., Leventhal, M. M. Oh, W. and Brando, J. L. Electron microscopy of cutaneous capillaries of newborn infants: Effects of placental transfusion. *Pediatrics*, **42** (1968), 678–683

17. Alpert, J. S., Coffman, J. D., Balodimos, M. C., Koncz, L. and Soeldner, J. S. Capillary permeability and blood flow in skeletal muscle of patients with diabetes mellitus and genetic prediabetes. *New Engl. J. Med.*, **286** (1972), 454

18. McFarlane, A. S. (1969) *Physiology and Pathophysiology of Plasma Protein Metabolism.* Pergamon Press, Oxford and New York, p. 87

19. Parving, H.-H., Rossing, N. and Jensen, H. E. Increased metabolic turnover rate and transcapillary escape rate of albumin in essential hypertension. *Circulation Res.* **35** (1974), 517–552

Discussion

TAKEDA

I would not argue these findings with respect to albumin escape rate in essential hypertension. However, in our studies of fibrinogen, the fractional transcapillary transfer rate was decreased by about 40%.

Now, with respect to plasma volume in essential hypertension, whether it is decreased or not depends on how it is expressed: the problem is that many of the hypertensive patients are obese. So that if the plasma volume is expressed in terms of per kg actual body weight, it would be decreased, but we thought that way is not accurate. Therefore, we expressed the plasma volume in terms of per kg ideal body weight, which is obtained from body height, and the plasma volume was found normal. We also found that plasma fibrinogen concentration was increased by about 40%. As a result, the plasma fibrinogen pool was increased by about 40%, and the transcapillary fibrinogen flux was about normal despite the decreased fractional transcapillary transfer rate. It is my understanding that the reason for high blood pressure in essential hypertension is the increased muscle tone in precapillary muscles. If this is correct, we should expect a decreased capillary flow, resulting in the decreased fractional transcapillary transfer. The accuracy or reliability of the measurement of fractional escape rate depends on the quality of tracer-labelled proteins and the analytical methods employed.

ROSSING

Why fibrinogen shows a decreased TER in hypertension, I do not know. The fact that it is not increased, however, is in harmony with the lack of fibrinogen among the proteins found in the urine of hypertensive patients. Concerning plasma volume, it was decreased, no matter how it was expressed in our hypertensive patients. The control group matched with respect to body weight.

Whether the capillary flow is decreased or not does not affect the results, since albumin is so macromolecular and the extravasation so slow that it is not determined by flow, but by the condition of the microvascular wall and, if filtration is involved, by the hydrostatic pressure.

What was found in *in vivo* microscopic studies in animals with acute hypertension was the 'sausage string' phenomenon in the arterioles, that is localised spasms between ballooned areas, where particles can be shown to ooze out through the distended endothelium[1].

REEVE

I wonder if an easier explanation of your findings would not be that you do have pinocytosis, because that would take care of all proteins. The only evidence for passage through holes in the circulation is in the gut where there are quite large fenestrae through which plasma proteins can pass. According to the studies of Palade and co-workers[2] the holes are too small elsewhere.

ROSSING

All right, but then you must make the assumption that pinocytosis is pressure dependent. Additionally you are not right in stating that plasma proteins only leak out in the gut. There are beautiful electron-microscopic demonstrations of myoglobin passing through the capillaries in the lung[3]. Furthermore, in hypertension it has been shown that these distended parts of the microcirculation are leaky not only to plasma proteins, but even to carbon particles[4].

References

1. Giese, J. (1966). *Pathogenesis of hypertensive vascular disease*, Munksgaard, Copenhagen
2. Reeve, E. B. and Chen, A. Y. (1970). Regulation of Interstitial albumin. In *Plasma Protein Metabolism*, M. A. Rothschild and T. A. Waldmann, eds., Academic Press, New York, pp. 89–109
3. Pietra, G. G., Szidan, J. P., Leventhal, M. M. and Fishman, J. P. Hemoglobin as a tracer in hemodynamic pulmonary edema. *Science*, **166** (1969), 1643–1646
4. Goldby, F. S. and Beilin, L. J. Relationship between arterial pressure and the permeability of arterioles to carbon particles in acute hypertension in the rat. *Cardiovasc. Res.*, **6** (1972), 384

29

Increased albumin catabolism in early uremia as a cause of protein depletion*

R. BIANCHI, G. MARIANI, A. PILO, F. CARMASSI and M. G. TONI

Introduction

The present study was undertaken with the aim of assessing acute changes of albumin turnover in the early phase of uremia, in order to explain the pathogenesis of protein depletion in that condition[1-3]. Previous turnover tracer experiments have shown that such a depleted state does not depend on protein restriction in the diet followed by the patients, assuming that the uremics were in a metabolic steady-state at the time of the study[1-3]. Therefore, it was supposed that some rapid transitory changes of albumin turnover occurred in the early phase of renal failure, during which catabolism exceeds production, without compensation by subsequent periods of relative excess production.

To demonstrate the occurrence of such acute changes, an original two-tracer method was developed, which allowed us to measure albumin catabolism during an experimental time of 6 hours only, in patients undergoing hemo- or peritoneal dialysis[4,5].

As previously discussed extensively[5], steady-state conditions are probably maintained during such a short interval, even if unbalances between synthesis and catabolism occur.

The method is based on the propensity of iodide to diffuse through the peritoneum or the cuprophan membranes of the artificial kidney, after injection of radioiodinated human serum albumin. Free iodide arising from protein degradation is measured on the dialysate, thus making it possible to sample at frequent intervals (e.g. every 20 min), and hence to obtain a much greater definition of the experimental functions in comparison with that usually obtained by urine collection.

*Partly supported by Public Health Service, U.S.A., Contract No. PH-43-68-707.

Material and methods

Patients

A total of 16 studies were carried out on 12 uremic patients undergoing maintenance hemo- or peritoneal dialysis treatment. Patients were divided into two main groups, the first group comprising 8 subjects in apparent clinical and metabolic equilibrium at the time of the experiment, the second one relating to 3 patients in the early phase of uremia or during a relapse from it. Subsequent turnover studies in two of these latter uremics and in one patient at the end of a relapsing uremia period were excluded from both groups and are considered separately. Patients of the second group showed at the time of the study clinical and hematochemical signs of 'hypercatabolism' (true rapid loss of body weight and of muscular mass; marked increase in blood urea nitrogen, serum creatinine and impairment of nutritional state)[6].

The experimental procedures, theoretical approach and reliability assessment of the method have been extensively presented and discussed elsewhere[5]. We give here only a brief account of the experimental protocol and of computational procedure.

Experimental

Electrolytically labelled ^{131}I-Human Serum Albumin (^{131}I-HSA)[7] and sodium ^{125}I-iodide were used as tracers (*see* figure 29.1). Particular care was given to

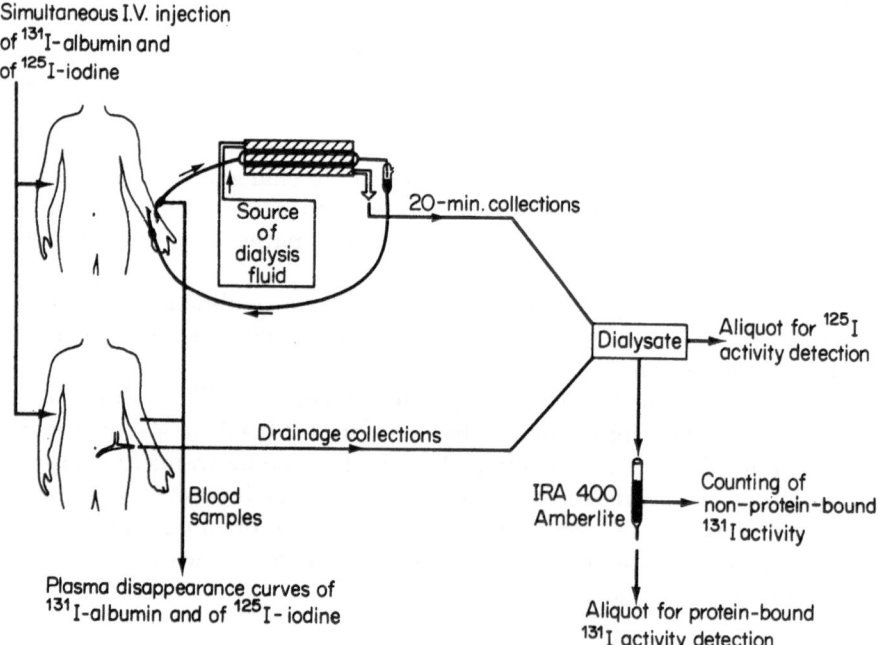

Figure 29.1. Experimental set up (see text).

remove any free iodide from labelled albumin, by dialysis against IRA 400 Amberlite and by gel filtration on Sephadex G-10.

After simultaneous i.v. injection of ^{131}I-HSA and ^{125}I-iodide into patients undergoing dialysis, the respective plasma disappearance curves up to the 6th hour were obtained by blood sampling, corrections being made for haematocrit changes during the experiment. Excreted activities of [131]I- and ^{125}I-iodide were measured in the diffusate, cumulatively collected up to 6 hours after the injection; the radioactive iodide released from albumin breakdown, present at very low concentration in the dialysate, was concentrated by passage through IRA 400 Amberlite columns.

The radioactivity detected in each diffusate sample was considered as 'excreted' during the interval time of dialysate collection (outflow from peritoneal cavity or from artificial kidney).

Computation

[125]I activity in diffusate $LI(t)$, represents the output of the iodide system after unitary injection of the tracer; that is, $LI(t)$ is the cumulative form of the transit time distribution function from the iodide system. The cumulative activity in diffusate corresponding to any input can then be predicted by convoluting this input with the function $LI(t)$.

In our case, the input to the iodide system is represented by the iodide released from albumin catabolism, that is the fractional catabolic rate of intravascular albumin (FCR) times the plasma activity at the same time, $P(t)/P(0)$. In the following equation the input to the iodide system convoluted with its transit time distribution function is equated to its output, $LA(t)$, that is the cumulative activity of iodide from the protein in the dialysate:

$$LA(t) = \text{FCR} \times P(t)/P(0) * LI(t)$$

where symbol $*$ means convolution product operation.

Since $P(t)$, $LI(t)$, and $LA(t)$ are experimentally known, the relation enables us to compute fractional catabolic rate, which is the only unknown.

Results

The results obtained for each patient are reported in detail in table 29.1, while figure 29.2 schematically represents the albumin turnover data found in both the first and second group of patients. Serum albumin concentration (4 ± 0.5 g 100 ml^{-1} in normal subjects) was slightly reduced in both the first and second group (3.4 ± 0.56 g 100 ml^{-1} and 3.57 ± 0.27 g 100 ml^{-1}, respectively), without significant difference between each other ($P < 0.01$ and $P < 0.1$, respectively, as compared with normal values). Plasma volume (42 ± 10 ml per kg body weight in normals) was significantly increased in both groups (60.3 ± 12.6 ml kg^{-1}, with $P < 0.001$, and 56.2 ± 4.9 ml kg^{-1}, $P < 0.02$,

Table 29.1. Results for various patients

Case	Dialysis	Age (years)	Body weight (kg)	Serum albumin concentration (g 100 ml⁻¹)	Haematocrit value (%)	Plasma volume (ml kg⁻¹)	Intravascular albumin (g kg⁻¹)	FCR (% day⁻¹)	CR (mg kg⁻¹ day⁻¹)	Body Weight changes (kg) A	B
C.C.	P	40	52.5	3.47	22.5	69.4	2.41	7.60	183.2	0	—
D.M.	P	36	69.0	4.20	26.4	59.0	2.48	10.69	265.0	0	0
V.A.	H	55	76.0	4.10	25.4	62.5	2.56	9.33	238.8	0	0
N.F.	H	20	60.5	3.50	28.3	74.3	2.60	12.74	331.2	0	0
P.M.	H	24	50.0	3.30	14.8	75.3	2.48	11.13	276.0	0	0
L.N.	P	54	61.0	3.33	23.4	42.4	1.41	13.17	185.9	0	0
G.E.	P	69	67.0	2.60	24.3	55.2	1.43	14.37	206.2	0	0
D.C.P.	P	46	60.0	2.74	21.8	44.3	1.21	14.78	178.8	0	0
D.A.M.	P	20	50.5	3.80	13.2	61.5	2.33	31.05	723.5	−3.5	—
L.D.*	P	47	56.0	3.28	18.5	55.4	1.82	30.67	558.2	−8.5	−3
M.M.*	P	46	77.0	3.64	34.5	51.8	1.89	35.65	673.8	0	−1
C.P.	H	41	73.0	2.90	19.5	66.5	1.93	15.65	302.0	0	0
L.D.**	H	47.5	53.0	2.70	24.5	76.0	2.05	30.05	617.0	0	0
L.D.***	H	48	57.0	2.60	25.6	66.9	1.74	12.66	220.3	+4	+1
M.M.**	H	46.3	78.5	3.60	39.0	54.3	1.95	11.35	221.3	+2.5	0
M.M.***	H	47	78.5	3.20	40.5	43.8	1.40	16.23	227.2	0	+1

P = Peritoneal dialysis
H = Hemodialysis
FCR = Fractional Catabolic Rate of intravascular albumin
CR = absolute Catabolic Rate of albumin

(∗) = first measurement
(∗∗) = second measurement
(∗∗∗) = third measurement
A = before study
B = after study

respectively, in the first and second group), mainly in relation to the low haematocrit value in all patients (*see* table 29.1). As a result of the enlarged plasma volume, the intravascular albumin pool was also increased in both groups (2.07 ± 0.6 g kg⁻¹, with $P < 0.01$, and 2.01 ± 0.28 g kg⁻¹, $P < 0.05$, respectively, as compared with 1.62 ± 0.35 g kg⁻¹ in normals), despite the lowered serum albumin concentration.

When considering only the values obtained in patients in apparent steady-state conditions (first group), the fractional catabolic rate of intravascular albumin was found within the normal range ($11.73 \pm 2.5\%$ day⁻¹, versus $12 \pm 2\%$ day⁻¹ in normals), while the absolute catabolic rate was at the upper limit of the normal range (233.14 ± 54.65 mg kg⁻¹ day⁻¹, with $P < 0.05$, normal value being 194 ± 49 mg kg⁻¹ day⁻¹), due to the slightly enlarged intravascular albumin mass.

In the three patients of the second group both fractional and absolute catabolic rates of albumin were conspicuously increased ($32.46 \pm 2.77\%$ day⁻¹

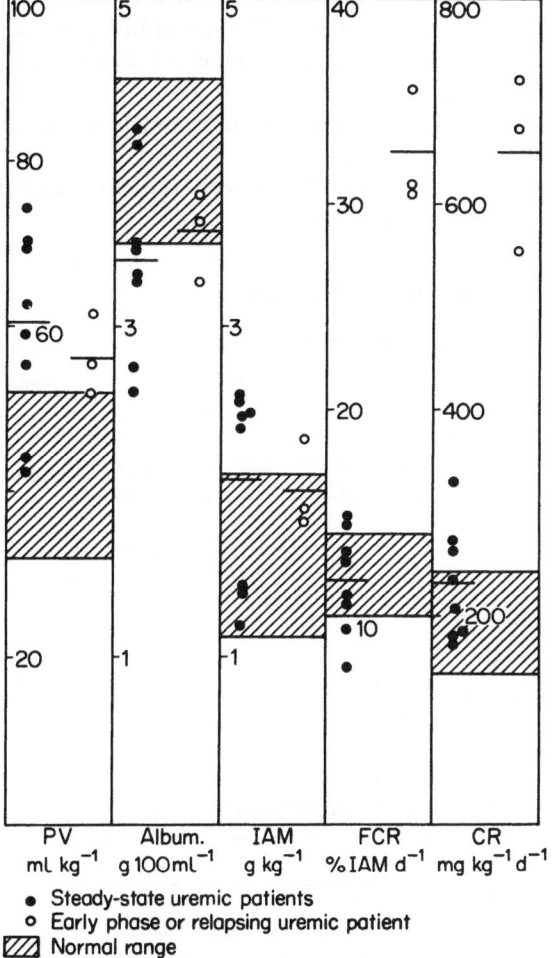

Figure 29.2. Albumin turnover results obtained from the uremic patients in
clinical steady-state (first group) and from patients in the early phase or during
relapse of uremia (second group).

and 651.8 ± 84.8 mg kg^{-1} day^{-1}, respectively), with high statistical significance in comparison with either normal values or values of the first group
($P < 0.001$).

Subsequent turnover changes of the two patients repeatedly submitted to
the study are reported in table 29.1.

Discussion

The only significant difference between the two groups of patients was found in respect of albumin catabolism, which was within the normal range in the 8 subjects in clinical steady-state conditions, and increased up to three times the normal in the patients in the early phase or during relapse of uremia; accordingly, these latter patients presented at the time of the study clinical features of 'hypercatabolism', a clinical definition which appears well founded on the basis of the present studies.

The results obtained in repeated turnover studies on 2 of these unsteady-state uremics are represented in figure 29.3.

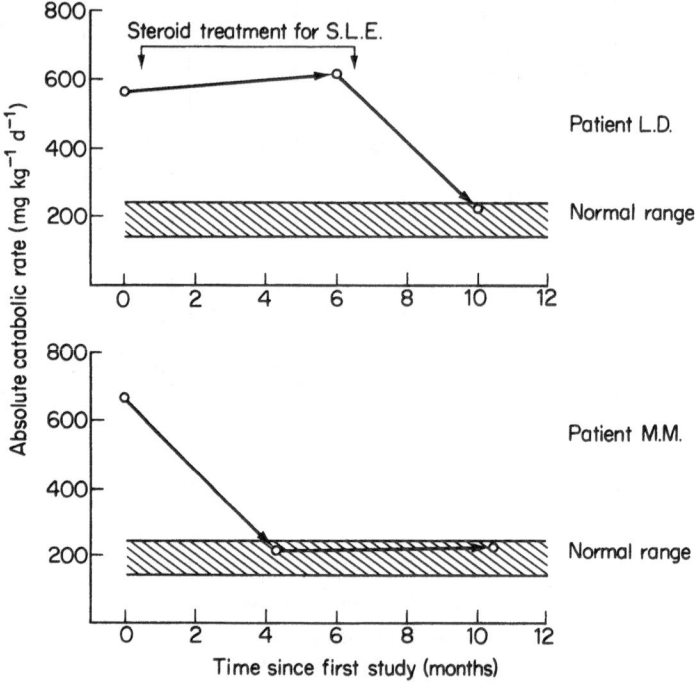

Figure 29.3. Absolute catabolic rates of albumin obtained by repeated turnover measurements in two patients from the second group.

As can be seen, albumin catabolism in patient L.D., with Systemic Lupus Erithematosus, returned to normal only when high dose steroid treatment was interrupted; no body weight changes were recorded during the second measure, when catabolism remained again increased. On the contrary, patient M.M. had a normal catabolic rate in the two studies subsequent to the first one.

All these findings suggest that highly increased catabolism in the early phase of the disease, not counterbalanced by a correspondingly increased synthesis is the cause of the albumin depletion in chronic uremia.

However, the actual origin of this increased protein breakdown still remains obscure.

Since previous studies in steady-state uremics[1-3] and the present one prove that increase of catabolism is only transitory, some inefficiencies of the protein production mechanisms are probably relevant as to the persistence of the metabolic deficiency in the protein pool; the lack of an inverted compensatory turnover pattern in uremia is probably due to the endotoxic factors peculiar to this disease. Further studies are planned to investigate if such factors are diffusable substances or not.

References

1. Bianchi, R., Mariani, G., Pilo, A. and Donato, L. (1971). Albumin metabolism in patients with chronic renal failure on low protein balanced diet. In *Protides of the Biological Fluids*, 19th ed., H. Peeters, ed., Pergamon Press, Oxford, p. 483
2. Bianchi, R., Mariani, G. and Pilo, A. (1972). Albumin metabolism in uremic patients on low-protein diet. In *Uremia*, R. Kluthe, G. M. Berlyne and B. T. Burton, eds., Georg Thieme Verlag, Stuttgart, p. 206
3. Bianchi, R., Mariani, G., Pilo, A., Toni, M. G. and Carmassi, F. Albumin depletion in uremic patients on conservative management. In this volume, p. 237
4. Bianchi, R., Mariani, G., Pilo, A. and Toni, M. G. (1974) Albumin catabolic rate during peritoneal dialysis in acute and chronic uremia. In *Protides of the Biological Fluids*, 21st ed., H. Peeters, ed., Pergamon Press, p. 441
5. Mariani, G., Bianchi, R., Pilo, A., Palla, R., Toni, M. G. and Fusani, L. Albumin catabolism measurement by a double tracer technique in uremic patients during a single dialytic treatment. *Europe. J. Clin. Invest.*, **4** (1974), 435
6. Monasterio, G. (1970) *Le Nefropatie Mediche*. USES, Firenze
7. Rosa, U., Pennisi, F., Bianchi, R., Federighi, G. and Donato, L. Chemical and biological effects of iodination on human albumin. *Biochim. Biophys. Acta*, **133** (1967), 486

Discussion

TAVILL

What mass of plasma protein was lost during dialysis from each patient?

MARIANI

The albumin loss did not exceed 3 g during the 6 h of the experiment. In one patient, G.E., with evident peritonitis at the time of the study, albumin loss was about 12 g, a half of this being lost into the first two dialysates, that is by exchange with the dialysis solution left in the abdomen from the preceding dialysis.

HOFFENBERG

How heavy was the proteinuria in these patients?

MARIANI

Patients were rather oliguric; in every case daily proteinuria did not exceed 0.5 g.

PART 7

Acute Phase Proteins

30
The acute phase plasma proteins

A. H. GORDON

Introduction

During the last four years increasing attention has been given to those plasma proteins which increase or decrease in concentration after trauma or during acute infections. As a result of such studies it has become apparent that additions must be made to the category of those which increase in concentration, usually referred to as 'acute phase proteins', or 'acute phase reactants' (APRs). Recent data on APRs have been summarised by Koj[1]. Thus information is now available that certain proteins present in plasma after injury which on injection into rats have anti-inflammatory effects should be classified as APRs (table 30.1). Similarly, kininogen, kininogenase, angiotensinogen[2] and those constituents of complement which are responsible for reactive lysis must also be considered to be APRs. Greatly increased concentrations of kininogen and kininogenase have recently been detected in the blood of rats after injury[3]. As shown in figure 30.1, the kininogen and kininogenase concentrations were found to be at a maximum after 2 days, as is typical of many other APRs.

On the other hand it has now become clear that not all of the plasma proteins which appear during neoplasia can be considered to be APRs. Thus, for instance, a distinction can be made between α-foetoprotein and those APRs which occur in embryo plasma, but disappear post-natally and only reappear in adult plasma as a result of neoplasia, trauma or infection. This is possible because the very low level of α-foetoprotein present in the plasma of adults does not increase as a result of trauma. The appearance of α-foetoprotein in adult plasma has been linked by Sell and Wepsic[4] with liver cell division. Especially high concentrations of this protein are associated with hepatomas.

The likelihood that control of the rate of synthesis of each APR is such that a specific response to different stimuli can occur was mentioned by Gordon[5]. Furthermore, it was also apparent at that time that the rates of synthesis of APRs may be affected by more than one factor. Since then, despite some new information on the factors controlling the synthesis of certain APRs, in particular transferrin and ceruloplasmin, no general picture of the control mechanisms responsible has yet emerged. On the other hand, attempts to understand

Table 30.1. Acute phase reactants

APR	Function *in vivo* or activity *in vitro*
Fibrinogen	Clot formation
Haptoglobin	Binds haemoglobin
Antitrypsin	
Antichymotrypsin	Prevention of damage after release of lysosomal enzymes
α_2-Macroglobulin*	
Kininogen	Precursor of bradykinin
Kininogenase	Liberator of bradykinin
Angiotensinogen	Precursor of angiotensin
Anti-inflammatory factor*	Reduction of Carrageenin oedema
C-reactive protein	Affects lymphocyte transformation *in vitro*
	Binds mucopolysaccharides in serum
Complement C3	Cell lysis or agglutination
Complement C4	Viral neutralisation
Complement C5̄6̄	Neutrophil chemotaxis
	Opsonisation leading to phagocytosis
Ceruloplasmin	Oxidase
Transferrin†	Fe transport
Thyroxine binding prealbumin‡	Thyroxine transport

*Constituent of rat plasma.

†Decreases in concentration after trauma due to excess catabolism over synthesis. Synthesis rate in these circumstances is increased.

‡Decreases in concentration after trauma.

how such mechanisms may be expected to operate have become more realistic as a result of investigations of the synthesis of ovalbumin, conalbumin, ovo-mucoid and lyzozyme in the chick oviduct. Thus Palmiter[6] has been able to show that stimulation by treatment *in vivo* with oestradiol brings about changes of three kinds. These involve an increased rate of initiation of protein synthesis, followed by an increased rate of elongation during synthesis of each protein molecule. Because in addition the concentration of mRNA was also found to be increased, the overall rate of protein synthesis increased by the multiple of all three factors. Especially relevant to studies of synthesis of APRs is the further finding of Palmiter[6] that changes in the relative rate of synthesis of ovalbumin to conalbumin are the result of alteration in concentration of the two relevant mRNAs.

After trauma, not only are there altered plasma concentrations of adrenal and other hormones but in addition substances released at the injury site enter the plasma and are carried into the liver. Thus the parenchymal cells in which the APRs are synthesised will be influenced by at least two classes of messenger substance. Furthermore, the possibility that nervous control of protein synthesis in these cells may also play a role has to be recognised. An important consequence of this complex situation may well be that stimulation by any given class of messenger substances cannot be expected to lead to a maximal response. For such to occur more than one factor may have to change simultaneously. In this regard special attention may be given to the work of Thomp-

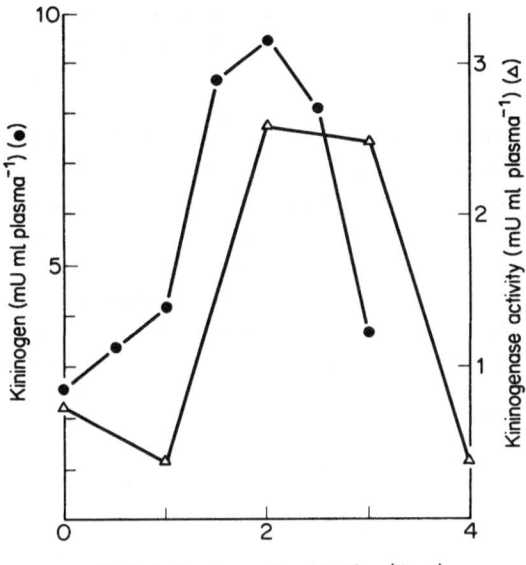

Figure 30.1. Plasma kininogen and kininogenase at various times after s.c. injection of turpentine into rats. Each point is an average value for 4–6 rats. Plasmas from each rat were estimated separately for kininogen by means of the guinea pig ileum. Different groups of rats were used for each time coordinate to avoid rebleeding. One milliunit (mU) of kininogen is the amount of substrate which will yield 1 nanomole of bradykinin equivalent in the presence of excess enzyme. One unit of kininogenase is the amount of enzyme necessary for liberation of 1 μg of bradykinin per hour.

son et al.[7] and Kampschmidt et al.[8–10] who have injected extracts of granulocytes into rabbits and rats. These extracts were prepared by the method originally developed for endogenous pyrogen but are referred to as LEM (leucocyte endogenous mediator) to emphasise that they lead to several different effects. Among these are increased concentrations of APRs, especially α_2AP-globulin, α_1-globulin, and ceruloplasmin. The nature of the several responses obtained with these extracts will be discussed in more detail below. Initially it is sufficient to note that the failure of LEM to lead to maximal rates of synthesis of the APR chosen as indicator substance does not necessarily prove that it plays a minor role in stimulation of the acute phase response. Alternative explanations may be either, that for a maximum response to be obtained, the stimulator should be supplied continuously instead of as a single injection, or as indicated above that its action has been tested in isolation.

Reasons for the use of liver slices for estimation of the rates of synthesis of albumin, α_1-globulin and transferrin

Despite the relatively low rates of synthesis of plasma proteins known to occur when liver slices are incubated in simplified media, this system has certain advantages for investigation of factors which control the synthesis of these proteins. Thus, because of its convenience, livers from rats which have been maintained on many different regimes can be employed. The rates of synthesis of several plasma proteins are fast enough to permit accurate measurement by radial immunodiffusion. Although as a result of the unphysiological conditions employed for incubation, many of the factors which operate *in vivo* to control protein synthesis rates, must be assumed to be non-functional; under such conditions the more stable metabolic relationships are likely to persist. An indication as to how far the low rates of synthesis of plasma proteins given by slices incubated in the manner employed in this work are proportional to those existing *in vivo* is the observed ratio for the rates of synthesis of transferrin and albumin. As shown in table 30.2 this is 0.37 compared with 0.21, the value found by Tavill et al.[11] for the isolated rat liver perfused with rabbit blood.

Table 30.2. Rates of synthesis by liver slices.

Pretreatment of rat	Strain of Rats	No of rats	μg protein g liver^{-1} h^{-1}		
			Albumin	Transferrin	α_1-globulin
None	H*	6	92 ± 18§	34 ± 5.7	L‡
	SD†	7	160 ± 15	76 ± 14	—
Fe-deficient	SD	2	121	66	—
Phenobarbitone	H	4	100 ± 12.4	109 ± 23	L
Turpentine	H	4	54 ± 10.5	63 ± 5.2	55 ± 17.7
Phenobarbitone + turpentine	H	3	50 ± 10.7	60 ± 5.9	33 ± 9.1

*H = rats of the hooded strain maintained at the National Institute for Medical Research, London. †SD = Sprague Dawley rats. ‡L = too low for estimation. § = standard error of the mean.

Note. After slicing, 1 g amounts were suspended in 10 ml BHK Eagles medium which had been equilibrated with 5% CO_2/95% O_2. The 25 ml flasks were then rapidly shaken in a water bath at 37° for 3 h with passage of 5% CO_2/95% O_2. The contents of all flasks were homogenised, ultra-sonicated and concentrated by pressure dialysis. The experiments using Sprague Dawley rats differed only in that 500 mg samples of liver slices were suspended in 3 ml of medium and the concentration stage was omitted

As will be explained below, the increase in this ratio is almost certainly due to a very low rate of synthesis of albumin by the liver slices rather than to an increased rate of synthesis of transferrin. In addition to normal rats, those at two days after subcutaneous (S.C.) injection of turpentine, others that had been maintained for 12 to 24 days on drinking water containing pheno-barbitone at 1 g litre^{-1} and others which had been maintained on an Fe-deficient diet have been investigated. S.C. injections of turpentine were used in the present experiment because this is known to be an effective means for bringing about increased synthesis of those plasma proteins usually referred to as APRs.

The dietary treatments were selected because of the possibility that one or both might lead to an increased rate of synthesis of transferrin. Previous data obtained by Tavill and Kershenobich[12] suggested that this would be likely to ensue as a result of iron deficiency. The effect of dieting with phenobarbitone was investigated because of the possibility of a similar response to that to carbon tetrachloride which, when given acutely, brings about increased plasma concentrations of transferrin[13].

The synthesis rates of albumin, α_1-globulin and transferrin were investigated because previous experiments had indicated that difference in respect to these proteins would be likely to follow appropriate pretreatments *in vivo*. α_1-globulin and transferrin were believed to be specially interesting in this respect because whereas the former can be considered to be a typical APR the rate of synthesis of which is determined mainly by trauma or tissue damage, synthesis of the latter, at least in the rat, is subject to multiple means of control. Thus as mentioned above, the synthesis rate of transferrin is affected by the Fe status of the rat. In addition, previous work[14] has shown that increased synthesis of transferrin may occur after injury.

Because estimation of the chosen plasma proteins was carried out by radial immunodiffusion[15] it was necessary to remove the plasma originally present in the livers as thoroughly as possible. This was done by perfusing for 3 min with cold Tyrode's solution before slicing. Incubation was at 37° in Eagle's medium (for details, *see* table 30.2).

Results of experiments using liver slices

In agreement with the results of Tavill and Kershenobich[12], the rates of synthesis of each protein was found to be linear for at least three hours. The results obtained with slices of livers from normal rats and those pretreated as already mentioned, given in table 30.2, may be summarised as follows:

Albumin
Injury by s.c. injection of turpentine led to reduced rates of synthesis of this protein both in previously normal rats and in those receiving phenobarbitone. In previously normal rats in which α_1-globulin synthesis was negligible, albumin synthesis expressed as a percentage of the total synthesis of the three proteins investigated in these experiments fell from 73% to 31% as a consequence of injury. For rats on the phenobarbitone diet the corresponding figures were 48% and 35%. While these decreased percentages must correspond to considerable reductions in the percentage of albumin to total proteins synthesised *in vivo* in rats which have been injured, the absolute rates of synthesis of albumin on which they are based (table 30.2) should be considered as minimal. This proviso is necessary because experiments with liver slices incubated in simplified media such as that used in the present experiments have indicated that albumin synthesis in such conditions is always reduced more than that of the other plasma proteins. Doubtless this finding explains the very high ratios found for the synthesis rate of α_1-globulin to that

of albumin when livers from injured rats are used. Unfortunately no explanation for the greatly decreased synthesis of albumin brought about by injury is yet available. Because other plasma proteins are synthesised more rapidly competition for intermediates in protein synthesis such as initiation factors may be of importance.

α_1-Globulin

As shown in table 30.2, the synthesis of this protein by liver slices from rats at 48 h after injury was found to be as rapid as that of albumin. On the other hand, livers from rats dieted with phenobarbitone synthesised no more than the minimal amounts of α_1-globulin characteristic of normal rats. The livers from rats on the phenobarbitone diet when tested at 2 days after injury synthesised nearly two-thirds as much α_1-globulin per g liver compared with that from livers from normal rats also at 2 days after injury. Because of their greatly increased size the total amount of α_1-globulin synthesised by livers from the phenobarbitone-dieted rats was much larger.

Transferrin

In consequence of the phenobarbitone diet the livers were enlarged and the rates of synthesis of transferrin were faster. As shown in table 30.2, this rate when expressed per g of liver was 3 times higher than that given by the livers of normal rats. Due to the extra size of these livers the total rate of synthesis of transferrin was increased approximately 6 times above normal.

A greatly increased rate of transferrin synthesis was also found when livers from rats that had been injected with turpentine were used. This rate, approximately double that given by normal livers, was the same whether or not the rats had also been dieted with phenobarbitone before the injection with turpentine. As shown in table 30.2, increases in the rates of synthesis of transferrin also occurred when the livers from two rats which had been maintained on an Fe-deficient diet were tested. However this increase was less than that found after phenobarbitone dieting or injection with turpentine.

Conclusions from liver slice experiments

The greatly increased rates of synthesis of transferrin observed in the present experiments as a result of Fe-deficiency, phenobarbitone dieting and injection of turpentine must firstly be considered in relation to previous knowledge of the conditions known to stimulate the rate of synthesis of this protein. Because Fe deficiency has already been shown[12] to lead to increased rates of synthesis of transferrin, the present results are confirmatory. Furthermore, the finding of an increased rate of synthesis of transferrin after injection of turpentine is in agreement with the report of Mutschler and Gordon[14] who observed increased synthesis of this protein by a perfused liver from a rat at 12 h after laparotomy.

Using a liver slice system and incorporating ^{3}H-lysine, Tavill and Kershenobich[12] obtained a similar increase, to 152% of the normal, at 48 h after laparotomy. The observation of O'Shea *et al.*[16]—using the same means for estimating the synthesis rate—of no change in respect to transferrin, may be due to a less severe means of stimulation and estimation at an earlier time after stimulus (24 h). Thus, only the earlier report of Gordon and Darcy[17], who employed liver perfusion at 48 h after injection of turpentine, fails to agree with the many experiments which suggest that the rate of synthesis of transferrin, at least in the rat, is subject to a similar control mechanism to that responsible for the synthesis of the plasma proteins usually classified as APRs. It is of particular interest that this type of control can function in conjunction with controls responsive to quite different metabolic conditions.

If it is assumed that increased transferrin synthesis is indeed induced by injury and/or inflammation, an explanation becomes necessary for the reduced plasma concentration of the protein known to occur under these conditions, e.g. O'Shea *et al.*[16]. Fortunately, this is provided by the observation of Jarnum and Lassen[18] that transition to the diseased state is characterised by a more rapid rate of catabolism than of synthesis of transferrin.

The existence of several very different conditions all of which lead to increased synthesis of transferrin raises the question as to whether the rates of synthesis of more typical APRs are sensitive to metabolic conditions other than those directly due to trauma. Thus it seemed to be important to ascertain whether a treatment causing an increased rate of synthesis of transferrin (phenobarbitone diet) would also lead to an increased synthesis of α_1-globulin. The outcome of the present experiments in regard to the specificity of the systems responsible for control of the synthesis of α_1-globulin and transferrin is clear. Thus whereas turpentine injection led to increased synthesis of both proteins, dieting with phenobarbitone led only to a large increase in transferrin synthesis without any effect on synthesis of α_1-globulin.

Since more than one type of stimulus can lead to an increased rate of synthesis of transferrin, the question also arises as to whether a common pathway may be involved. That such may be true for turpentine injection and phenobarbitone dieting, both of which lead to increased transferrin synthesis, seems probable because the increases produced in these two different ways are not additive.

In view of the fact that Fe-deficiency has been shown[12] to lead to increased synthesis of transferrin, the possibility, originally envisaged by Tavill, cannot be ignored that phenobarbitone has the same effect because it brings about a reduction in the concentration of intracellular iron. In view of the increased amounts of cytochrome P. 450 synthesised in such livers this may be true. Evidently further information as to the effect of phenobarbitone on both the soluble and insoluble or storage Fe of the liver parenchymal cells will be valuable. However, because dieting with phenobarbitone for 50 days (phenobarbitone at a concentration of 1 g l^{-1} in the drinking water) as used in the present experiments has been found to lead to a doubling of plasma Fe, and an even greater increase (to 286% of normal) of plasma TIBC, any Fe deficiency must be intracellular only. If a localised Fe deficiency can be shown to exist and to stimulate transferrin synthesis the increased plasma Fe and transferrin concentrations found in this condition may then be considered to be the result of a compensatory mechanism.

Leucocytes as a source of stimulators of the acute phase response

The many attempts to test extracts of cells known to accumulate at injury sites to ascertain whether they show properties which would qualify them to be considered as stimulators of the acute phase response are difficult to interpret. Thus only in the absence of local inflammation at the injection site can the results of tests involving s.c. injection be considered to be positive. Also there is the possibility that the acute phase response may be brought about indirectly. If so, substances originating from leucocytes may act on the surrounding tissue to release stimulators which in turn could be carried to the liver. As will be described below efforts have been made to avoid such criticisms. Thus Darcy[19] who observed increased concentrations of α_1-globulin in rats after injection of extracts of leucocytes obtained from the same species was careful to record the presence or absence of local inflammation at the injection site. Eddington *et al.*[20] adopted a different approach by comparison of the acute phase responses obtained after intraperitoneal (i.p.) injection of leucocyte endogenous mediator (LEM) with that given by the supernatant from granulocytes which had been homogenised.

Before consideration of these results, the quantitative relationship between granuloma size and the acute phase response may be mentioned. In this regard Darcy[21] reported that the AP response is much greater per unit of stimulator substance in larger rats. A standard s.c. injection of 0.2 ml turpentine was used. In the larger rats both the granuloma size and the AP response were much larger, the granuloma size increasing most. A possible explanation for these findings would appear to be that because more leucocytes were available in the larger rats more of these cells accumulated at the lesion and thus a greater quantity of a stimulator substance was able to enter the blood stream. Because both the granuloma size and the AP response were found to be more than proportional to the size of the rats the likelihood is increased that the response is brought about by factors originating in the granuloma. In support of this proposition, there is evidence that when both granulocytes and macrophages are present in an injury site they secrete some part of their lysosomal content. Initiation of this process is believed to be due to interaction between auto-antibodies to tissue components and proteins which have been altered to some degree in the injury site. In rats the presence of such complement fixing autoantibodies has been demonstrated in the serum of normal adults[22]. Such autoantibodies, if complexed with denatured proteins, are likely to affect both granulocytes and macrophages. Thus in experiments with granulocytes from normal human plasma Weissmann *et al.*[23] were able to demonstrate the release *in vitro* of the enzymatic contents of the granules. Later Cardella *et al.*[24] using mouse peritoneal macrophages reported a similar phenomenon in respect to the lysosomal hydrolases.

Some support for a theory of this kind can be derived from the finding of Willoughby *et al.*[25] that complement levels are depressed after tissue damage. Also relevant is the existence of a strain of mice (Webster Smith BRVS), the Kupffer cells of which are specially sensitive to the presence of antigen–antibody complexes. In these mice at 5 mins after anaphylactic shock Santos-Buch and Treadwell[26] described a virtual disintegration of the Kupffer cells. At least in this strain a consequence of anaphylactic shock must be that the parenchymal cells of the liver are subjected to the influence of the contents of the Kupffer cell granules.

Effects produced by injection of extracts of leucocytes

The nature of the stimulatory substances which are obtainable from granulocytes present in peritoneal exudates of rats and rabbits have been studied in great detail. Although most attention has been given to the endogenous pyrogen present in such extracts, Eddington *et al.*[20] have successfully demonstrated that a typical acute phase reaction usually follows when extracts obtained from rabbit granulocytes are injected into rats. The effects of these extracts differ both according to whether the granulocytes have themselves been stimulated by the simultaneous presence in the peritoneum of a mild irritant such as glycogen or of a microorganism, according to whether they are obtained from rats or rabbits and also according to the species into which they are injected for testing. Not all possible combinations have been examined but certain facts have been established.

1. Extracts from rabbits if tested in rats lead to the following changes: decreased concentration[27] of plasma Fe and Zn; release of neutrophils from the bone marrow; shift of amino acids from muscle to liver and small increases in at least 3 APRs (α_2-AP-globulin, α_1-globulin and ceruloplasmin[28]). When injected into rats the endogenous pyrogen or some other constituent of these extracts leads to a slight decrease in body temperature.

2. Extracts from rats if tested in rats bring about most or possibly all of the above responses, at least if the granulocytes have been stimulated *in vivo*. Information based on measurement of plasma Fe concentration indicate that, by this criterion, extracts from rabbit and rat granulocytes are equally active.

Rat serum taken four hours after injection of *D. pneumoniae* also leads to a rapid fall of plasma Fe if injected i.p. into rats[8].

Since all of the metabolic changes mentioned above are known to occur during infections or as a consequence of trauma, the possibility that release of soluble substances from granulocytes may be an important initiating factor in the acute phase response deserves serious consideration.

Preliminary attempts to characterise the active substances released from rabbit granulocytes have indicated the presence of complex mixtures. Indeed, as shown by Kampschmidt *et al.*[9] most of the activities are associated with substances of rather similar properties. No definite separation of pyrogenity from the factors controlling the levels of plasma Fe and Zn have so far been achieved. Unfortunately, the most highly purified materials so far obtained[9] produced only slight increases in the concentration of plasma α_2-AP-globulin when injected into rats. Evidently, although injection of extracts derived from granulocytes and other leucocytes leads to metabolic consequences closely resembling those existing during the acute phase response, the central problem as to whether the response is thus mediated must remain in doubt. Furthermore, before the physiological significance of these findings can be properly evaluated, they must be considered in relation to the existence, at least occasionally, of an acute phase response in the absence of granulocytes. Since a few cases have been reported of almost complete depletion of these cells and at the same time a normal acute phase response, it is probable that initiation of an increased synthesis of the APR group of plasma proteins can be brought about in more than one way.

Physiological significance of APRs

As shown in table 30.1, certain of the functions *in vivo* of 13 of the APRs are now known. Furthermore, suggestions as to the possible function of certain of the remaining members of this essentially rather diverse group of plasma proteins, based on their biological activities *in vitro*, can also be made, e.g. for the anti-inflammatory factors of plasma, C-reactive protein and ceruloplasmin.

Antitrypsin and antichymotrypsin

Following surgery, the plasma concentration of antitrypsin more than doubles[29], and that of antichymotrypsin also increases[30]. Recent work has shown that milder forms of trauma including inhalation of cigarette smoke also causes statistically significant increases of antitrypsin. Thus Elson *et al.*[31], who examined 32 cigarette smokers, found the concentration of this APR to be nearly 20% higher than in a comparable group of non-smokers. These findings are of special interest because emphysema has been produced in dogs forced to smoke cigarettes by inhalation[32].

A relationship between smoking and emphysema has also been demonstrated in humans by Auerbach *et al.*[33]. Cases of inherited α_1-antitrypsin deficiency are especially at risk because the rate of synthesis of this APR fails to respond normally during inflammation and early emphysema often follows. In this situation lysosomal enzymes are released and, as a result of an inadequate level of α_1-antitrypsin, severe tissue damage ensues.

Complement components

Both C3 and C4 and the complex formed from $C\overline{56}$—known as activated reactor—must be recognised as APRs[30]. Evidence concerning increased plasma concentrations of C3 in several disease states has been provided by Hornung and Arquembourg[34], Alper[35], Kushner et al.[36] and Harveit et al.[37]. Although the increased concentration of C3 in disease is rather slight, (122% of the pre-operative value[29]) its physiological importance is undoubted. Thus the efficiency of many aspects of the immune system are tremendously enhanced. For example, antibody-mediated phagocytosis is greatly potentiated.

As shown by Thompson and Rowe[38], much larger increases in reactor activity may appear in post-operative serum. The phenomenon known as reactive haemolysis has been shown to require the presence of the complex of $C\overline{56}$ (activated reactor) and C7. Activated reactor is not present as such in serum but appears after certain treatments, e.g. Zymozan at 37° C for 1 hr. Thompson and Rowe[38] found that 19% of healthy subjects showed reactor activity and that after various forms of trauma and in severe disease states the proportion of reactors was much higher. Physiochemical data obtained by Lachmann and Thompson[43] show $C\overline{56}$ to be a complex of M.W. 360000 which is destroyed by heating at 56° C for 20 min.

Activated reactor is capable of complexing with C7 and it is only in this state with C5, C6 and C7 present in equimolecular proportions that the ability to lyse cells is finally achieved.

Since in inflammatory states and after trauma two of the components which under appropriate conditions unite to form the active complex increase in concentration in the plasma, they may, as already stated, be classified as APRs. However, it is necessary to add that formation of activated reactor does not always follow the stimuli which are effective in most individuals.

Anti-inflammatory factors

Evidence has been obtained by Billingham et al.[39,40] for the existence of proteins with anti-inflammatory activity in the plasma of rats suffering acute inflammation. The protein nature of this factor and the fact that it is synthesised in the liver have been established by Billingham et al.[39]. Preliminary evidence suggests that it belongs to the α_1-group of plasma proteins. The anti-inflammatory properties of this factor have been demonstrated by measurements of the paw volume of rats suffering from carrageenin oedema. More recently Van Gool et al.[41] have provided evidence that the α_2-AP-globulin of rats has the same property. The same test has been used by Persellin et al.[42] to demonstrate the presence of an anti-inflammatory factor in human pregnancy serum.

Conclusion

As it becomes possible to identify the physiological roles of additional members of the APR group of plasma proteins their association with the metabolic changes produced by tissue damage becomes clearer. However, much further work is required before any conclusion on the physiological importance of the plasma level of any of these proteins can be reached. Similarly, knowledge as to how the rates of synthesis of these proteins are controlled is as yet very slight.

References

1. Koj, A. (1975). Acute phase reactants. In *Plasma Proteins*, Allison, ed. Plenum, London, p. 73
2. Nasjletti, A. and Masson, G. M. C. Studies on angiotensinogen formation in a liver perfusion system. *Circulation Research*, **31**, Suppl. 2, (1972), 187
3. Borges, D. R. and Gordon, A. H. (1975). Kininogen and kininogenase synthesis by the livers of normal and injured rats. *J. Pharm. and Pharmacol.*, in press.
4. Sell, S. and Wepsic, H. T. (1974). In *The Liver, the Molecular Biology of its diseases*. F. Becker, ed., Marcel Dekker, New York
5. Gordon, A. H. (1970). The effects of trauma and partial hepactectomy on the rates of synthesis of plasma proteins by the liver. In *Plasma Protein Metabolism*, M. A. Rothschild and T. Waldmann, eds., Academic Press, New York, p. 351
6. Palmiter, R. D. Regulation of protein synthesis in chick oviduct. *J. Biol. Chem.*, **247** (1973), 6770
7. Thompson, W. L., Pekarek, R. S., Powanda, M. C. and Wannemaker, Jr. R. W. LEM induced alterations in hepatic RNA synthesis—possible regulatory mechanism for synthesis of acute phase serum globulins. *Fed. Proc.*, **33**(3), (1974), p. 696, Abstract 2747
8. Kampschmidt, R. F., Pulliam, L. A. and Upchurch, H. F. Sources of leucocytic endogenous mediator in the rat. *Pr. Soc. Exp. Biol. and Med.*, **144** (1973), 882
9. Kampschmidt, R. F., Upchurch, H. F., Eddington, C. L. and Pulliam, L. A. Multiple biological activities of a partially purified leucocytic endogenous mediator. *Am. J. Physiol.*, **224** (1973), 530
10. Kampschmidt, R. F. and Upchurch, H. F. Effect of leucocytic endogenous mediator on plasma fibrinogen and haptoglobin. *Proc. Soc. Exp. Biol. and Med.*, **146** (1974), 904
11. Tavill, A. S., East, A. G., Black, E. G., Nadkasni, D. and Hoffenberg, R. (1973). Regulatory factors in the synthesis of plasma proteins. In *CIBA Foundation Symposia Protein Turnover*, 9 (new series), G. E. W. Wolstenholme and M. O'Connor, eds., Elsevier, Amsterdam, p. 155
12. Tavill, A. S. and Kershenobich, D. (1972). Regulation of transferrin

synthesis. In *Protides of the Biological Fluids*, Proc. 19th Congress, H. Peeters, ed., p. 489

13. Loh, T. T. and Juggi, J. S. Tissue iron in acute and chronic liver damage from carbon tetrachloride. *Austr. J. Exp. Biol. and Med. Sci.*, **49** (1971), 493

14. Mutschler, L. E. and Gordon, A. H. Plasma protein synthesis by the isolated perfused regenerating rat liver. *Biochim. Biophys. Acta*, **130** (1966), 486

15. Mancini, G., Carbonara, A. O. and Heremans, J. F. Immunochemical quantitation of antigens by single radial immunodiffusion. *Immunochem.*, **2** (1965), 235

16. O'Shea, M. J., Kershenobich, D. and Tavill, A. S. Effects of inflammation on iron and transferrin metabolism. *Brit. J. Haem.*, **25** (1973), 707

17. Gordon, A. H. and Darcy, D. A. Production of α_1-globulins by the perfused rat liver. *Brit. J. Exp. Path.*, **48** (1967), 81

18. Jarnum, S. and Lassen, N. A. Albumin and transferrin metabolism in infectious and toxic diseases. *Scand J. Clin. Lab. Invest.*, **13** (1961), 357

19. Darcy, D. A. Polymorphonuclear cell fractions which stimulate increase of an acute phase protein in the rat. *Brit. J. Exp. Path.*, **49** (1968), 525

20. Eddington, C. L., Upchurch, H. F. and Kampschmidt, R. F. Effect of extracts from rabbit leucocytes on levels of acute phase globulins in rat serum. *Proc. Soc. Ex. Biol. and Med.*, **136** (1971), 159

21. Darcy, D. A. Granuloma weight and the α_1-acute phase protein response in rats injected with turpentine. *Brit. J. Exp. Path.*, **51** (1970), 59

22. Elson, C. J. and Weir, D. M. Development of anti-tissue antibodies in rats. *Clin. and Exp. Immunol.*, **4** (1969), 241

23. Weissmann, G., Zurier, R. B., Spieler, P. J. and Goldstein, I. M. Mechanisms of lysosomal enzyme release from leucocytes exposed to immune complexes and other particles. *J. Exp. Med.*, **134** (1971), 149A

24. Cardella, C. J., Davies, P. and Allison, A. C. Immune complexes induce selective release of lysosomal hydrolases from macrophages. *Nature*, **247** (1974), 46

25. Willoughby, D. A., Coote, E. and Turk, D. A. Complement in acute inflammation. *J. Path.*, **97** (1969), 295

26. Santos-Buch, C. A. and Treadwell, P. E. Disruption of Kupffer cells during systemic anaphylaxis in the mouse. *Am. J. Path.*, **51** (1967), 505

27. Kampschmidt, R. F. and Upchurch, H. F. Lowering of plasma iron concentration in the rat with leucocytic extracts. *Am. J. Physiol.*, **216** (1969), 1287

28. Pekarek, R. S., Powanda, M. C. and Wannemacher, Jr. R. W. The effect of leucocyte endogenous mediator (LEM) on serum copper and ceruloplasmin concentrations in the rat. *Pr. Soc. Exp. Biol. and Med.*, **141** (1972), 1029

29. Minchin Clarke, H. G. and Freeman, T. Quantitative immunoelectrophoresis of human serum proteins. *Clin. Sci.*, **35** (1968), 403

30. Ganrot, K. Plasma protein pattern in acute infectious disease. *Scand. J. Clin. Lab. Invest.*, **34** (1974), 75

31. Elson, L. A., Betts, T. E. and Darcy, D. A. α_1-antitrypsin in cigarette smokers. Supp. to *Excerpta Medica*. Characterisation of Human Tumours. *Int. Cong. Series*, **321** (1974), 151

32. Auerbach, O., Cuyler Hammond, E., Kirman, D. and Garfinkel, L. Emphysema produced in dogs by cigarette smoking. *J. Am. Med. Assoc.*, **199** (1967), 241

33. Auerbach, O., Cuyler Hammond, E., Garfinkel, L. Benante, C. Relation of smoking and age to emphysema. *New Eng. J. Med.*, **286** (1972), 853

34. Hornung, M. and Arquembourg, R. C. β_{1C}-globulin, an 'acute phase' serum reactant of human serum. *J. Immunol.*, **94** (1965), 307

35. Alper, C. A. (1970) Regulation and metabolism of the third component of complement (C3). In *Plasma Protein Metabolism*, M. A. Rothschild and T. Waldmann, eds., Academic Press, New York, p. 393

36. Kushner, I., Edgington, T. S., Trimble, C., Lieni, H. H. and Muller Eberhard, U. Plasma hemopexin homeostasis during the acute phase response. *J. Lab. Clin. Med.*, **80** (1972), 18

37. Hartveit, F., Børve, W. and Thunold, S. Serum complement levels and response to turpentine inflammation in mice. *Acta Path. Microbiol. Scand.*, Sect. A. Suppl. **236** (1973), 54

38. Thompson, R. A. and Rowe, D. S. Reactive haemolysis—a distinctive form of red cell lysis. *Immunology*, **14** (1968), 745

39. Billingham, M. E. J., Gordon, A. H. and Robinson, B. V. The role of the liver in inflammation. *Nature New Biology* **231** (1971), 26

40. Billingham, M. E. J. and Robinson, B. V. Separation of irritancy from the anti-inflammatory component of inflammatory exudate. *Br. J. Pharmac.*, **44** (1972), 317

41. Van Gool, J., Schreuder, J. and Ladiger, N. C. J. J. Inhibitory effect of foetal α_2-globulin, an acute phase protein, on carrageenin oedema in the rat. *J. Path.*, **112** (1974), 245

42. Persellin, R. H., Vance, S. E. and Peery, A. Effect of pregnancy serum on experimental inflammation. *Brit. J. Exp. Path.*, **55** (1974), 26

43. Lachman, P. J. and Thompson, R. A. Reactive lysis: the complement mediated lysis of unsensitised cells. *J. Exp. Med.*, **131** (1970), 643

31

Sequential changes in plasma proteins in various acute diseases

C.-O. KINDMARK

The increase in the erythrocyte sedimentation rate (ESR) has long been used as a measure of the intensity of the systemic effect of inflammation. As fibrinogen, the immunoglobulins and β-lipoprotein influence the height of the ESR, the clinician requires more specific information. Since the introduction of methods for determination of specific plasma proteins, attention has been focussed on the concentration of only one, or at most a few, of these proteins in various acute diseases. In Professor Laurell's laboratory in Malmö we have tested a battery of specific proteins in a series of diseases to study its potential clinical diagnostic value. A question that has interested us is whether the inflammatory response varies in intensity with the tissue affected. Because of the limited time available, I will confine myself to the proteins, which are the topic of today's session, i.e. the acute phase proteins. Here I include only those which increase rapidly in acute conditions, i.e. C-reactive protein (CRP), antichymotrypsin, α_1-antitrypsin (α_1-AT), orosomucoid, haptoglobin and fibrinogen.

With but one exception, haptoglobin, which was measured as a peroxidase activity, the plasma proteins studied were generally determined immunochemically (by electroimmunoprecipitation assay), which permits determination of plasma proteins in concentrations down to 0.6 mg l^{-1}. CRP within the normal range was determined by radioelectroimmunoassay, which permits determination down to 0.01 mg l^{-1}.

The following conditions were studied: surgical trauma, myocardial infarction, hepatitis A and B, peritonsillitis, influenza A, serous meningoencephalitis.

Surgical trauma has attracted special attention because it is a standardised injury of varying intensity and affects one or more organs. We studied cholecystectomy and cholecystectomy plus cholodocholithectomy in a number of patients sufficient to warrant conclusions. Figure 31.1 shows the response of the 'classical' acute phase reactants. As you see, a substantial increase occurred within 24 hours and reached a maximum after 3 days. To obtain better infor-

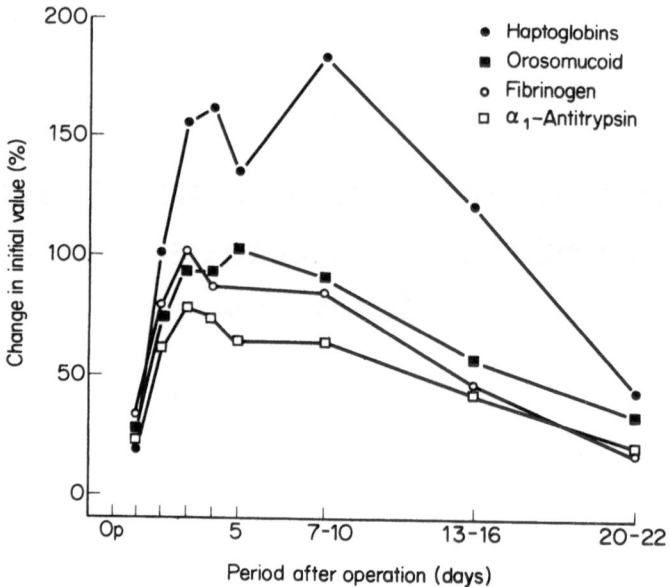

Figure 31.1. Changes in plasma concentration of haptoglobins, orosomucoid, fibrinogen and α_1-AT during the first 3 weeks after cholecystectomy.

mation about the course of events during the initial postoperative period, blood was obtained from 6 patients at 4 hour intervals on the first postoperative day. The results are given in figure 31.2. Gravimetric values were used for C-reactive protein, which shows a much larger relative increase than the others. They are presented as percentage deviation from the initial level. The chronological order of the changes in the protein pattern was the same in all the subjects. From the fourth hour antichymotrypsin and C-reactive protein increased rapidly; the α_1-AT, orosomucoid and haptoglobin not before 12 to 20 hours.

Of the acute phase reactants, C-reactive protein and antichymotrypsin deserve special attention because their reaction is recognisable in plasma already within 8 hours. The biological function of antichymotrypsin is not known and its name is misleading since other protease inhibitors of plasma have a higher affinity for chymotrypsin. The prompt increase of CRP is of special interest in the light of its proposed function as a phagocytosis-promoting factor.

Haptoglobins, fibrinogen, orosomucoid and α_1-AT increase by roughly the same relative amount and at the same relative rate. The relative increase in concentration of the haptoglobins is larger than that of the others, but this seems to mirror the same relative increase in rates of synthesis, as there is no reason to assume that the haptoglobin drain, through ineffective erythropoiesis, is changed by the inflammation. The responses of these four proteins

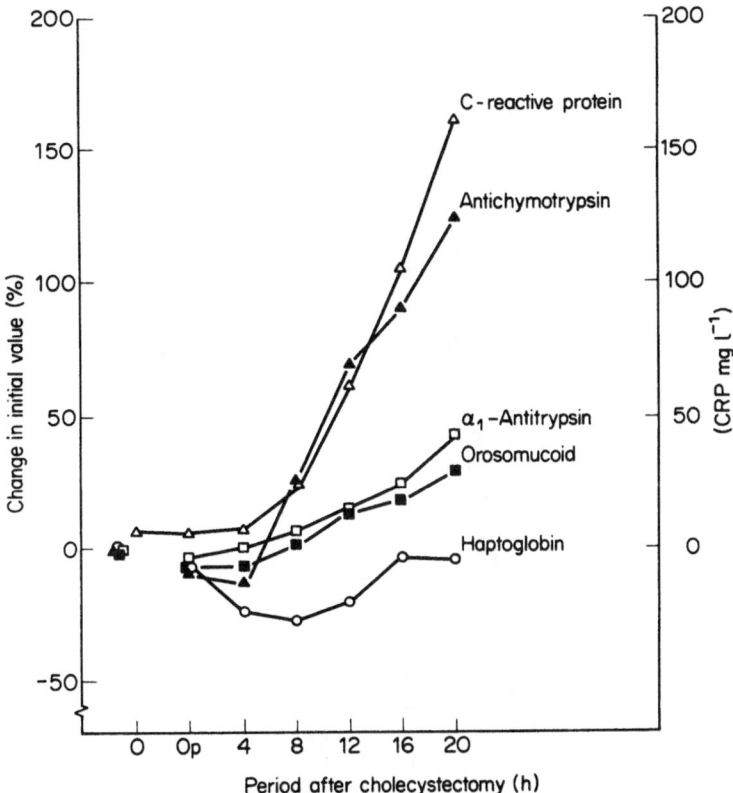

Figure 31.2. The acute phase reactants during the first day after cholecystectomy. The results are given as means from blood samples of 6 patients taken every fourth hour after the operation. C-reactive protein is given as mgl^{-1} while the percentual deviation from the initial level has been used for the other variables.

seem to be so similar that it is tempting to assume that the increased synthesis is stimulated by some common trigger. The leucocytosis reached its maximum within 24 hours and preceded the plasma protein shift. On the other hand, these cells did not seem to be responsible for the increased synthesis, as the same pattern of inflammatory response is seen in patients with agranulocytosis[1].

In the study of myocardial infarction the most prominent changes were shown by haptoglobins, followed in order and magnitude by antichymotrypsin, fibrinogen, orosomucoid and α_1-AT (figure 31.3). No initial decrease was seen. C-reactive protein also followed this general pattern although it increased more rapidly and, furthermore, the mean relative increase was some 10 times larger (figure 31.4). This does not imply that these changes are related to similar biological functions or handled by the same mechanisms, but it is

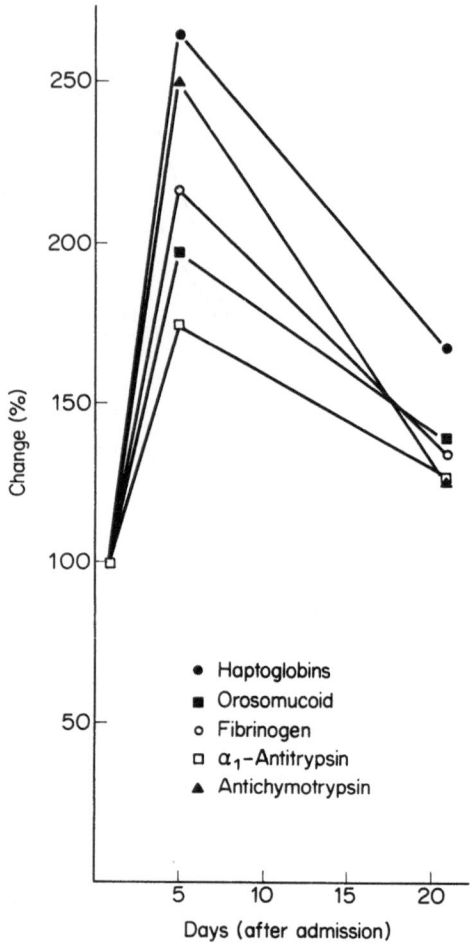

Figure 31.3. Mean changes in protein levels expressed as a percentage of levels obtained on admission (day 1).

conceivable that better knowledge of reaction patterns of a large number of individual proteins in well defined clinical situations, such as myocardial infarction, might produce evidence of biological interest.

The cause of the protein changes is an intriguing problem. Increased biosynthesis of certain proteins, mainly in the liver, and an increased capillary permeability most probably play an important role in the cause of the changes. Despite the lack of confirmatory evidence it is tempting to assume that lysosomal proteins released from neutrophil leucocytes might play a role.

Apart from studies of ceruloplasmin, haptoglobin and β_{1C}-globulin studied separately in acute infections and a study of infectious hepatitis, which I will return to, there appears to be no available survey of the plasma protein pattern

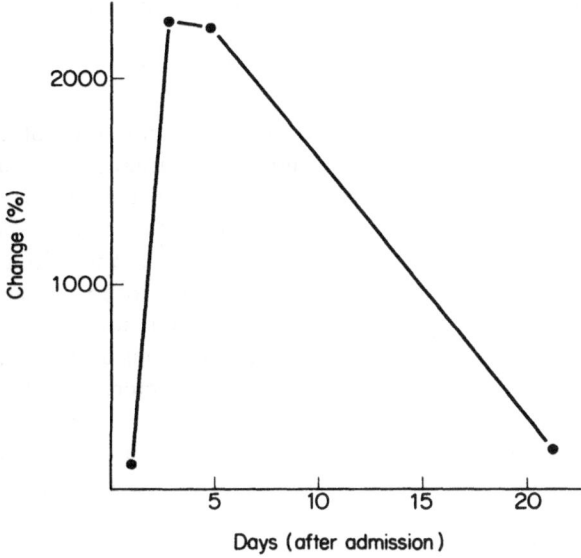

Figure 31.4. Changes in concentration of C-reactive protein expressed as a percentage of mean levels on admission (day 1).

in acute infectious diseases. To shed light on the plasma protein response, we chose[2] 3 diseases, *viz* peritonsillitis, serous meningoencephalitis and influenza A. The concentrations of haptoglobin, orosomucoid, fibrinogen, antichymotrypsin, α_1-AT and C-reactive protein on the fourth day of disease were compared with the values found after recovery of the individual. In peritonsillitis and meningoencephalitis the pattern was largely the same and showed a substantial increase in all the proteins examined. The levels found in influenza A were compatible with those in meningoencephalitis. This holds also for CRP, which was about 3 times as high in influenza A as in meningoencephalitis.

These proteins changed much more in peritonsillitis than in either of the other two diseases. This was rather remarkable since the general condition of the patients with peritonsillitis is, if anything, better than that in the other two diseases. The difference in strength of the reaction of the acute phase proteins might perhaps be ascribed to the difference in the cause of the disease, i.e. bacterial or viral. The strong reaction in peritonsillitis might, perhaps, also be due to a more extensive local distruction of the tissue.

In these three diseases, then, the reaction patterns were largely uniform except in their intensity. Thus CRP was increased in all three diseases, but most in peritonsillitis. The differences in level of the concentrations of the acute phase proteins in the three diseases were not large enough to be of any differential diagnostic value in a given case. Most investigators have failed to demon-

strate CRP in viral diseases, probably because the methods used were not
sensitive enough.

So far the pattern of the acute phase reactants in the acute stage of various
inflammatory conditions, induced or not, as well as in acute bacterial or viral
diseases has been rather uniform, but the day-to-day study of samples sent to
the laboratory for investigation has shown that inflammation affecting some
organs of the body produce a particular plasma protein pattern of differential
diagnostic importance. This is analogous with alterations in the plasma protein homeostasis caused by some hormones. One example of this is the sequential changes in the plasma protein pattern seen in 23 consecutive cases of
inoculation hepatitis cared for at the department for infectious diseases in
Malmö. 90% of these patients admitted intravenous abuse of drugs with the
use of non-sterile equipment. The day when the icterus was first noticed was
taken as the day of onset of the illness.

Judging from the values found for serum bilirubin and aminotransferases,
the intensity of the liver disease varied from mild to severe during this study,
and this made the material less suitable for statistical treatment. Individual
observations were therefore plotted as well as mean values. The relation
between plasma protein values in individual patients resembled the mean
curve, except in the few cases with a relapse.

The acute inflammatory response in most febrile conditions, i.e. increase in
the acute phase reactants, is missing in hepatitis B. An α_1-AT increase was
demonstrable from the onset and it persisted for one month. Two exceptions
were heterozygotes for α_1-AT deficiency. The highest α_1-AT values were
found in sera from women using oral contraceptives.

The orosomucoid values crowded around the normal mean throughout the
course of the disease. On the other hand, the haptoglobin concentrations were
low or normal at the beginning of the disease and generally remained fairly
constant during the first month. The CRP determinations were occasionally
above the upper limit of the normal range in the first month. The mean curve
for antichymotrypsin persisted within the normal limits. The mean fibrinogen
values varied but little during the disease.

In hepatitis B, then, the pattern of the inflammatory response is characterised by an increase in α_1-AT and a decrease of haptoglobin and a normal
orosomucoid level. The regularity of this pattern is evident from individual
cases plotted in figure 31.5, in which scales were so chosen that the values for
the three proteins of healthy individuals—if connected—will give vertical lines
around 100, and those for patients with bacterial diseases, vertical lines displaced to the right. Judging from our experience, lines leaning to the right
are characteristic of this type of hepatitis. This pattern may be explained by
increased α_1-AT synthesis and a normal synthesis of haptoglobin and orosomucoid. The last two usually accompany each other closely if the red cell turnover is normal. The tendency of the haptoglobins to decrease in hepatitis may
be ascribed to the retarded flow through the splanchnichus, with consequent

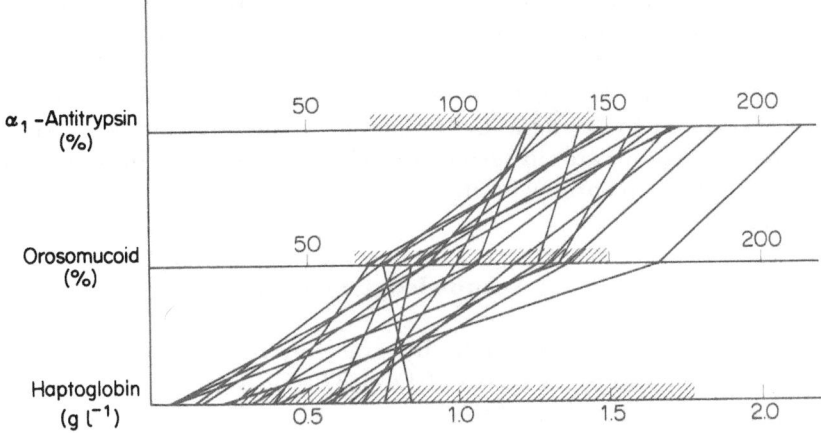

Figure 31.5. Relation between individual values of α_1-AT, orosomucoid and haptoglobin during the first week of inoculation hepatitis.

reduction of red cell survival. This relation between the levels for these three acute phase reactants appears to persist during the icteric phase of hepatitis B, but the absolute levels vary with the intensity of the response. This constellation is also regularly seen during spells of jaundice in patients with cirrhosis, and contrasts with the findings in cholecystitis and biliary occlusion, where the conventional inflammatory response is most obvious during febrile periods. Owing to the increased α_1-AT values, already at the onset of jaundice the pattern resembles that found in pregnancy and hepatosis induced by oestrogen-containing contraceptive pills. These findings support the view that the homeostasis of α_1-AT and the pair haptoglobin and orosomucoid are regulated by independent mechanisms.

On epidemiological and clinical grounds it is obvious that the two types of hepatitis, A and B, are different clinical entities. In Sweden an epidemic of hepatitis A is not very common. Nevertheless, a few months ago an epidemic broke out in a large factory on the west coast of Sweden and we were fortunate to receive blood samples from these patients. The whole battery of plasma proteins has not yet been investigated, but it is already evident that in this study of hepatitis A we did not find the pattern of changes in plasma proteins typical of hepatitis B and the icteric phase of cirrhosis. Thus, in hepatitis A we found a simultaneous increase in α_1-AT, orosomucoid and haptoglobin. In other words, in this type of viral hepatitis the acute phase reactions show the same harmonic increase as in other febrile conditions.

Although it is beyond the scope of my paper, I would like to add that we have also found substantial differences in immunoglobulin level between hepatitis A and B, as might be expected from the literature dealing with this subject. Thus the IgM is much higher in hepatitis A than in hepatitis B. This is also true for IgG, though the difference is not so large.

Summarising, I feel that the results of our investigations lend support to the hypothesis that the concentrations of the acute phase reactants, orosomucoid and haptoglobin, are regulated by some common mechanism: α_1-AT does not belong to the same group. In acute inflammatory conditions and in some infections caused by bacteria and viruses the qualitative alterations in the acute phase reactants are changed in much the same way. On the other hand, the reaction pattern varies with the site of the inflammation.

Acknowledgement

This investigation was supported by grants from the Swedish Medical Research Council (Project Nr B75-03X-4137-03), Svenska Läkarsällskapet fonder and Alfred Osterlunds Stiftelse.

References

1. Kindmark, C.-O., Möller, H. and Neumann, E. Ultraviolet light inflammation and C-reactive protein in a case of agranulocytosis with erysipelas. *Acta Dermatovener.*, (Stockholm), **51** (1971), 210
2. Ganrot, K. (1974). On plasma proteins in inflammation, University of Malmö (Thesis)

Discussion

JOHANSSON

Many proteins in plasma have been designated acute phase reactants. What is the reason for your picking out only C-reactive protein, antichymotrypsin, α_1-antitrypsin, orosomucoid and fibrinogen?

KINDMARK

Your question is very relevant. Among the so called acute phase proteins only a few have been included in my discussion and different authors define them in different ways. My opinion is that only those proteins that increase rapidly after the onset of inflammation should be called acute phase proteins. Nearly all proteins in plasma are affected by inflammation to some extent. Some proteins increase very little and others decrease. I cannot see any basis for including any more proteins among the acute phase reactants than those mentioned, especially from a clinical point of view.

32
Albumin turnover in burns and trauma

J. W. L. DAVIES

Some of the complex alterations of normal metabolism resulting from injury to skin by burning, or to the skeleton and surrounding tissues by fractures have been studied with iodine labelled serum albumin[1-7]. While the unstable metabolic state resulting from injury limits the deductions which can be made even from detailed studies, it has proved possible to confirm the beneficial effects of a warmer than usual (30–32°C) environmental temperature on the metabolic responses of patients with a variety of injuries[1,8-13].

In this review observations made on 24 adult patients with burns of widely ranging severity and 15 adult patients with fractures are used to indicate the problems involved in the studies. Some of them are technical, others are inherent in studying an unstable metabolic state. In spite of the problems, valid indicators are available for at least some of the complex metabolic changes which affect the turnover of albumin.

Materials and methods of analysis

Albumin used in all studies was obtained from liquid plasma taken from donors who appeared to be free of hepatitis. The albumin solution was isolated and purified by passage through a column of G200 Sephadex prior to labelling with [131]iodine or [125]iodine using the iodine monochloride method[14]. The albumin solution was labelled with an average of less than one iodine atom per protein molecule. Extreme care was taken in the preparation of the labelled albumin solution to prevent denaturation of the protein since a pilot study has shown that the rate of catabolism of labelled albumin in injured patients is similar to that which would be observed after giving a partly denatured albumin preparation to a normal individual. At the time of injection the labelled solution contained less than 1% of the total radioactivity as free iodide. All patients received sufficient non-radioactive iodide to limit thyroid uptake of liberated radioactive iodide.

In the analysis of the *in vivo* behaviour of each batch of injected labelled albumin the whole body content of radioactivity of each patient was measured at daily intervals using a sensitive whole body counter incorporating two large (15 cm diameter) scanning scintillation crystals enclosed in a steel room with lead lined walls[15]. Urine was collected over 24-hour periods and blood samples were taken at daily intervals. A multichannel analyser and sensitive scintillation counter was used to assay unit volumes of both urine and plasma for the content of ^{131}I or ^{125}I or a mixture of both isotopes. Apparently pure albumin was isolated from aliquots of plasma using 1% trichloracetic acid in 94% ethanol as previously described[9]. From the radioactive iodine content of the isolated albumin it was possible to calculate values for the specific radioactivity of the protein.

The plasma volume was estimated at the start and end of the studies with one iodine isotope and often about half way through the study with the other iodine isotope. The haematocrit of each blood sample was measured and the plasma volume calculated assuming a constant blood volume. These calculated plasma volumes were interpolated between the measured volumes for the daily estimate of the plasma albumin pool. The plasma albumin concentration was determined each day by electrophoresis of plasma on cellulose acetate, followed by elution of the appropriate stained fraction and colorimetric estimation of the dye concentration.

The distribution of albumin between the intra- and extravascular spaces was determined from the daily estimates of intravascular radioactivity (counts per ml plasma × plasma volume) and the whole body content of radioactivity measured directly with the whole body radioactivity counter. The extravascular content of radioactivity was calculated by subtraction of the daily total plasma content of radioactivity from the appropriate estimate of total body content.

The intravascular albumin pool was calculated from the plasma albumin concentration and the estimates of plasma volume calculated from the observed haematocrit values.

The absolute catabolic rate was calculated from the specific radioactivity of each sample of pure albumin isolated from each daily plasma sample divided into the appropriate 24-hour urine content of radioactivity. The specific radioactivity measurements were made at the midpoint of each urine collection period. This method of calculation of the absolute catabolic rate gave more consistent values than use of the electrophoretic albumin concentration and the counts per ml plasma (to give an indirect estimate of specific radioactivity) and the urine content of radioactivity.

The fractional catabolic rate was calculated from the 24-hour urine radioactivity content divided by the calculated total plasma content of radioactivity at the midpoint of the urine collection period.

Particular care was taken in designing the clinical studies so that each preparation of labelled albumin was tested in a patient before it was used to indicate the response to surgical injury. Aliquots of the various albumin preparations which had been shown to have a satisfactory *in vivo* behaviour were then used in patients with more severe accidental injury.

The behaviour of satisfactory labelled albumin preparations in patients with minor surgical injuries

Two paraplegic patients were admitted to hospital for the excision and repair with tissue grafts, of extensive areas of necrotic tissue. Albumin labelled with [125]Iodine was injected between 7 and 10 days before the excision and grafting operation. The results before and after the operative repair are shown in

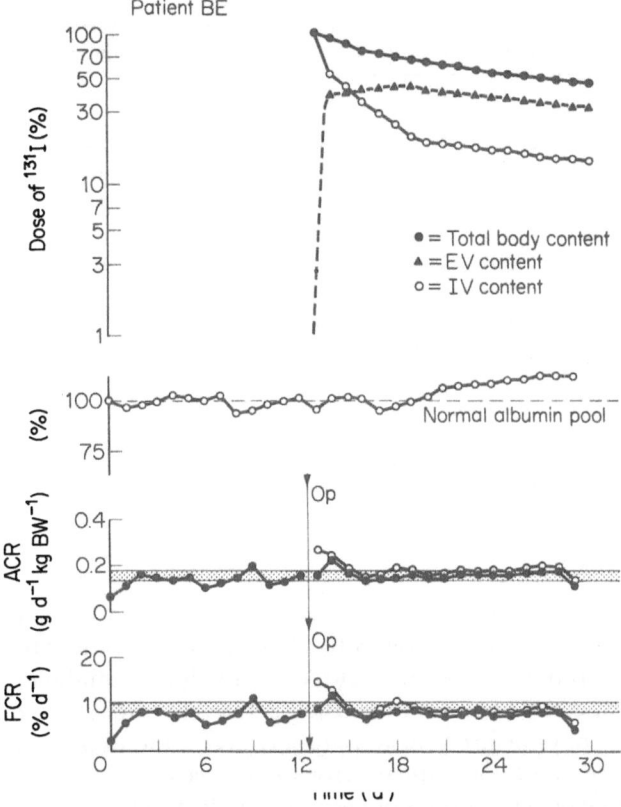

Figure 32.1. A paraplegic patient admitted for repair of decubitus ulcers covering 1–2% of the body surface. A.C.R. = absolute catabolic rate (g day^{-1} kg body weight^{-1}) F.C.R. = fractional catabolic rate (% plasma pool day^{-1}) Catabolic rates measured with ^{131}I albumin (○) and ^{125}I albumin (●). The shaded bands denote the expected normal range of values for the fractional and absolute catabolic rates. Op. = time of operation. EV = extravascular content of radioactivity. IV = intravascular content of radioactivity.

figure 32.1 for one of the patients. The same batch of albumin labelled with
[131]Iodine was used for the postoperative injection. Preoperatively, both the
fractional and absolute catabolic rates were very close to the range of rates
observed in normal individuals. After the operation there was a transient
increase in both catabolic rates lasting about 24 hours. During the postoperative period the two different labelled albumin preparations showed
very similar changes in both the absolute and fractional catabolic rates. The
plasma albumin pool remained close to the expected normal value. The total
amounts of radioactivity in the whole body, in the extravascular tissues and in
plasma decreased at abnormal rates (table 32.1). There appears to be a third
exponential rate of decrease added to the normal pattern. This third rate may
represent the combined effects of a transiently increased rate of catabolism and
slower mixing in the extravascular pool.

Table 32.1. Biological half-lives of radioactivity in the whole body, the plasma and the
extravascular tissues of patients with non-burn injuries

Patient	Whole body		Plasma		Extravascular	
	period (day)	half-life (day)	period (day)	half-life (day)	period (day)	half-life (day)
BE	—	—	0–1	0.7	—	—
	1–6	1.6	1–6	1.5	—	—
	6–20	22.0	6–20	18.0	6–20	24.0
R.LI	—	—	0–1	0.7	—	—
	1–7	3.4	1–7	1.4	—	—
	7–20	16.0	7–20	20.0	3–20	14.5
NI	—	—	0–1	0.7	—	—
	1–5	1.5	1–8	1.0	—	—
	5–20	15.5	8–20	14.5	2–20	17.5
GR 1st study	0–2	0.8	0–2	0.8	—	—
	2–15	14.0	2–15	9.7	3–15	19.0
GR 2nd study	0–3	1.2	0–3	0.9	—	—
	3–20	12.5	3–20	13.5	3–20	12.6

The results in the second paraplegic patient are shown in figure 32.2. The
vascular albumin pool was consistently about 20% below the expected normal
value. Preoperatively, the absolute and fractional catabolic rates were
within the expected normal range. Postoperatively, they showed a transient
increase for 2 days before returning to just above the range of preoperative
values. The amounts of radioactivity in the various compartments again
showed multiexponential rates of decrease (table 32.1).

The behaviour of satisfactory labelled albumin preparations in patients with more severe injuries

Two other paraplegic patients had fractures of the tibia and fibula as well as

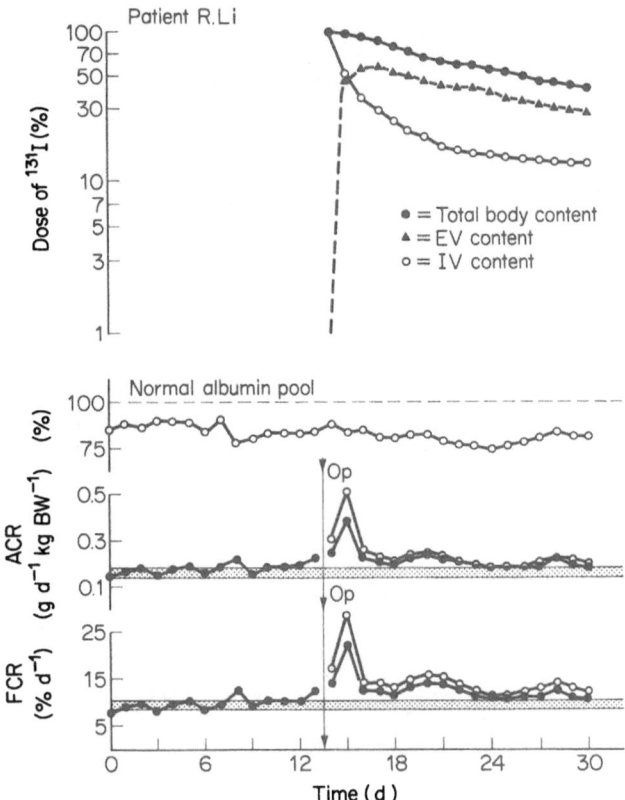

Figure 32.2. A paraplegic patient admitted for repair of decubitus ulcers covering 3–4% of the body surface. Other data as in figure 32.1.

their spinal injury. After about a week in hospital to stabilise their clinical condition the fractured tibiae were repaired by compression plating. Albumin turnover studies were started soon after admission to hospital with [125]Iodine albumin and continued after reparative surgery using [131]Iodine albumin. The results in one of the two patients are shown in figure 32.3. Again both the fractional and absolute catabolic rates observed during the preoperative period were close to the expected normal range. In contrast, postoperatively both labelled albumin preparations (which were different from those used in either of the patients described in figures 32.1 and 32.2) showed raised fractional and absolute catabolic rates persisting for at least two weeks after the operation. The plasma albumin pool was essentially constant during the preoperative period; it decreased for a 2-day period immediately after operation and then returned to values similar to those found during the preoperative period. Note again the different rates of disappearance of radioactivity from the total body, the extravascular space and the plasma (table 32.1).

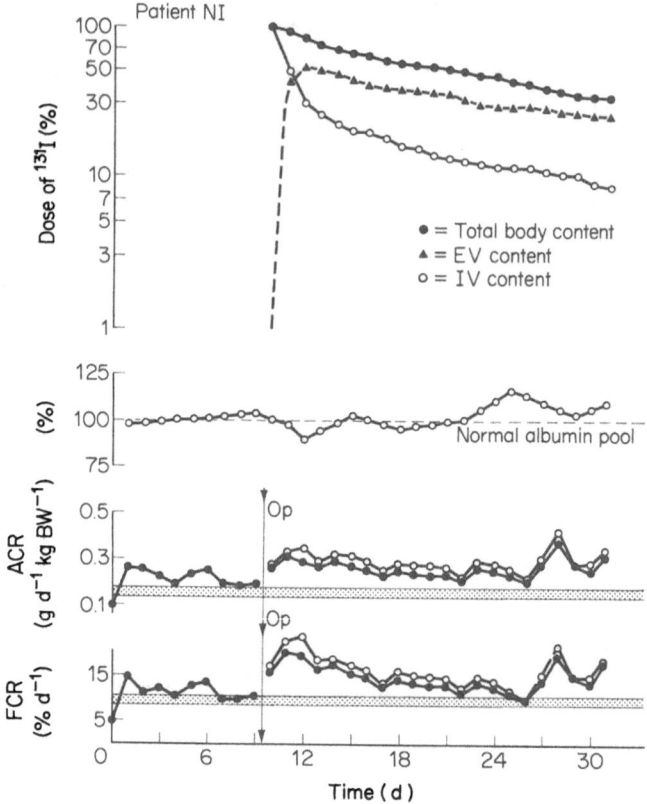

Figure 32.3. A paraplegic patient admitted for stabilisation of clinical condition and subsequent repair of the fractured tibia and fibula by compression plating. Other data as in figure 32.1.

The pattern of changes observed in non-paralysed injured patients may be seen in figure 32.4. This patient had chronic non-union of a fractured tibia and fibula which could only be treated with a bone graft and arthrodesis of the knee joint. Preoperative studies with ^{131}Iodine albumin showed a normal fractional catabolic rate and a slightly elevated absolute catabolic rate. During this time the plasma albumin pool was steady and about 15% above normal. Postoperatively, after a further injection of the same batch of labelled albumin, there was a threefold increase in both the fractional and absolute catabolic rates lasting about 5 days. The intravascular albumin pool dipped by 20% for a few days postoperatively and then returned to the preoperative values. Multiexponential rates of change of the content of radioactivity in the total body, the extravascular and plasma pools were again observed (table 32.1).

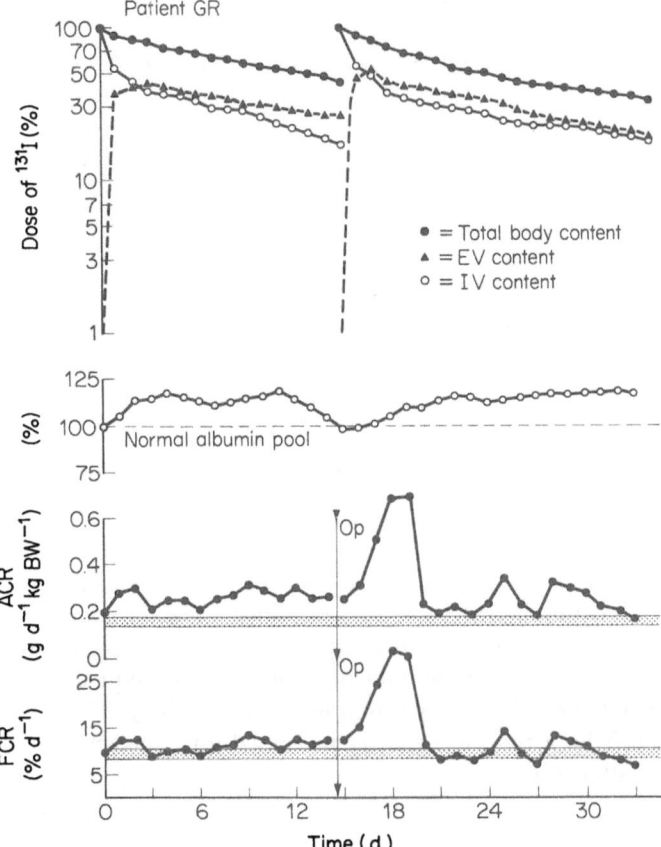

Figure 32.4. A patient admitted for bone grafting and arthrodesis of chronic non-united fractures of tibia and fibula. Other data as in figure 32.1.

The behaviour of satisfactory labelled albumin preparations in patients with severe accidental injury

The labelled albumin was given about 24 hours after injury to these patients who sustained either major fractures of the pelvis, femur or fractures of two or more long bones. The effect of injury alone on the metabolism of labelled albumin was determined in two patients with fractures of the pelvis who were treated without any operative repair of the fractures. It was not possible to make whole body radioactivity measurements in either of these patients or in those with more extensive multiple injuries described below.

Both patients with pelvic fractures (see an example of the results in figure 32.5), showed subnormal plasma albumin pools, and raised fractional and

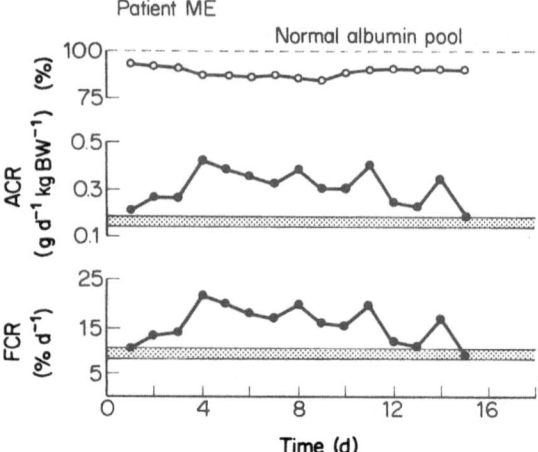

Figure 32.5. Changes in the plasma albumin pool and the absolute and fractional catabolic rates of albumin in a patient with a fractured pelvis. Other data as in figure 32.1.

absolute catabolic rates of albumin during the first 10 days following injury. Twice normal catabolic rates were observed for a few days after injury.

The observations in figure 32.6 are typical of those observed in three patients with multiple fractures. The plasma albumin pool was reduced to about 75% of the expected normal value within a few days of injury. The fractional and absolute catabolic rates were greatly increased and persisted for between 7 and 14 days after injury.

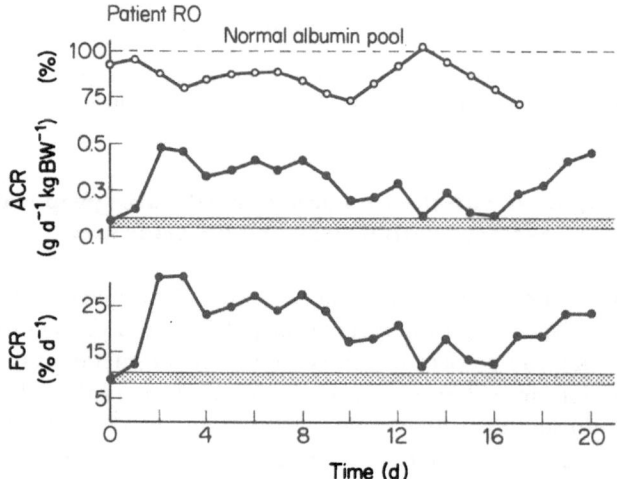

Figure 32.6. Changes in the plasma albumin pool and the absolute and fractional catabolic rates of albumin in a patient with multiple fractures. Other data as in figure 32.1.

The patient with the fractured femur also had a vascular lesion which made it necessary to amputate the injured limb late on the fourth day (figure 32.7). He had a very high catabolic rate of albumin prior to the mid-thigh amputation. By the end of the day after operation the four times increased rate of catabolism was reduced to a near normal value. The 30% reduction in the plasma albumin pool observed preoperatively showed little further change after the operation had been completed.

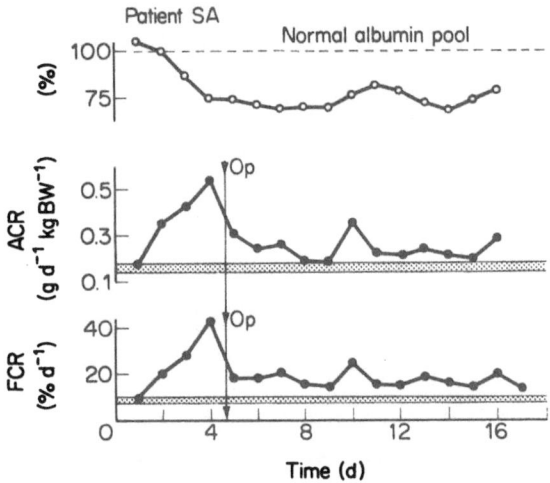

Figure 32.7. Changes in the plasma albumin pool and the absolute and fractional catabolic rates of albumin in a patient with a badly fractured femur with vascular lesions which required mid-thigh amputation on the 4th day after injury. Other data as in figure 32.1.

This obvious trend of increasingly abnormal rates of catabolism as the severity of injury increases is also observed in patients with burns. The following four patients illustrate this direct relationship.

A patient with a partial thickness skin loss burn covering 25% of the body surface

In this patient all the burned area healed without the need for skin grafting. As may be seen in figure 32.8, the fractional and absolute catabolic rates were within the expected normal range during the 3-week period of study. During this time the vascular albumin pool increased from normal to about 20% above the expected normal value. The radioactivity in the plasma, the extravascular tissues and in the whole body appeared to have exponential rates of decrease during the second and third weeks after burning (table 32.2). The raised extravascular content of albumin will be discussed more fully below.

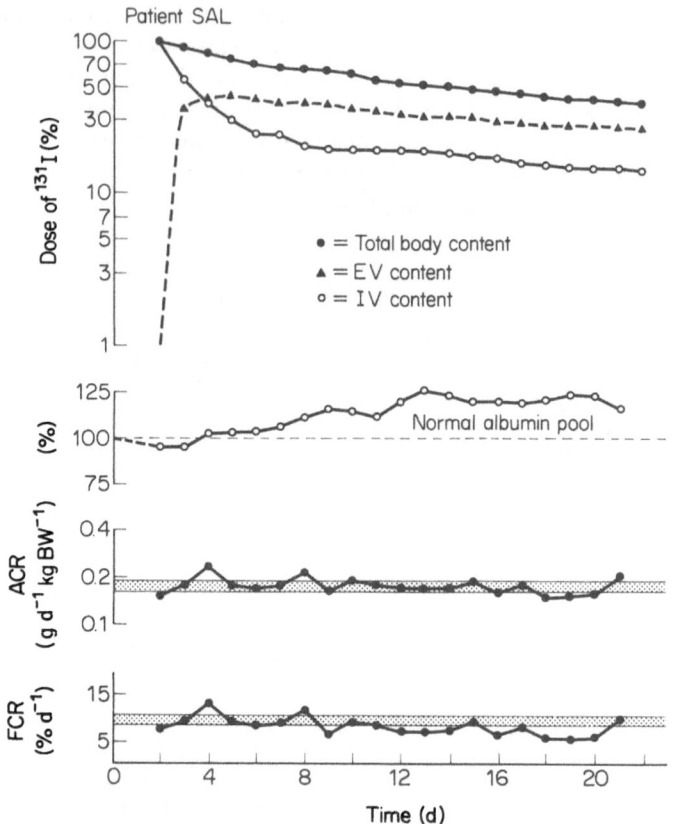

Figure 32.8. Changes in the distribution of injected radioactivity, the plasma
albumin pool and the absolute and fractional catabolic rates of albumin in a
patient with partial thickness skin loss burns covering 25% of the body surface.
None of the burn was full thickness skin loss. The patient was treated in a warm
(32°C) environment. Other data as in figure 32.1.

A patient with a full thickness skin loss burn also covering about 25% of the body surface

As may be seen in figure 32.9, this patient showed a much more abnormal
pattern of results during the whole period of study. Almost none of the burned
area healed during this period. The fractional catabolic rates were more than
twice the normal value during the first week after burning and were still
abnormally high during the second and third weeks. The absolute catabolic
rates were only abnormally high during the first week after burning, during
which time the plasma albumin pool was only about 60% of normal. During
the second and third weeks after burning the plasma albumin pool slowly
increased to 80% of the expected normal value. Measurements of the distribution of radioactivity between the plasma and extravascular tissues revealed a

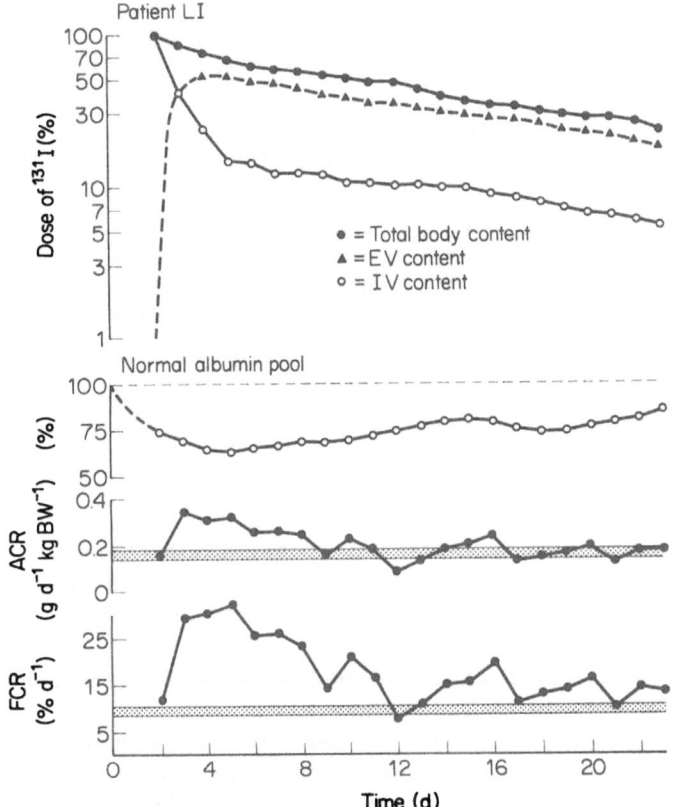

Figure 32.9. Changes in the distribution of injected radioactivity, the plasma albumin pool and the absolute and fractional catabolic rates of albumin in a patient with a full thickness skin loss burn covering about 25% of the body surface treated in the warm environment. Other data as in figure 32.1.

marked increase in the extravascular content. The various rates of disappearance of radioactivity from the whole body, the plasma and the extravascular tissues are shown in table 32.2.

A patient with a 50% burn, half the area of which was full thickness skin loss

These severe burns caused a more abnormal pattern of results than observed in any of the patients with less extensive burns or those with fractures. As may be seen in figure 32.10, the absolute and fractional catabolic rates were five times normal during the first week, decreasing to between 2 and 3 times normal during the second week after burning. The plasma albumin pool decreased to about 55% of normal soon after burning and then slowly increased to about 75% of normal by the end of the third week. The amount of radioactivity in

Table 32.2. Biological half-lives of radioactivity in the whole body, the plasma and
the extravascular tissues of patients with burns

Patient	Whole body		Plasma		Extravascular	
	period (day)	half-life (day)	period (day)	half-life (day)	period (day)	half-life (day)
SAL	—	—	0–1	0.7	—	—
	0–3	1.2	1–5	1.0	—	—
	3–20	16.5	5–20	19.5	5–20	16.0
LI	0–3	1.3	0–3	0.75	—	—
	3–20	10.5	3–20	10.5	3–20	9.8
LJ	0–1	0.4	0–1	0.25	—	—
	1–8	2.5	1–3	1.7	2–8	2.4
	8–20	7.3	3–20	10.0	8–20	6.7
ER	0–3	0.75	0–3	0.4	—	—
	3–10	3.7	3–20	4.3	3–10	3.7
	10–20	12.5			10–20	13.0

the extravascular tissues appeared to be many times that found in the plasma.
The various rates of disappearance of radioactivity from the whole body, the
plasma and the extravascular tissues are shown in table 32.2.

**A patient with full thickness skin loss burns covering 80% of the body
surface**
The results in this patient were the most abnormal ever measured (figure
32.11). The fractional and absolute catabolic rates were at least five times the
expected normal values. The plasma albumin pool decreased to about 50%
of normal and the distribution of radioactivity between the extravascular and
intravascular tissues was so abnormal that there appeared by the end of the
study to be at least ten times as much labelled albumin outside the plasma as
in it. The various rates of disappearance of radioactivity from the whole body,
the plasma and the extravascular tissues are shown in table 32.2.

Uses of a whole body radioactivity counter[16]

The whole body counter has improved the accuracy with which the exudate
losses of albumin from the tissues of burned patients can be estimated. In an
earlier study[3] losses of albumin in exudate from burned tissues were determined by analysis of the radioactive iodine content of dressings and soiled
bed linen. The availability of the whole body counter made it unnecessary to
collect dressings and soiled bed linen since the difference between the rate of
loss of radioactivity from the whole body—measured immediately after
removing soiled dressings—and that measured in complete 24-hour urine
samples indicated the non-renal losses. In the absence of diarrhoea these

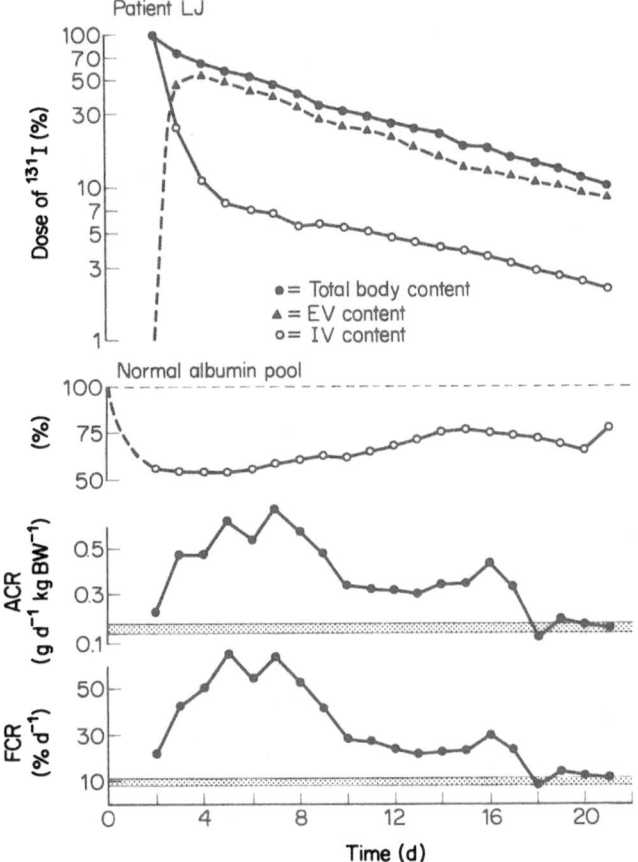

Figure 32.10. Changes in the distribution of injected radioactivity, the plasma albumin pool and the absolute and fractional catabolic rates of albumin in a patient with a total burned area covering 50% of the body surface. Half of the burned area was full thickness skin loss burn. The patient was treated in the warm environment. Other data as in figure 32.1.

non-renal losses were assumed to be in exudate. The magnitude of these exudate losses in patients with burns of differing severity are shown in figure 32.12. It was not unusual, particularly in the patients with the more extensive burns, for the daily exudate losses of radioactivity to be almost as great as those in urine.

These exudate losses of radioactivity may be only minimum values since proteolytic bacteria, which sometimes infect the burned tissues have been shown to catabolise iodine labelled albumin. It seems not unreasonable to suggest that the breakdown products of such catabolism may be readsorbed through the burned tissues and excreted by the kidneys. It still remains to be determined how much of the increased rate of catabolism of albumin during

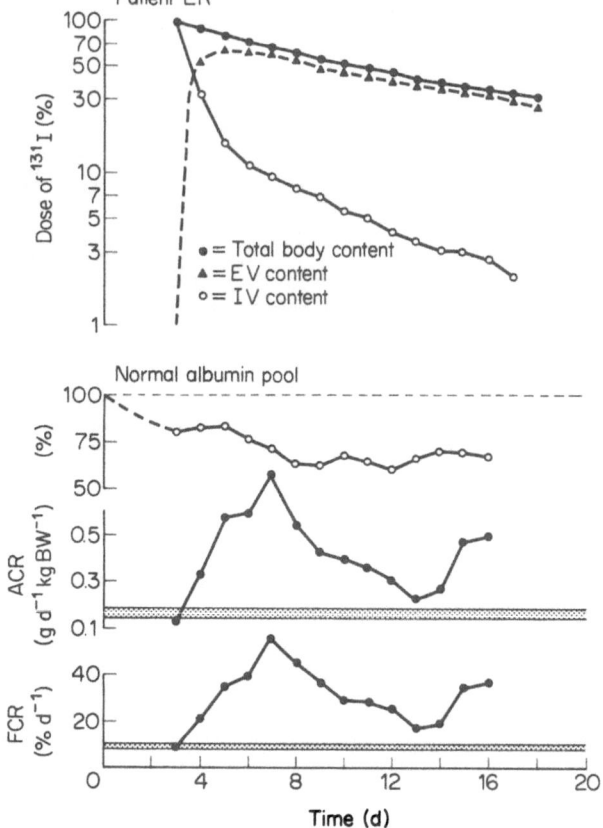

Figure 32.11. Changes in the distribution of injected radioactivity, the plasma albumin pool, and the absolute and fractional catabolic rates of albumin in a patient treated in the warm environment with a burned area covering 85% of the body surface, most of which was full thickness skin loss. Other data as in figure 32.1.

the first week or so after burning could be the result of bacterial decomposition of serum albumin in the burned tissues. It seems probable that this effect will only be small since many of the patients in this study had relatively little bacterial infection of large areas of the burn.

The whole body counter also provided more accurate estimates of the distribution of labelled albumin between the plasma and extravascular tissues[17]. The earlier studies[3] using indirect estimates of the whole body content of radioactivity are considered to be inaccurate, since the whole body content was deduced from cumulative losses in urine and exudate subtracted from the dose of radioactivity injected. The cumulative losses tend to become increasingly underestimated due to small but probably persistent losses in faeces and

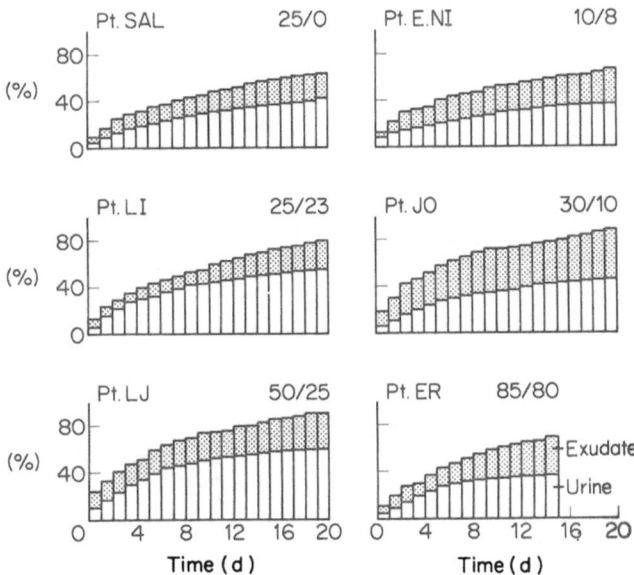

Figure 32.12. Cumulative losses of ^{131}I radioactivity in urine and exudate from 6 patients with burns of differing severity. The losses are expressed as percentages of the injected dose of radioactivity.

sweat. Calculated values for the extravascular to intravascular ratios of radioactive content were available on all 6 burned patients and in three of the patients with relatively minor fracture injuries (figure 32.13). The three patients with major injuries had indwelling urinary catheters, from which complete urine collection gave an estimate of the total body content of radioactivity.

An increased extravascular content of albumin was observed in 10 of the 12 patients. In 5 of the 6 patients with skeletal injuries a two-fold increase in extravascular albumin content was observed. Similar raised values have been reported in another series of studies on patients with fractures[1]. The raised extravascular content of albumin in these patients was unexpected particularly when the injury was relatively minor. It may reflect the lack of muscular activity in the paraplegic patients due to the spinal injury, and in the more severely injured patients due to immobilisation by traction or plaster casts. The lack of muscular activity may limit the ability of the lymphatic system to return to the vascular system the protein which normally passes through the walls of capillaries under the influence of arterial pressure. The low plasma albumin concentrations frequently observed at this time confirm the findings made in other injured patients[18,19].

In 5 of the 6 patients with burns the extravascular labelled protein content was often more than three times that found in normal persons. Only the patient with the least severe burn showed a near normal pattern of distribu-

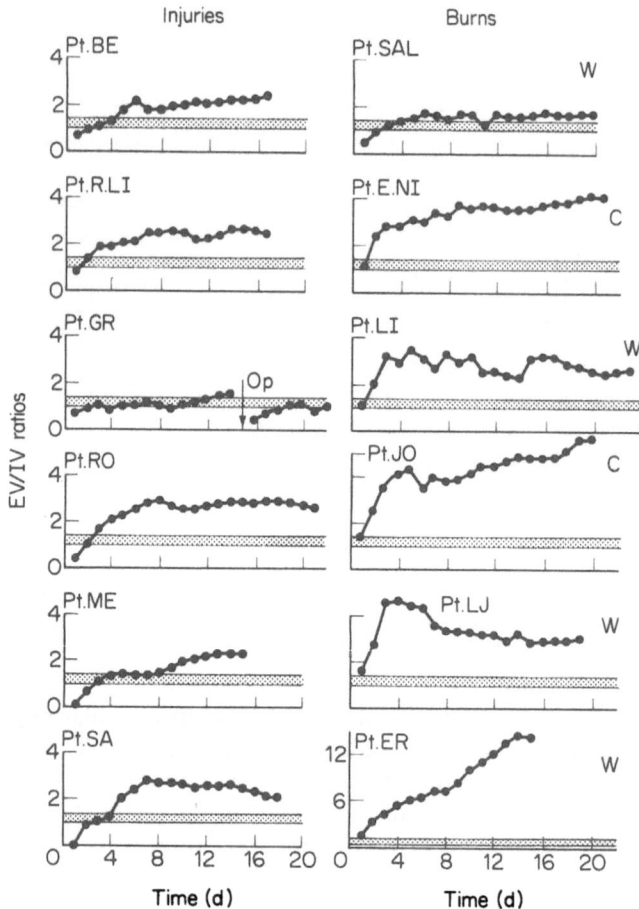

Figure 32.13. Estimates of the distribution of radioactivity between the extra-vascular (EV) and intravascular (IV) compartments of 6 patients with fractures and 6 patients with burns. The shaded bands denote the EV/IV ratios observed in normal persons. W (warm) and C (cool) indicate the temperature of the environment in which the burned patients were treated.

tion. The high values observed in most of the patients no doubt arise from the altered permeability of the heat-damaged capillaries. While in patients with skeletal injuries the tissue damage will be mainly around the site of the fracture, the patients with burns have a generalised oedema, with the altered capillary permeability being more widespread than the actual burned area.

The temperature of the environment in which the burned patients are nursed has þeen shown to alter the distribution of albumin between the extra- and intravascular tissues[9]. In the warmer environment the extravascular content of protein is significantly less than that found in a comparable group of patients treated in a cooler environment. Examples of these differences are shown in figure 32.14 for two comparable patients with relatively minor burns (Patients E.NI and SAL), and for two comparable patients with more severe burns.

All the patients with fractures were treated in a relatively cool (22°C)

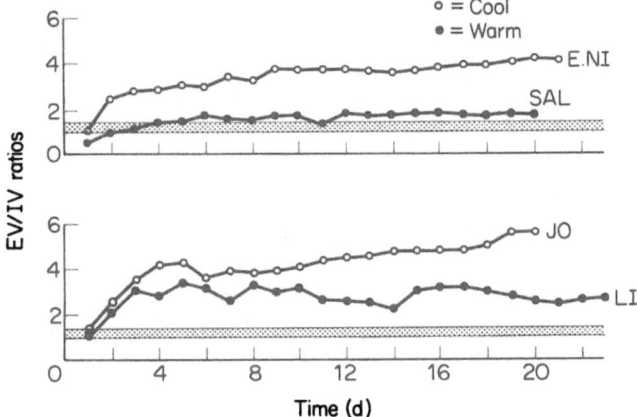

Figure 32.14. Estimates of the effect of environmental temperature on the distribution of radioactivity between the extravascular and intravascular compartments in four patients with burns. Patients E.NI. and SAL had similar severity burns. Patients JO and LI had comparable burned areas of greater severity. The shaded bands indicate the range of EV/IV ratios observed in normal individuals.

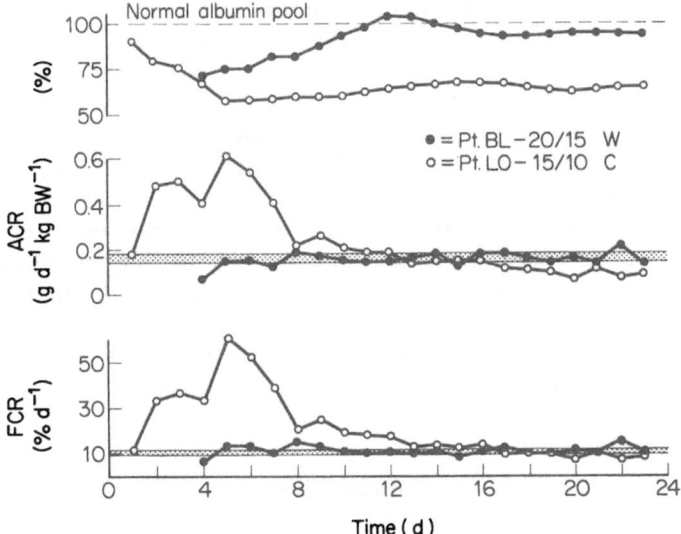

Figure 32.15. The effect of differing environmental temperatures on the plasma albumin pools and the fractional and absolute catabolic rates of albumin in two patients with comparable severity burns.

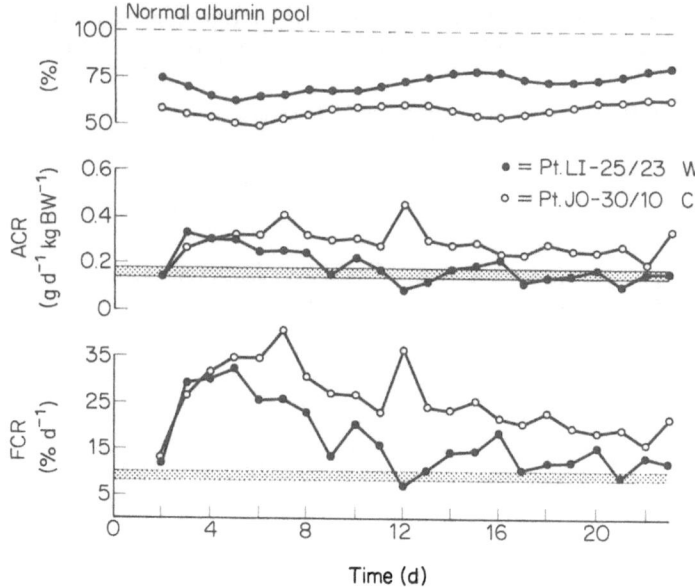

Figure 32.16. The effect of differing environmental temperatures on the plasma albumin pools and the fractional and absolute catabolic rates of albumin in two patients with more severe burns of comparable severity.

environment. Evidence from studies on another group of patients[1] indicates that raising the environmental temperature to 30° C would probably not have altered the distribution of protein between the extra- and intravascular compartments compared with that observed in the cooler environment.

It has also been shown in burned patients that raising the environmental temperature in which they are nursed from 22° C to 32° C reduces the catabolic rate of albumin[9,12,13]. The absolute and fractional catabolic rates shown in figure 32.15 are those measured in two patients of comparable age and body size, and with burns of relatively minor severity. The raised rates of catabolism were observed when the environment was relatively cool; they were within normal limits at the raised environmental temperature. The vascular albumin pool was nearer the expected normal value in the patient treated in the warmer environment.

Comparable patients with more severe burns (figure 32.16) also showed nearer normal rates of catabolism and albumin pools in the warmer environment.

Conclusions

1. These studies have shown that the increased rate of catabolism of albumin following injury is confirmed and is not an artefact due to the administration of partially denatured labelled albumin.

2. In contrast to at least one other report[1] the available evidence indicates that both the fractional and absolute catabolic rates are raised after moderate or severe injury.

3. In patients with fractures the rate of catabolism is directly related to the severity of injury.

4. Inactivity resulting from paraplegia or from surgical immobilisation may cause an increase in the amount of albumin in the extravascular tissues of injured patients.

5. Patients with extensive burns show remarkably increased rates of catabolism of albumin which may persist for at least 3 weeks after burning.

6. Increased permeability of heat-damaged capillaries is the most likely cause of the greatly increased extravascular albumin content in burned patients.

7. The problems associated with complete collection of urine in all patients, and the difficulty of measuring exudate losses of albumin from the tissues of burned patients makes the use of a sensitive whole body radioactivity detector obligatory.

8. The unsteady metabolic state of injured patients is associated with multiexponential rates of decrease of labelled albumin from the whole body, the extravascular tissues and the plasma.

9. The catabolism of albumin in patients with burns and other forms of injury can be determined most reliably by expressing the rate of urinary loss of radioactivity in terms of that remaining in either the plasma or the whole body.

Acknowledgements

Grateful thanks are due to Prof. S-O. Liljedahl of the Department of Surgery, Regional Hospital, Linköping, Sweden, for permission to describe some of the studies made on patients in his clinical care.

References

1. Ballantyne, F. C. and Fleck, A. The effect of environmental temperature (20° and 30°) after injury on the catabolism of albumin in man. *Clinica Chimica Acta*, **46** (1973), 139–146

2. Birke, G., Liljedahl, S-O., Plantin, L. O. and Wetterfors, J. Albumin catabolism in burns and following surgical procedures. *Acta Chir. Scand.*, **118** (1960), 353–366

3. Davies, J. W. L., Ricketts, C. R. and Bull, J. P. Studies of plasma protein metabolism I. Albumin in burned and injured patients. *Clin. Sci.*, **23** (1962), 411–423

4. Davies, J. W. L., Liljedahl, S-O. and Reizenstein, P. Metabolic studies

with labelled albumin in patients with paraplegia and other injuries. *Injury*, **1** (1970), 271–278

5. Mouridsen, H. T. and Faber, M. Accumulation of serum albumin at the operative wound site as a cause of postoperative hypoalbuminaemia. *Lancet*, **2** (1966), 723–725

6. Mouridsen, H. T. Turnover of human serum albumin before and after operations. *Clin. Sci.*, **33** (1967), 345–354

7. Sterling, K., Lipsky, S. R. and Freedman, L. S. Disappearance curve of intravenously administered ^{131}I-tagged albumin in the postoperative injury reaction. *Metabolism*, **4** (1955), 343–349

8. Cuthbertson, D. P., Smith, C. M. and Tilstone, W. J. The effect of transfer to a warm environment (30° C) on the metabolic response to injury. *Brit. J. Surg.*, **55** (1968), 513–516

9. Davies, J. W. L., Liljedahl, S-O. and Birke, G. Protein metabolism in burned patients treated in a warm (32° C) or cool (22° C) environment. *Injury*, **1** (1969), 43–56

10. Davies, J. W. L. and Liljedahl, S-O. (1970). Protein catabolism and energy utilisation in burned patients treated at different environmental temperatures. In *Energy Metabolism in Trauma* Ciba Foundation Symposium, R. Porter and J. Knight, eds., Churchill, London, pp. 59–78

11. Davies, J. W. L. Protein metabolism following injury *J. Clin. Path.*, **23** Suppl (Coll. Path), **4** (1970), 56–64

12. Davies, J. W. L. and Liljedahl, S-O. (1971). Metabolic consequences of an extensive burn. In *Contemporary Burn Management*, H. C. Polk and H. H. Stone, eds., Little, Brown and Co., Boston, Chapter 10

13. Davies, J. W. L. and Liljedahl, S-O. The effect of environmental temperature on the metabolism and nutrition of burned patients. *Proc. Nutr. Soc.*, **30** (1971), 165–172

14. McFarlane, A. S. Efficient trace labelling of proteins with iodine. *Nature*, **182** (1958), 53

15. Reizenstein, P. and Karlsson, H. Å. Clinical whole body counting— whole body scanner with two crystals. *Acta Radiol.*, **4** (1966), 209–220

16. Mouridsen, H. T., Bojsen, J. and Faber, M. The application of whole body counting to studies of the turnover of ^{131}I labelled plasma proteins. *Scand. J. Clin Lab. Invest.*, **23** (1969), 379–390

17. Mouridsen, H. T. The extravascular retention of serum albumin in the operative wound. *Acta Chir. Scand.*, **134** (1968), 417–424

18. Ballantyne, F. C. and Fleck, A. The effect of environmental temperature (20° and 30°) after injury on the concentration of serum proteins in man. *Clinica Chimica Acta*, **44** (1973), 341–347

19. Mouridsen, H. T. The extravascular retention of albumin in wound tissue and its contribution to the postoperative hypoalbuminaemia in rabbits. *Clin. Sci.*, **37** (1969), 431–441

33
Increased fibrin formation with tumours and its genesis*

J. J. FRANKS, S. G. GORDON, B. KAO, T. SULLIVAN and D. KIRCH

O'Meara and his associates[1-5] and Clarke[6] showed over ten years ago that malignant tissues produce a lipid containing coagulative factor which can diffuse out of cells, initiate coagulation in and around tumour tissue and lay down fibrin, particularly at the tumour margin. They speculated that such fibrin deposition could enhance the invasiveness of tumours, both locally and at metastatic sites, by providing a lattice work for new growth. Other workers using histopathologic[7,8] and radioisotopic techniques[9-14] have confirmed that fibrin deposition does occur in many animal and human neoplasms. Whether there is also a relationship between a tumour-produced coagulative factor and the increased incidence of vascular thrombosis seen in cancer patients is not known but the clotting system is certainly disturbed in many of these patients who have high levels of fibrinogen, platelets, factors V and VIII, and fibrin split products as well as other abnormalities indicative of a 'hypercoagulable state'[15-19].

We have been interested in this relationship between neoplasia and certain clotting proteins and have approached the problem in two different ways. First, in an effort to obtain quantitative information about fibrinogen metabolism and fibrin formation in cancer patients, we have studied the behaviour of radioiodinated fibrinogen in a number of patients and animals with large tumours. Secondly, we have made progress in characterising a tumour-produced coagulative factor originally proposed by O'Meara. Our initial studies were carried out in 14 patients, 10 men and 4 women, whose ages ranged from 21 to 63, with most patients in their early fifties. Nine patients had cancer of the lung, two cancer of the pancreas, one oesophageal cancer,

*This work is supported by Veterans Administration Research Allocation 554-01-2920.01-170,01; grants RR-51 from the General Clinical Research Centers Program of the Division of Research Resources, National Institutes of Health; Research Career Development Award AM 19578; a grant from the Milheim Foundation for Cancer Research, Denver, Colorado; U.S. Public Health Service General Research Assistants Grant No. S01-RR-05357; and grant R01-CA14408 from the National Cancer Institute, National Institutes of Health.

one carcinoma of the colon, and one a primary hepatoma. All of these patients were studied in the Clinical Research Center at the University of Colorado Medical Center where excellent laboratory and nursing support was available and, in particular, where meticulous urine collections could be made. Each patient received his own fibrinogen, labelled with either ^{131}I or ^{125}I iodide. Plasma and urine samples were collected daily or more often, and the isotopic data were interpreted in terms of a standard four compartment model (figure 33.1, solid lines) in which x represents the intravascular radioactive fibrinogen, y the interstitial radioactive fibrinogen, z the radioactive breakdown products and u the urinary compartment into which radioactivity is excreted and in which it accumulates. Normally, as shown by Atencio *et al.*[20], fibrinogen is catabolised to its constituent amino acids with no significant or, at least, no long-lasting conversion to fibrin. In other words it appears that fibrin formation is not a detectable intermediate step in normal fibrinogen catabolism.

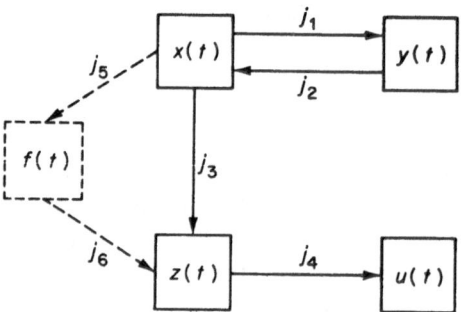

Figure 33.1. Model(s) of fibrinogen metabolism.

As is well known, this model predicts, and the data confirm, that plasma radioactivity decays in an exponential fashion. By appropriate curve-fitting procedures, numerical estimates can be obtained from the plasma data for the rate constants, and most important for j_3. The product of j_3 and the total amount of native fibrinogen in the vascular compartment provides a good estimate of the absolute catabolic rate in mg day^{-1} kg body weight^{-1}. Figure 33.2 shows mean values for the total amount of intravascular fibrinogen in two groups of patients. On the left is the mean from twelve normal subjects, studied by Takeda several years ago[21]. These subjects averaged 127 mg of fibrinogen per kg body weight. In marked contrast, our patients had greatly expanded fibrinogen masses, almost 3 times normal. When we compared j_3, the fractional catabolic rate in the two groups (figure 33.3), we found that j_3 was significantly increased in the tumour patients, averaging 25% more than normal. This moderate increase in j_3, combined with a markedly expanded plasma fibrinogen pool, resulted in a more than three-fold increase in the absolute catabolic rate, averaging 107 mg day^{-1} kg body weight^{-1}, compared with a normal of 31 mg day^{-1} kg^{-1}. Figure 33.4 shows a semi-log

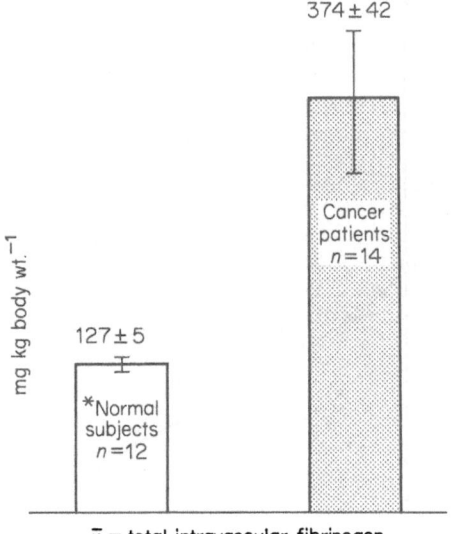

Figure 33.2. Intravascular fibrinogen.

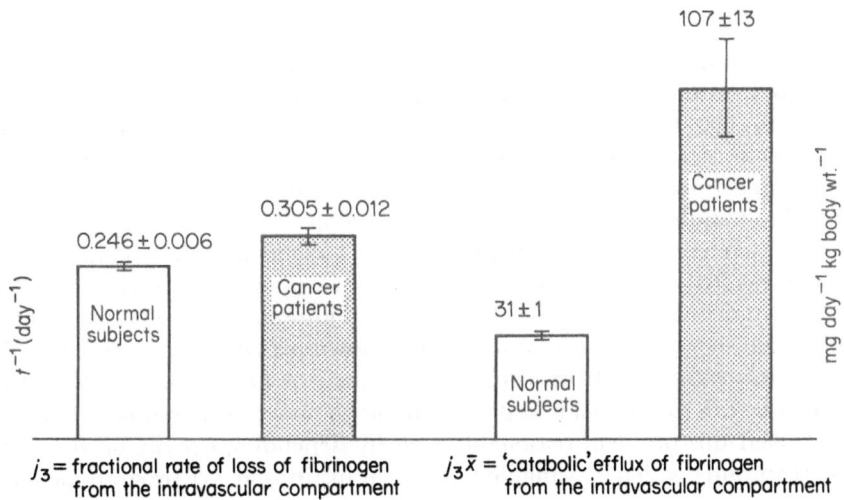

Figure 33.3. Fractional and absolute fibrinogen catabolic rates.

plot of plasma and urinary data from a patient with a pancreatic carcinoma. The lower curve indicates plasma radioactivity. Here we have a good fit of our data with a double exponential function, just as expected. However, when we look at total body radioactivity, estimated by subtracting the amount of radioactivity excreted each day, we find a marked discrepancy, with total

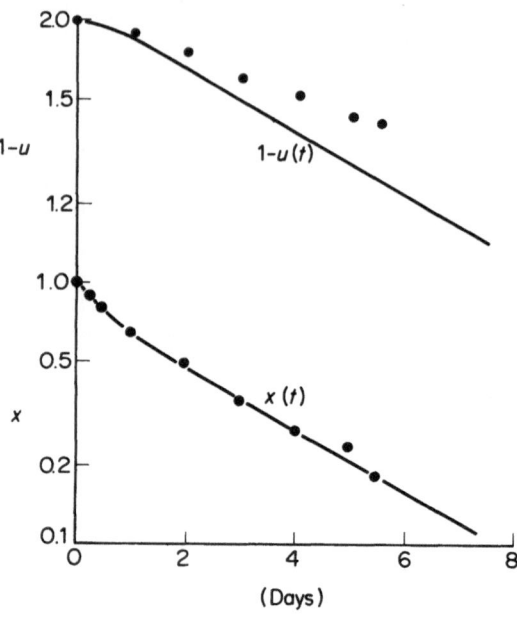

Figure 33.4. Fibrinogen metabolism in a female with pancreatic cancer.

body loss of radioactivity lagging behind the plasma loss. Similar plasma and excretory patterns were seen in all of our other cancer patients. This discrepancy suggested to us that labelled fibrinogen, perhaps as fibrin, was being retained in these patients' bodies outside the vascular and interstitial fibrinogen compartments. We considered a number of modifications of the standard model to handle the concept of fibrin formation. We settled on the five compartment linear model shown in figure 33.1 (solid and dotted lines). This model was first described by Reeve and his associates about 10 years ago[20]. It differs from the standard model in that an additional compartment is interposed between the vascular and the breakdown products compartments. We tentatively have called this compartment the f or fibrin compartment. The difficulty with this model is apparent; namely, how can we estimate j_5 and j_6. We solved this problem reasonably well by depending heavily on the CDC 6400 computer which is available to us at our institution. We programmed into this computer the analytical solutions for the differential equations which describe this five compartment model. A least-squares fit of the plasma data provides us with numerical values for the slopes and intercepts of the plasma curve and for the rate constants j_1, j_2 and the sum of j_3 and j_5 which we have called j_7. For j_4 we used[21] 2.0 day^{-1}. Armed with this information the computer program proceeds to generate a series of whole body radioactivity curves based on different values for j_5 and j_6. The computer can handle a 50 × 50 matrix of values for j_5 and j_6 and can thus test 2500 different pairs of numerical values for these two rate constants in a single run. For each of these 2500 pairs the program computes the sum of the squares of the differences between the predicted whole body radioactive curve for that pair and the experimental data. It then chooses the j_5–j_6 pair which gives the minimum value for the sum of the squares of the differences out of all the 2500 pairs examined. The program can be repeated using smaller and smaller increments between the various values of j_5 and j_6 until the sum of the squares of the differences can be minimised no more. The obvious traps of multiple minima for the sum of the squares and for physiologically impossible values for the rate constants, for example, negative values, have not been seen. It would appear that this method at least has the virtue of working and giving reasonable values for the rate constant estimates. Figure 33.5 shows data from a patient with lung cancer with total body radioactivity curves predicted by the old four compartment model and by the new model shown here as a dotted line. The fit to the whole body radioactivity data with the 5 compartment model is, of course, good because the program is designed to make a good fit: it is quite different from the predictions of the 4 compartment model. The lower curve on this figure shows a computer predicted plot of our new f-compartment into which, after a few days, is collected nearly one tenth of the total injected radioactivity. Figure 33.6 shows a comparison of j_3, the fractional catabolic rate, in normal subjects and in our patients with the use of the new five compartment model. There is no real difference in these two values,

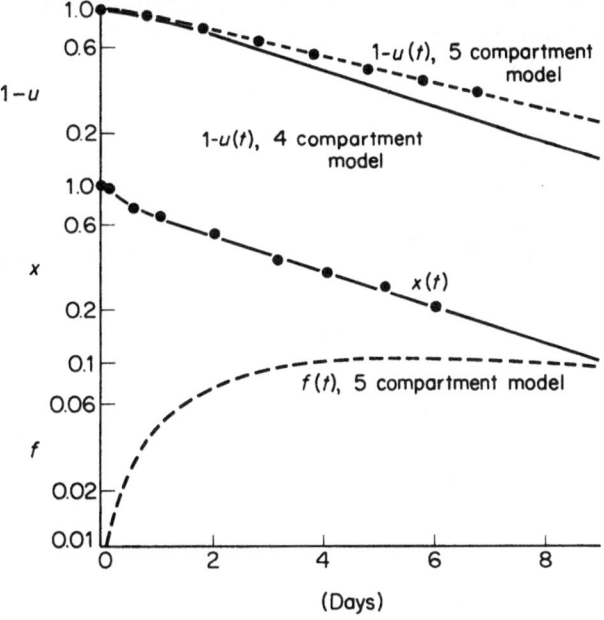

Figure 33.5. Fibrinogen metabolism in a male with lung cancer.

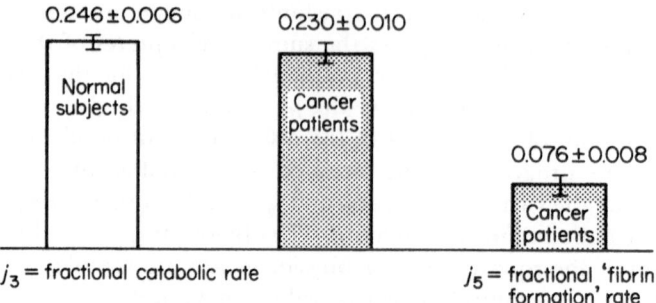

j_3 = fractional catabolic rate j_5 = fractional 'fibrin formation' rate

Figure 33.6. Revised fractional rates based on five compartment model.

namely, 0.246 and 0.230. The discrepancy between plasma loss rates and whole body loss rates in the tumour patients is instead accounted for by the diversion of some of the plasma fibrinogen into a fibrin compartment indicated here by j_5. j_5 averaged about one-third of j_3 in our 14 patients.

Now, does j_5 really represent conversion of fibrinogen to fibrin? There is certainly considerable support for the idea in the literature. A number of investigators have shown by injecting high specific activity radiofibrinogen or radioantifibrin that fibrinogen or fibrin accumulates in neoplastic tissue. More direct microscopic techniques lend further support to this idea. To test this possibility more directly we did repeat studies in 5 of our 14 cancer patients. During this second study they received heparin continuously by slow intra-venous drip. We attempted to maintain their Lee White coagulation times be-tween two and two and one half times their control values. If we could show that heparin reduced j_5 to near zero we would have reasonably good evidence that j_5 represents a fibrin formation rate. If j_5 is reduced to zero by heparin infusion, the predictions of the five compartment model become exactly the same as the four compartment model, just as if the f-compartment did not exist. Figure 33.7 shows that this is true. The total body radioactivity curves generated by the two models are virtually superimposable. Table 33.1 shows pre-heparin and heparin values for j_5 which we now call the fractional rate of

Table 33.1. Fractional rate of 'fibrin formation'*

Patient	Fractional rate (day^{-1}) No Heparin	Heparin
L.G.	0.070	0.041
C.K.	0.062	0.002
C.M.	0.069	0.045
M.T.	0.061	0.010
L.W.	0.106	0.003
Mean	0.074	0.020
S.E.	0.008	0.009

*$p < 0.01$

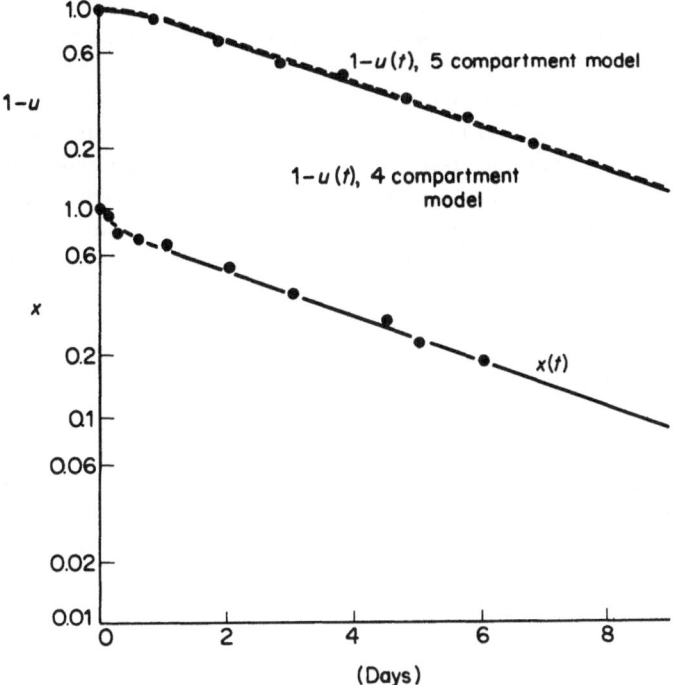

Figure 33.7. Fibrinogen metabolism in a man with lung cancer receiving continuous intravenous heparin.

fibrin formation. In every case j_5 was reduced by treating the patient with heparin and in three patients it was reduced to 0.01 or less. It was interesting to us that the more modest reductions were obtained in the two patients whose clotting times were the most difficult to maintain in the therapeutic range.

We felt that these studies in man provided convincing evidence that labelled protein—most likely fibrin—was being retained abnormally in these patients. What we could not determine was whether fibrin was being retained specifically in the tumour or as part of a generalised, low grade, chronic process of disseminated intravascular coagulation. We attempted to scan the tumours of some of our patients after giving them ^{131}I-fibrinogen but the levels of activity which we were able to use were too low to provide sufficient contrast. For this and for other reasons we decided to resort to animal studies, provided we could find a suitable tumour model. Thanks to the kindness of Dr Sumner Wood, we were able to start a small colony of rabbits carrying the V2 carcinoma[22]. The rabbit is particularly suitable because of our long experience with it in plasma protein studies and because much information regarding fibrinogen metabolism in the rabbit is available in the literature. The V2 carcinoma, originally derived from a rabbit papilloma, but much modified by repeated passage, has been used experimentally for many years. The tumour is passed from animal to animal by injecting a small amount of minced tumour into the recipient. We chose to inject into the muscles of the upper hind leg where growth of the tumour could be followed easily by direct measurement and where radioactive scanning would be relatively easy. Our studies are still in progress but the results so far are quite striking and confirm our results in man. Figure 33.8 shows, for purposes of comparison, plasma and whole body count data from a normal rabbit. The upper curve is the predicted total body radioactivity based on plasma data parameters if $j_5 = 0$, that is if there is no flow into the f-compartment. As has been shown before by others, the predictions of the four compartment model fit the data in a normal animal very well.

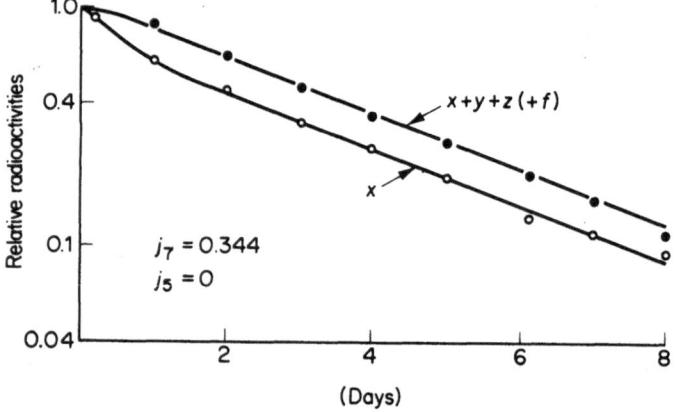

Figure 33.8. Fibrinogen metabolism in a rabbit (B3).

Figure 33.9 shows the same kind of data in an animal with a large and rapidly growing V2 carcinoma. Note first that the plasma disappearance rate is greatly accelerated, with a calculated metabolic loss rate of 0.84, about twice normal. This calculated metabolic loss rate, j_7, represents the sum of j_3 and j_5, that is loss into the f-compartment plus loss by catabolism. Further, just as in man, the total body radioactivity declines much more slowly. The line drawn through these whole body count points is again obtained by choosing values for j_5 and the outflow constant, j_6, which give the best least-squares fit according to our program. The value of j_5 computed from this program for this animal is 0.32, indicating that a substantial proportion of this rabbit's plasma fibrinogen is being transferred to the fibrin compartment each day. Table 33.2 summarises the results of animal studies completed so far. At the top are two normal animals whose plasma fibrinogen levels were less than 4 mg ml^{-1}. The total intravascular fibrinogen in these two animals was 105 and 120 mg kg body weight^{-1}, the plasma fractional loss rate 0.35 and 0.34, and the absolute catabolic rate 0.37 and 0.41 mg day^{-1} kg body weight^{-1}. These results agree well with the published values of other workers. In contrast, the plasma

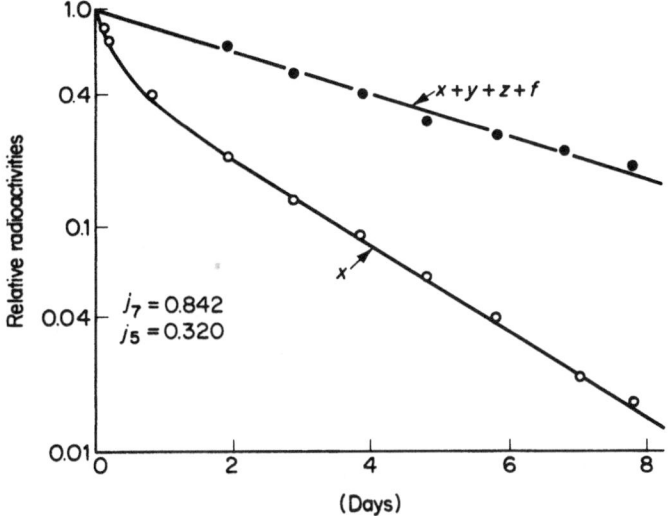

Figure 33.9. Fibrinogen metabolism in a rabbit (F32) with a large V2 carcinoma

Table 33.2. Fibrinogen data in normal and tumour-bearing rabbits

Rabbit	Condition	Plasma Fibrinogen (mg ml^{-1})	$\bar{x}$	j_7	j_5	$j_7 x$	$j_5 \bar{x}$
B-2	Normal	3.48	105	0.35	0	37	0
B-3	Normal	3.95	120	0.34	0	41	0
F-8	Small Growing Tumour	5.23	261	0.67	0.19	175	49
F-9	Small Growing Tumour	4.43	210	0.71	0.37	149	78
F-32	Large Growing Tumour	10.65	383	0.84	0.32	320	103
F-21	Large Growing Tumour	10.59	381	0.80	0.13	310	40

$\bar{x}$ = Total intravascular fibrinogen in mg kg^{-1}. j_7 = Fractional loss rate from plasma in day^{-1}. j_5 = Fractional fibrin formation rate in day^{-1}.

fibrinogen concentration in tumour-bearing animals was increased—modestly in animals with small tumours and markedly in animals with large tumours. The fractional loss rate in the four animals with tumours was increased two- to three-fold and the absolute catabolic rate three to eight times normal. In addition there was significant transfer of fibrinogen to the f-compartment. In these it varied from 0.13 to 0.37 per day, with absolute 'fibrin formation rates' of 40 to 103 mg day^{-1} kg body weight^{-1}.

We were also able to image most of these animals, using a Searle Radiographics pho/Gamma III high performance Anger Camera with videotape data store and area-of-interest playback capability. Figure 33.10 shows how the animals were positioned. The bladder, usually a site of intense radioactivity, was centred over the crystal and the hind legs were extended laterally

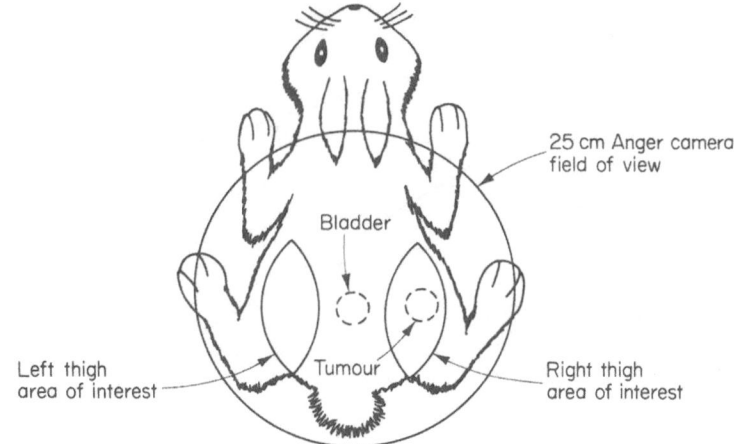

Figure 33.10. Rabbit positioned for radionuclide imaging.

and posteriorly. The areas of interest which we focussed on were the outlined upper leg areas. This scintagraphic system permits us to record counts exclusively from the areas of interest during tape playback. Figure 33.11 shows the results of a typical imaging study done on day 4 of the experiment

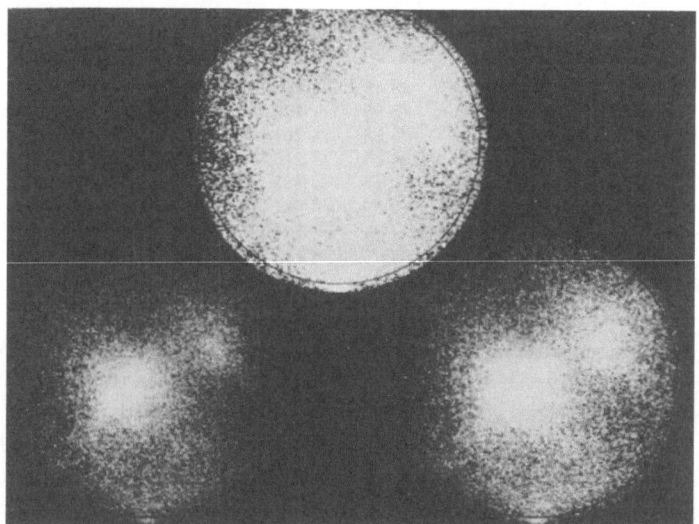

Figure 33.11. Scintagraphic image of a tumour-bearing rabbit. A rabbit which had received ^{131}I-fibrinogen four days previously was positioned over the Anger Camera detector as indicated in figure 33.10. Three photographs of the oscillo-scope screen are shown, taken at different lens apertures. Note the rather intense radioactivity arising from the bladder near the centre of this photograph. In addition there is a clearly defined hot-spot on the upper right outlining the tumour.

in a rabbit with a rapidly growing tumour. This figure is comprised of three photographs of the oscilloscope screen at three different camera apertures. The photograph at the lower right shows the best contrast. Note the rather intense radioactivity arising from the bladder near the centre of this photograph. In addition there is a clearly defined hot-spot on the right which outlines the tumour. This localisation of radioactivity by tumour appears early in the study, usually during the second day, and it persists throughout the study, as shown in figure 33.12. Here we have plotted, against time, the radioactivity arising from the tumour bearing area and from the corresponding area of the opposite leg. Not only is there concentration of radioactivity in the tumour, but the loss of activity from the tumour lags behind the opposite side, as expected.

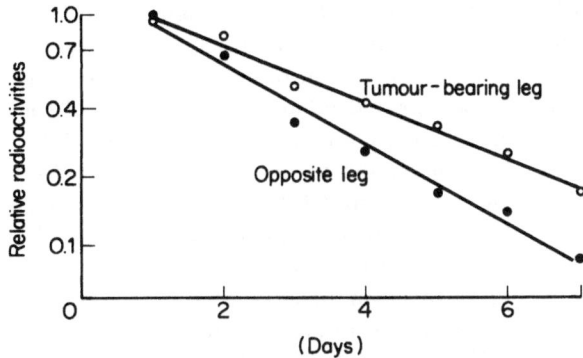

Figure 33.12. Scintascan radioactivity of a rabbit (F8).

No such difference in the uptake and loss of radioactivity was seen in normal animals. This characteristic time course of radioactivity, obtained by scanning tumours after radioiodinated fibrinogen has been injected, was reported and extensively studied by the group at Pisa several years ago[23-27].

Finally, I would like to tell you briefly about our efforts to purify the tumour coagulative factor initially proposed by O'Meara and our efforts to determine its mode of action[28,29]. Normal tissue and tumour tissue were obtained at surgery from patients with a variety of cancers and from rabbits carrying V2 carcinomas. A crude extract of these tissues was obtained by placing them in cold Tris buffer at a pH of 8.5 for several hours. We assayed the procoagulant activity of these crude extracts by their effect on the coagulation time of re-calcified citrated human and bovine plasma, with rabbit brain thrombo-plastin as a standard. Table 33.3 shows a comparison of the specific coagula-tion promoting activity, in equivalent rabbit brain thromboplastin activity, of normal and tumour tissue from six patients. The normal tissue was taken from sites adjacent to the tumour and, with the exception of the liposarcoma, from the same histologic source as the tumour. In every case the procoagulant activity of the crude extracts of tumour tissue was greater than the extracts of

Table 33.3. Procoagulant activity in crude extracts of normal and malignant tissue from the same source

| Tissue | Tumour type | Specific activity (equiv RBT* mg protein^{-1}) | | Tumour / Normal |
		Tumour	Normal	
Lung	Carcinoma	0.163	0.093	1.8
Colon	Adenocarcinoma	3.8	1.9	2.0
Muscle	Liposarcoma	0.113	0.081	1.4
Breast	Carcinoma	0.068	0.007	9.7
		0.067	0.025	2.7
Kidney	Carcinoma	0.0163	0.0096	1.8

*RBT = Rabbit Brain Thromboplastin

the corresponding normal tissue, with ratios ranging from 1.4 to 9.7. We have been able to concentrate the procoagulant activity of crude extracts of both human and animal tumours by factors of more than a thousand by a purification sequence involving repeated ethanol precipitation of the crude extract followed by filtration through an XM-100 ultrafilter. This concentrated material shows up as a single band on polyacrylamide gel electrophoresis and has a molecular weight of about 62000. When normal tissue extracts are handled in the same way, procoagulant activity disappears. Furthermore, no electrophoretic band is seen with normal tissue extracts.

The mode of action of this tumour-associated procoagulant, which we call cancer thrombogenic factor or CTF, is currently under study. We have found that the activity of CTF is inhibited by DFP (diisopropylfluorophosphate), both in crude and in partially purified extracts. However, DFP does not inhibit the procoagulant activity of crude extracts of normal tissue. This suggests that the activity in normal tissue is due to tissue thromboplastin, in contrast to the tumour tissue. As further evidence that CTF is not a thromboplastin and that it does not act through the extrinsic pathway, we find that CTF promotes coagulation in factor VII depleted plasma. On the other hand, we have also found that CTF is active in factor VIII deficient plasma, so it does not act through the intrinsic pathway either. The best evidence at hand suggests that CTF acts at a site in the coagulation cascade distal to both the intrinsic and extrinsic pathways, perhaps by activating either factor X or prothrombin or, less likely, by a direct effect on fibrinogen.

In summary, we have studied fibrinogen metabolism in patients with cancer and in rabbits with experimental tumours. Both the fractional and the absolute rates of catabolism were increased significantly. Total body radioactivity loss lagged behind plasma and interstitial fibrinogen compartments. A model with an added fibrin compartment is compatible with the data. Fibrin formation was prevented or reduced in human subjects by heparin infusion. Both human and animal tumours contain an extractable tumour-specific coagulative protein which has been partially purified and which appears to act late in the clotting pathway.

References

1. O'Meara, R. A. Q. and Jackson, R. D. Cytological observations on carcinoma. *Irish J. Med. Sci.*, **391** (1958), 327
2. Boggust, W. A., O'Meara, R. A. Q. and Thornes, R. D. The coagulative factors of normal human and cancer tissue. *Biochem. J.*, **80** (1961), 32
3. O'Meara, R. A. Q. and Thornes, R. D. Some properties of the cancer coagulative factor. *Irish J. Med. Sci.*, **423** (1961), 106
4. Boggust, W. A., O'Meara, R. A. Q. and Thornes, R. D. The coagulative factors of normal human and human cancer tissue. *Irish J. Med. Sci.*, **447** (1963), 131–144
5. Boggust, W. A., O'Meara, R. A. Q. and Fullerton, W. W. Diffusible thromboplastins of human cancer and chorion tissue. *Europ. J. Cancer*, **3** (1968), 467
6. Clarke, N. Intracellular location of tissue thromboplastin and possible relation to fibrin deposits in human neoplasms. *Nature*, **205** (1965), 608
7. Hiramoto, R., Bernecky, J., Jurandowski, J. and Pressman, D. Fibrin in human tumours. *Cancer Research*, **20** (1960), 592
8. Ogura, T., Tatsuta, M. and Yamamura, Y. Localisation of fibrinogen in the tumour tissue. *G.A.N.N.*, **58** (1967), 403
9. Day, E. D., Planinsek, J. A. and Pressman, D. Localisation *in vivo* of radio-iodinated anti-rat-fibrin antibodies and radioiodinated rat fibrinogen in the Murphy rat lymphosarcoma and in other transplantable rat tumours. *J. Natl. Cancer Inst.*, **22** (1959), 413
10. Day, E. D., Planinsek, J. A. and Pressman, D. Localisation of radio-iodinated rat fibrinogen in transplanted rat tumors. *J. Natl. Cancer Inst.*, **23** (1959), 799
11. Dewey, W. C., Bale, W. F., Rose, R. G. and Marack, D. Localisation of antifibrin antibodies in human tumors. *Acta Union Int. Cancer*, **19** (1963), 185
12. Shaeffer, J. R. Interference in localisation of I^{131} fibrinogen in rat tumours by anticoagulants. *Am. J. Phys.*, **206** (1964), 573
13. Spar, I. L., Bale, W. F., Marrack, D., Dewey, W. C., McCardle, R. J. and Harper, P. V. ^{131}I-labelled antibodies to human fibrinogen: Diagnostic studies and therapeutic trials. *Cancer*, **20** (1967), 865
14. Riccioni, N. Diagnosis of malignant lesions of the liver by radiocolloid and ^{131}I fibrinogen. *J. Biol. Nucl. Med.*, **13** (1969), 160
15. Fumarola, D. and DelBuono, G. The blood coagulation pattern in malignancies. *Prog. Med. Napoli*, **14** (1958), 327
16. Amundsen, M. A., Spittell, J. A., Thompson, J. H. and Owen, C. A. Hypercoagulability associated with malignant disease and with the postoperative state: Evidence for elevated levels of antihemophilic globulin. *Ann. Intern. Med.*, **58** (1963), 608
17. Miller, S. P., Sanchez-Avalos, Stefanski, T. and Zukerman, L. Coagula-

tion disorders in cancer I. Clinical and laboratory studies. *Cancer*, **20** (1967), 1452

18. Waterbury, L. S. and Hampton, J. W. Hypercoagulability with malignancy. *Angiology* **18** (1967), 197

19. Davis, R. B., Theologides, A. and Kennedy, B. J. Comparative studies of blood coagulation and platelet aggregation in patients with cancer and nonmalignant diseases. *Ann. Intern. Med.*, **71** (1969), 67

20. Atencio, A. C., Bailey, H. R. and Reeve, E. B. Studies on the metabolism and distribution of fibrinogen in young and older rabbits I. Methods and models. *J. Lab. & Clin. Med.*, **66** (1965), 1

21. Takeda, Y. Studies of the metabolism and distribution of fibrinogen in healthy men with autologous ^{125}I-labelled fibrinogen. *J. Clin. Invest.*, **45** (1966) 103

22. Wood, S., Jr. Experimental studies of the intravascular dissemination of ascitic V2 carcinoma cells in the rabbit, with special reference to fibrinogen and fibrinolytic agents. *Bulletin Swiss Acad. Med. Sci.*, **20** (1964), 92

23. Monasterio, G., Becchini, M. F. and Riccioni, N. Radioiodinated (I[131] and I[125]) fibrinogen for the detection of malignant tumours in man. *Med. Radioisotope Scanning*, **2** (1964), 159

24. Monasterio, G., Becchini, M. F. and Riccioni, N. Detection of tumours in man by means of I[131]-fibrinogen. *Excerpta Med. Internat. Cong. Ser.*, No. 105 (1965), 1270

25. Riccioni, N., Becchini, M. F. and Aloisi, M. Differentiation of malignant neoplasms from other lesions using radioactive fibrinogen. *Minerva Nucleare*, **9** (1965), 451

26. Riccioni, N., DeRenzi, G. Becchini, M. F. and Bartorelli, A. The time course of radiofibrinogen uptake in the I.R.E. sarcoma and in acute inflammatory process as induced in rats. *Minerva Nucleare*, **9** (1965), 250

27. Riccioni, N., Becchini, M. F., Vitek, F. and Donato, L. (1966). Analog computer study of the kinetics of extravascular distribution of ^{131}I-labelled plasma proteins in normal and tumoural tissues. Proc. Conf. on problems connected with the preparation and use of labelled proteins in tracer studies, Pisa, Italy, January 17–19, 193–198.

28. Gordon, S. G. Partial characterisation of cancer thrombogenic factor (CTF) activity. *Fed. Proc.*, **33** (1974), 209 (Abstract)

29. Franks, J. J. and Gordon, S. G. Some properties of cancer thrombogenic factor (CTF). *Fed. Proc.*, (1974), 209 (Abstract)

Discussion

DONATO

I have been very pleased to hear Dr Frank's paper, which was particularly interesting to me, because it was in 1964 that my former chief, Monasterio, with my colleague Becchini, started to use, to study and to investigate fibrinogen behaviour in tumours[1].

The results showed that tumours pick up fibrinogen in a very peculiar way, which makes them quite different for instance from the inflammatory lesions which also concentrate fibrinogen. The special behaviour of tumours is that they pick up fibrinogen and if you count over the lesion it stays there for a very long time, for several days. This behaviour can be enhanced if you give ε-aminocaproic acid to the patient, and can be suppressed completely if you heparinise the patient. Therefore these data completely agree with what you showed. Then in 1965 they demonstrated in animals with experimental tumours, both spontaneous and induced, that the ratio of activity in the tumours compared with normal was very high[2].

It was in 1966 that Dr Riccioni, with Dr Vitek and myself, tried to explain the very strange kinetics of fibrinogen in tumours, which did not fit with any ordinary models[3]. It was only by introducing a fibrin compartment at the tumour site that we could explain those data. Similarly, if you take the behaviour of an inflammatory lesion in a patient treated previously with ε-aminocaproic acid, you mimic the tumour behaviour. If you heparinise the patient, the kinetics returns to normal. You understand why I was so glad to see your data, because they confirm completely a trend that we well know. By the way, we routinely use fibrinogen for the specific diagnosis of tumours. We have examined not less than 400 patients by now with liver, brain and bone lesions. Labelled fibrinogen is, because of its particular behaviour, one of the best agents we have for tumour diagnosis.

FRANKS

Dr Donato, I am very pleased to learn that our observations rest on such a wide base of clinical and laboratory observations. One interest of ours is to see if there are long-term therapeutic advantages in the use of anticoagulants in cancer.

Have you attempted to use heparin or a fibrinolytic agent in the treatment of tumours in a large group of patients?

DONATO

No, we have not.

MILLER

At Rochester in our department, Dr William F. Bale has worked over a number of years in connection with the localisation of fibrinogen in tumours and became very interested in that because of its potential use in therapy. He has in fact literally cured a variety of rat tumours by preparing very potent iodinated antifibrinogen antibodies and having them localise in tumours and deliver a dose of radiation sufficient to lead to complete involution and disappearance of the tumour[4]. In connection with your work and with your observations, Dr Donato, I think the work that Bale did emphasised the fact that all tumours are obviously not the same, that the phenomena which you have observed and studied vary, that some tumours are characterised by a necrosis, sterile or otherwise, and show this phenomenon to a maximum degree. Others are quite cellular and vascular and do not localise fibrinogen anywhere near that extent. Have you any occasion to classify your observations on that kind of basis?

FRANKS

Every tumour we have looked at exhibits this phenomenon, but we have limited our studies to solid tumours and have excluded lymphomas. I should say that animal tumours, as has been known for some years, can be cured by a variety of manoeuvres, including heparin therapy. It is a more difficult problem in man.

DONATO

As a comment to Dr Miller's question, we have observed this kind of fibrinogen behaviour in all types of human tumours that we have examined.

DAVIES

I would like to comment on some of the work that Dr Franks reported. Particularly, I would have appreciated his computer program when we were studying fibrin formation in injury. In doing fibrinogen turnover studies we postulated that there was a fibrin pool, but we did not have the computer facilities to prove this.

Both in burn patients, where there is a great accumulation of fibrin in the burned area, and in patients with fractures, where fibrin collects around the site of the healing wound, there appear to be extra fibrin compartments. I would have liked to have had your program to sort out these problems. We have also attempted to study the effects of various anticoagulants, such as heparin, on the turnover of fibrinogen and the response of the fibrin pool to the administration of these materials. I wonder if you can tell me how much heparin you gave. Because of the possibilities of haemorrhage, we experienced difficulties with the clinicians, who wished to give less heparin than we would have preferred. Similarly when phenindione, they would not give enough drug to affect the turnover, because of the fear of haemorrhage following the accident or following surgery.

FRANKS

I divorced myself from the clinical care of these patients. We did however persuade the people who were taking care of them to let us give enough heparin to prolong the coagulation time to between 2 and 2.5 times the normal. We argued that if the tumour was going to erode into a major vessel, say in a patient with cancer of the lung, heparin was not going to make much difference. Furthermore, these patients are prone to develop thrombophlebitis and heparin helps to prevent its occurrence.

REGOECZI

I was very interested to see your table (33.3) in which you show the procoagulant activity of various tumours. Some years ago we became interested in a factor in mucin-producing adenocarcinomas having a procoagulant effect. Pineo and I extracted glycoproteins from mucin from such tumours and found an activating protein which would directly activate factor X. Quite interestingly, its activity was dependent on its high carbohydrate content. That was just a comment. I would like to ask whether this accumulation of fibrin around tumours reflects a natural event, with the plasma proteins passing through the circulation as usual? Then, if there happens to be a tumour in the way with a procoagulant factor in it, the fibrinogen might clot on its way through the tumour. Is it that, or do we have a net accumulation of fibrinogen by some specific attracting mechanisms? Can you, from your computations say something about it?

FRANKS

I do not know of any fibrinogen attracting mechanisms of tumours and I think there are several arguments against the idea that there is much net accumulation. For one thing, you do not see that much fibrin around most tumours. Secondly, if you carry out the experiments longer than I showed, you see a gradual disappearance of label from the fibrin compartment. The other bit of evidence is perhaps not all that pertinent, but I think it is interesting.

Dr Genton at our institution has been forming large thrombi in the venae cavae of dogs by putting phenol into a portion of the vessel. He has found first of all that if there is already labelled fibrinogen in the animal's body, there is an accumulation, a peak, and then a fall off of radioactivity in the thrombus, so that there is turnover in these fairly stable clots in the venous system. Alternatively if he adds labelled fibrinogen a 'cold' clot becomes 'hot' in time[5]. Dr P. W. Straub reported a similar phenomenon in a patient with an aortic aneurysm containing a large thrombus[6].

References

1. Monasterio, G., Becchini, M. F. and Riccioni, N. (1964). Radioiodinated (^{131}I and ^{125}I) fibrinogen for the detection of malignant tumours in man. In *Medical Radioisotope Scanning*, Vol. II, IAEA, Vienna, p. 159
2. Riccioni, N., De Renzi, G., Becchini, M. F. and Bartorelli, A. The time course of radiofibrinogen uptake in the I.R.E. sarcoma and in acute inflammatory process as induced in rats. *Min. Nucl.*, 9 (1965), 250
3. Riccioni, N., Becchini, M. F., Vitek, F. and Donato, L. (1966). Analog computer study of the kinetics of extravascular distribution of ^{131}I labelled plasma proteins in normal and tumoural tissues. In *Labelled Proteins in Tracer Studies*, L. Donato, G. Milhaud and J. Sirchis, eds., EURATOM 2950.d,f,e, Brussels, p. 193
4. Bale, W. F., Spar, I. L. and Goodland, R. L. (1960). Experimental radiation therapy of tumours using ^{131}I-carrying antibodies to fibrin. University of Rochester Atomic Energy Project Report, UR-567

5. Genton, E. Personal communication
6. Straub, P. W. Chronic intravascular coagulation. Clinical spectrum and diagnostic criteria, with special emphasis on metabolism, distribution and localisation of ^{131}I fibrinogen. *Acta Med. Scand.*, Supp., 1 (1971), 526

34
Hormonal regulation of net haptoglobin biosynthesis in the isolated perfused rat liver

L. L. MILLER

Haptoglobin, discovered and first described by Polonowski and Jayle[1] has been extensively studied over the last three decades. However, it was not until 1960[2] that Murray and Connell noted that subcutaneous injection of turpentine in rabbits was followed within 48 h by a marked rise in the serum haptoglobin level. Later observations of puromycin treated rats were consistent with the view that the increased level of haptoglobin was largely the result of increased synthesis in overall response to the subcutaneous inflammation. The pattern of increase in plasma haptoglobin level in response to the experimental injury has since been observed to occur in a variety of higher animals including the rat, dog, and human in response to diverse kinds of injury[3]. So characteristic has been this stereotyped response that haptoglobin is now reckoned among the acute phase plasma proteins.

Although it might have been deduced that the liver was the site of haptoglobin synthesis from our early demonstration[4] in the rat of the dominant role of the liver in the synthesis of all of the plasma proteins with the exception of the γ-globulins, Krauss and Sarcione[5] first showed labelled amino acid incorporation into haptoglobin by the isolated perfused rat liver.

Alper et al.[6] presented strong evidence based on labelled amino acid incorporation and immunofluorescence studies for the liver as the site of haptoglobin synthesis in the dog. The first demonstration of net haptoglobin synthesis in vitro was that of John and Miller[7]; they used the isolated perfused normal rat liver in experiments 12 hours in duration to define some nutritional, hormonal and temporal factors which had been thought to play a role in the regulation of plasma protein biosynthesis in the intact animal. They demonstrated net synthesis of the five plasma proteins albumin, fibrinogen, α_1-acid glycoprotein, α_2-(acute phase) globulin and haptoglobin by the isolated liver and for the first time presented evidence for the direct action on the liver of the hormones cortisol and insulin in the induction of increased synthesis of the acute phase proteins.

It is the purpose of this report to review our observations on the actions of hormones on the net synthesis of haptoglobin by the isolated rat liver. It will become clear that in the normal liver the glucocorticoid cortisol is essential for the induction of increased haptoglobin synthesis, that the quantitative magnitude of the increased synthesis can be modulated not only by the nutritional state of the liver donor but also by the perfusate pH, and by the thyroid status of the liver donor. The complexity of hormone interactions on haptoglobin synthesis by the liver is manifest in the response of the livers of totally hypophysectomised rats to the four hormones cortisol, insulin, growth hormone and triiodothyronine; only in the presence of all four hormones is the quantitatively low level of haptoglobin synthesis restored almost to normal.

Methods

Details of the methods used in these studies have been published, as follows: perfusion technique including the operative procedure for the isolated liver[8]; preparation of hypothyroid and hyperthyroid rats[9]; use of totally hypophysectomised rats[10]; methods of isolation and purification of rat haptoglobin, the preparation of anti-rat haptoglobin antiserum in rabbits and its use in measurement of haptoglobin in perfusion samples by the single radial immunodiffusion technique[7]. It is to be emphasised that in all experiments hormones were not only added to the perfusate at the beginning of perfusion, but were also added by continuous infusion along with the solution of glucose and amino acids throughout the entire 12 or 24 hours of each experiment.

Results and discussion

Effects of cortisol supplementation

As we described in 1969[7], livers from normal fed rats responded to continuous infusion of cortisol only after 4 to 6 hours with a two- to three-fold increase in the rate of synthesis of haptoglobin. Further addition of either growth hormone and/or insulin failed to enhance the already increased rate of synthesis in response to cortisol as shown in figure 34.1. We emphasise that it is the effect of cortisol which is manifest here, neither insulin nor growth hormone, alone or together, produced this effect in the absence of added cortisol.

Effect of perfusate pH on response to cortisol plus insulin

In 1971 we compared the effect of perfusate pH 7.10 with pH 7.40[11] on the biosynthesis of the acute phase proteins in response to the hormone combination of cortisol plus insulin. Figure 34.2 summarises our observations with respect to haptoglobin and indicates that at pH 7.10, which may be likened to a state of acidosis, the induction of increased haptoglobin synthesis in response

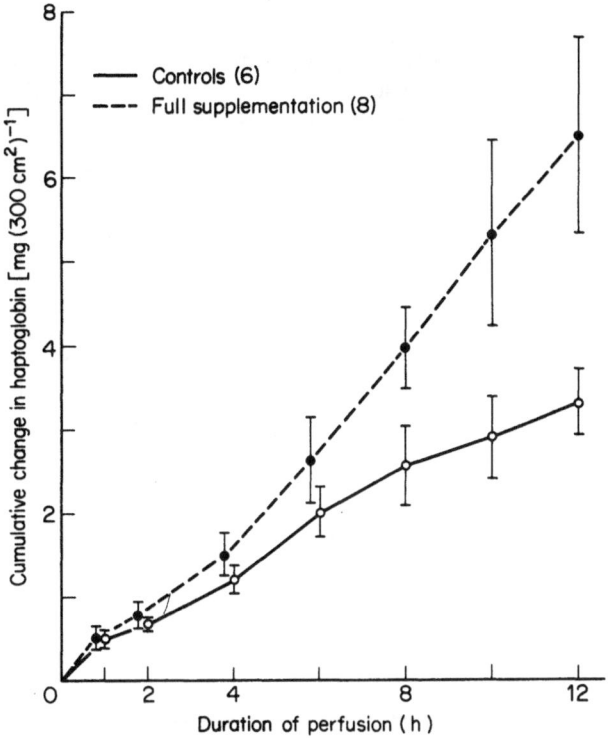

Figure 34.1. Cumulative average net change in rat haptoglobin in perfusate per 300 cm² body surface area of liver donor rat. Taken from John and Miller[7]. (For the sake of comparison with other workers' data, cumulative protein synthesised on the basis of a per 10 gm weight of fresh liver from a fed normal rat is usually within 10% of the values estimated on basis of 300 cm² body surface area). In this and subsequent figures bars indicate standard deviations for each group. Unless otherwise indicated, controls received a continuous infusion of Ringer solution totalling 18 ml and containing only 500 mg of glucose plus 3000 units of penicillin and 3.0 mg streptomycin over the 12 h of each experiment. 'Full Supplementation' indicates the continuous infusion contained not only glucose, 500 mg, but also 320 mg of the complete amino acid mixture plus insulin 6.8 units, bovine growth hormone 1.0 mg, and cortisol as the sodium hemisuccinate 5 mg. In addition, a priming dose of hormones were added to the perfusion reservoir at the start of perfusion, *viz.* 5.1 units of insulin, 5.0 mg cortisol hemisuccinate, and growth hormone 0.5 mg.

to the hormones is about 50 per cent greater than at pH 7.40. If one may presume that the increased synthesis of acute phase proteins serves a biologically useful purpose, this pH effect may be a reflection of the fact that clinical and experimental states of infection and injury are commonly associated with ketosis. It is of interest that increased synthesis of haptoglobin at pH 7.10 was paralleled by increased rates of synthesis of α_1-acid glycoprotein and α_2-(acute phase)-globulin but not of fibrinogen[11].

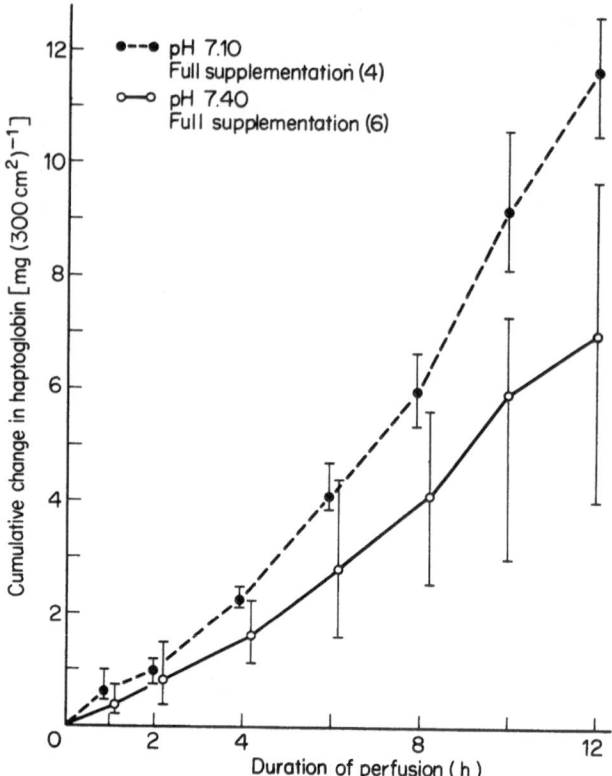

Figure 34.2. Infusion used in both groups of experiments was identical except that pH was maintained at 7.10 or 7.40 by regulated infusion of 0.75 M NaHCO$_3$ as monitored by a radiometer pH stat coupled with a radiometer autoburette.

Effects of nutritional state of the liver donor on the response to cortisol plus insulin

As revealed in figure 34.3, taken from a previous publication[12], the induction of increased haptoglobin synthesis by cortisol plus insulin is significantly but not seriously impaired by an 18 h fast of the liver donors; however, a 6 day fast of the liver donor not only decreases haptoglobin synthesis to about 25% that of livers from fed rats, but also virtually eliminates the induction of increased biosynthesis after 4 to 6 hours. This effect of starvation on the response of the liver to hormones is in harmony with observations of effects of prolonged starvation and protein depletion on acute phase protein synthesis in animals and humans.

Lack of effect of deletion of L-tryptophan and L-threonine from amino acid mixture on haptoglobin synthesis in response to cortisol plus insulin

Rat haptoglobin is a sialoglycoprotein which has been carefully purified and

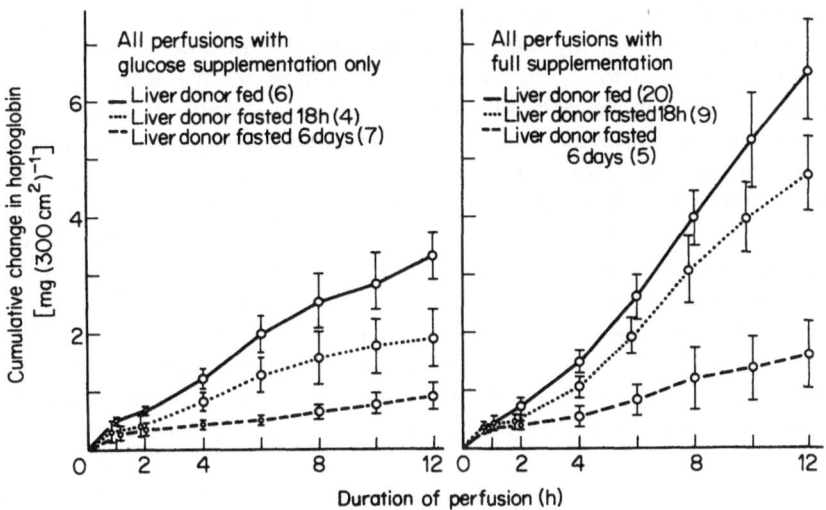

Figure 34.3. Taken from Miller and John[12].

analysed by Lombart *et al.*[13] who reported it to contain 2.98% L-tryptophan and 5.41% L-threonine. It was reported by Fleck *et al.* and Sidransky *et al.* that 18 h starvation or feeding rats or mice an amino acid mixture complete excepting for a total lack of L-tryptophan or L-threonine resulted in a profound

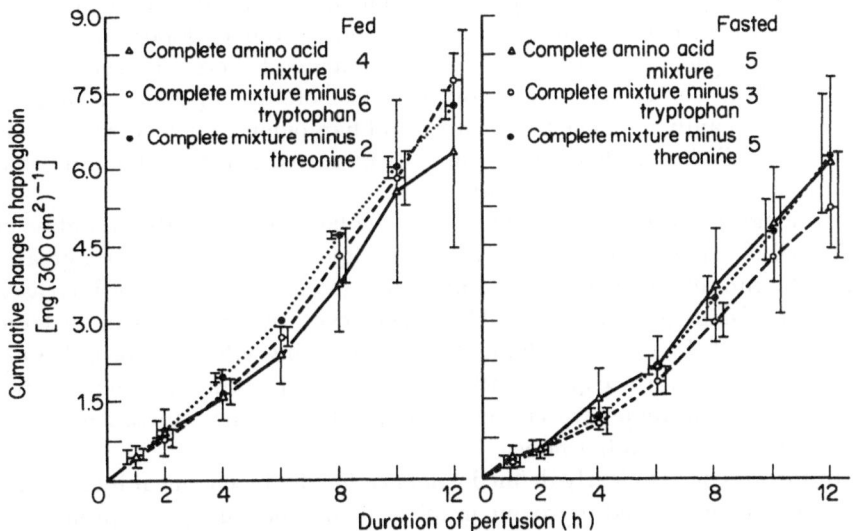

Figure 34.4. Taken from Miller and Griffin[16]. Conditions were similar to those of experiments in figure 34.1, excepting that perfusion pH was controlled at 7.4 and 'Full Supplementation' did not include growth hormone.

alteration of liver polysome profiles[14,15]. Furthermore, non-specific protein synthesis as measured by labelled amino acid incorporation by such altered polysome preparations was substantially depressed. We were very much surprised to find in the isolated perfused rat liver that *neither* 18 h fasting of the liver donor *nor* supplementation with an amino acid mixture (nutritionally complete excepting for omission of L-tryptophan or L-threonine) significantly altered the biosynthesis of five plasma proteins as typified by haptoglobin. Figure 34.4, taken from a report by Miller and Griffin[16], reveals that net haptoglobin synthesis and its increase by cortisol plus insulin are unchanged by either of the amino acid deficiencies. One must conclude that the intact perfused liver is capable of obtaining the tryptophan and threonine necessary for the increased synthesis of haptoglobin (and the other acute phase proteins) from catabolism of its own cell proteins or catabolism of haemoglobin (from haemolyzed red cells) or bovine serum albumin of the perfusate.

These kinds of observations of increased net synthesis of specific plasma proteins are difficult to reconcile with the gross alterations observed in poly-some profile preparations and their loss of capacity to incorporate labelled amino acids into trichloroacetic acid insoluble precipitates.

Modulation of the action on the isolated liver of cortisol plus insulin by thyroid status of the liver donor and by tetraiodothyronine (T_4) supplementation

This phenomenon has been recently described by Griffin and Miller[9] and is exemplified by the curves of figure 34.5. The profound differences in hapto-globin synthesis between the livers from hypothyroid liver donors and hyper-thyroid liver donors represent a three-fold range of increasing synthesis of haptoglobin. The relatively physiological dose of 4 μg T_4 added to normal livers significantly enhanced the synthesis of haptoglobin; however, the same dose of T_4 failed to further enhance already increased synthesis of haptoglobin by livers from hyperthyroid donors.

Although not presented in figure 34.5, we have observed that the dose of 1/5 cortisol, 1/10 insulin used in group 8 (figure 34.5) is five times greater than the minimal dose of cortisol necessary to elicit induction of increased synthesis of haptoglobin in normal livers; however, the admixture of the large 22 μg dose of T_4 virtually eliminated the inducing effect of cortisol. One may conclude that the experiments of figure 34.5 represent an example of the interactions of hormones with the effect on specific protein synthesis dependent on the dose of thyroid hormone.

In the absence of clearly defined evidence for the biochemical mechanism of action of cortisol and thyroxine in the induction of acute phase proteins in general and haptoglobin in particular, it is at best speculative to offer at present an interpretation of the manner in which thyroid hormone acts to modulate the response to cortisol.

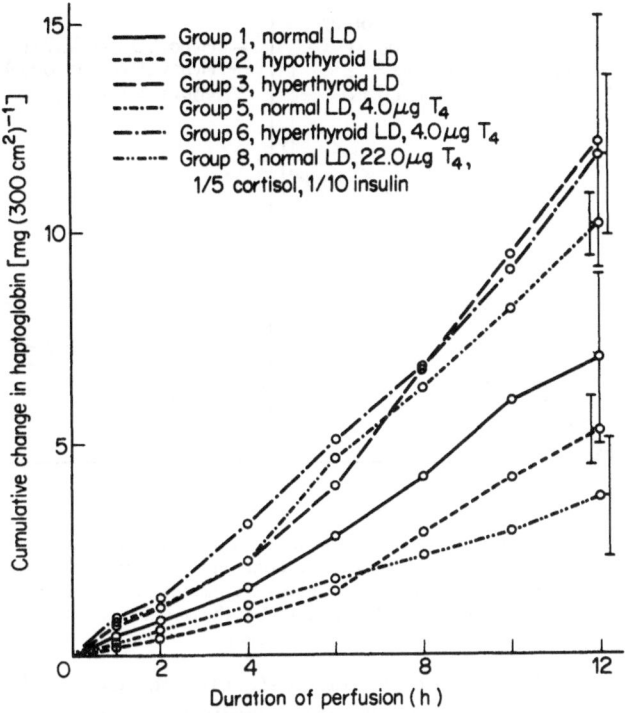

Figure 34.5. Taken from Griffin and Miller[9]. 'Full Supplementation' conditions without growth hormone were used throughout.

Effects of hypophysectomy of liver donors and addition of the hormones cortisol, insulin, triiodothyronine and bovine growth hormone on synthesis of haptoglobin by the isolated liver

These effects are summarised in figure 34.6. In brief, this figure taken from as yet unpublished data[10] makes clear that the liver of the fed totally hypophysectomised rat does not respond to cortisol unless it is accompanied by *all* three hormones, *viz.* insulin, T_3, and growth hormone. Any two of the three did not improve haptoglobin synthesis. However, such deficient combinations permitted the perfused liver to function for 24 h; without any hormone supplementation the perfusions failed after 12 h.

Although we have failed to demonstrate a requirement for, or an effect of, added growth hormone during perfusions of livers from normal[7] fed or 18 h fasted rats, the response of the hypophysectomised rat's liver to growth hormone in the presence of cortisol, insulin and T_3 is unequivocal and emphasizes the view that hormonal responses in intact animals are complex and that an experiment in which only a single hormone is studied may lose more in the nature of its response than it gains in the simplicity of its design.

On the basis of our studies utilising the isolated perfused rat liver it appears that the biosynthesis of haptoglobin, like that of the acute phase proteins

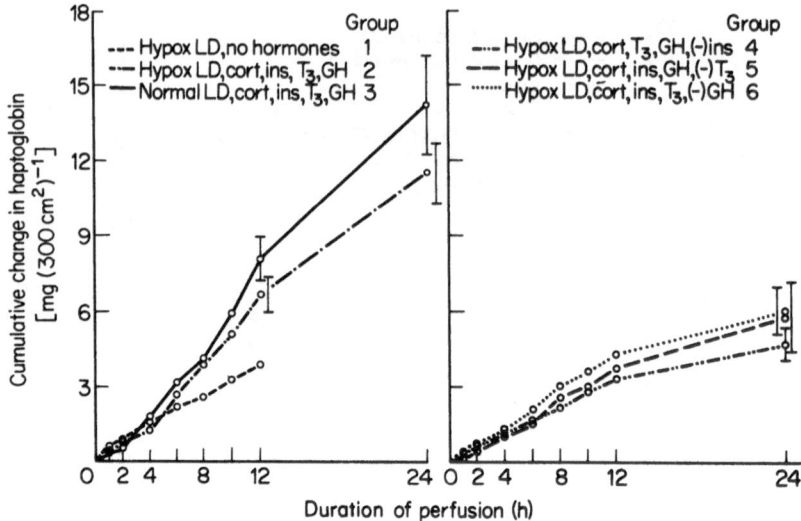

Figure 34.6. Taken from Griffin and Miller[10]. Amino acids and glucose were infused continuously as in 'Full Supplementation' in all experiments. Hormones were added in the doses indicated previously for insulin, cortisol, and growth hormone. When added, 2 μg of T_3 was added at the outset plus a total of 2 μg added to the infusion.

fibrinogen, α_1-acid-glycoprotein and α_2-(acute phase)-globulin, is most responsive to glucocorticoid and that the magnitude of the response to cortisol by livers from otherwise normal rats may be critically modulated by the nutritional state of the liver donor and by the thyroid status of the liver donor as well as by the level of thyroid hormone.

Influence of added insulin and bovine growth hormone on the synthesis of haptoglobin was demonstrable in livers from totally hypophysectomised rats but not in livers from normal rats.

We have observed that the totally adrenalectomised rat responds to the trauma of the operation with a short-lived elevation in haptoglobin levels. After return to normal values the haptoglobin level increases in response to the injury of subcutaneously injected turpentine 7 to 14 days post-adrenalectomy. Although the intensity and duration of this response is significantly less than that observed in normal rats, it implies that haptoglobin synthesis may be stimulated by agents other than glucocorticoid. The use of the isolated perfused rat liver makes practical a further search for the identification of other hormonal factors involved in the regulation of acute phase protein synthesis.

Acknowledgements

This paper is based on work partially performed under contract with the United States Atomic Energy Commission at the University of Rochester

Atomic Energy Project and partially by Grant 1-R01-AM11029 MET from the National Institutes of Health, United States Public Health Service. It has been assigned Report No. UR-3490-546.

We are indebted to Donna M. Eddy, Constanza Perez del Cerro, Janice W. Vergo, and Drusilla Wemett for their invaluable technical assistance, to Leon Schwartz and Gerald Cooper for their preparation of the illustrations, and to Rebecca Wilferth for her skilled secretarial assistance.

References

1. Polonowski, M. and Jayle, M. F. *Compt. Rend.*, **129** (1938), 457
2. Murray, R. K. and Connell, G. E. *Nature*, **186** (1960), 86
3. Glenn, E. M., Bowman, B. J. and Koslowske, T. C. *Biochem. Pharm. Supplement*, (1968), 27–49
4. Miller, L. L. and Bale, W. F. *J. Exp. Med.*, **99** (1954), 125
5. Krauss, S. and Sarcione, E. J. *Biochim. Biophys. Acta*, **90** (1964), 301
6. Alper, C. A., Peters, J. H., Birch, A. G. and Gardner, F. H., *J. Clin. Invest.*, **44** (1965), 574
7. John, D. W. and Miller, L. L. *J. Biol. Chem.*, **244** (1969), 6134
8. Miller, L. L. (1973). In *Isolated Liver Perfusion and Its Application*, Bartosek, I., Guaitani, A. and Miller, L. L., eds., Raven Press, New York, pp. 11–52
9. Griffin, E. E. and Miller, L.L. *J. Biol. Chem.*, **248** (1973), 4716
10. Griffin, E. E. and Miller, L. L. *J. Biol. Chem.*, **249** (1974), in press
11. Miller, L. L. and Griffin, E. E. (1973). In *Isolated Liver Perfusion and Its Application* Bartosek, I., Guaitani, A. and Miller, L. L., eds., Raven Press, New York, pp. 139–145
12. Miller, L. L. and John, D. W. (1970). In *Plasma Protein Metabolism*, Rothschild, M. A., and Waldmann, T., eds., Academic Press, New York, pp. 207–222
13. Lombart, C., Dautrevaux, M. and Moretti, J. *Biochim. Biophys. Acta*, **97** (1965), 270
14. Fleck, A., Shepherd, J. and Munro, H. N. *Science*, **150** (1965), 628
15. Sidransky, H., Bongiorno, M., Sarma, D. S. R. and Verney, E. *Biochem. Biophys. Res. Comm.*, **27** (1967), 242
16. Miller, L. L. and Griffin, E. E. *Am. J. Clin. Nutr.*, **24** (1971), 718

Discussion

DONATO

A very naive question to Dr Miller: What is the role of growth hormone in protein synthesis?

MILLER

Growth hormone in protein synthesis is not a naive question. I attempted to get an answer when Prof Doughaday came to Rochester. You are aware of the fact that at the present time there is a distinction between growth hormone and somatomedin. It is believed that in some mysterious and as yet unrevealed way growth hormone is converted by the liver to somatomedin, or stimulates hepatic synthesis of somatomedin. If it were so, then I have insisted that adding growth hormone directly into liver perfusion ought to lead to very high concentrations of somatomedin. We have never been able to demonstrate a truly significant effect of bovine growth hormone on the liver with exception of experiments on livers from hypophysectomised animals. You might simply say, the normal liver is already under the influence on growth hormone, so that you cannot elicit a further effect. I might remind you of the observation of Salaman and Korner, who claimed that growth hormone could be shown to have an effect in liver perfusion within a period of an hour or two[1,2].

We have never been able to reproduce those results in quite a few experiments. Their published data have no confidence intervals. They are simply two curves with some points on them, the two curves do not exactly coincide and they concluded that the one with growth hormone was higher than the other.

ROSSING

Just concerning the question of growth hormone, I can report that we had the opportunity to investigate albumin turnover in 6 acromegalic patients. They had a significantly increased catabolic rate and, considering that they were presumably in steady state conditions, we must conclude that they had also an increased rate of albumin synthesis. The albumin mass was quite normal, and they had an increased fractional catabolic rate too, and an increased transcapillary escape rate.

REEVE

I want just to say that Dr Miller does very long-term perfusion studies, which many workers were unable to do, and this is perhaps one reason why his results have gone further than the other investigators. One short question: How does the rate of synthesis of haptoglobin, which you are reporting here, compare with the rate of *in vovo* in rats?

MILLER

We have not attempted to make any precise calculations. Incidentally, some of you may take issue with the manner in which we expressed our results, mg per 300 cm² of body surface area. You can make a quick conversion by equating that to milligrams of protein synthesised per 10 grams wet weight liver, and you will have an answer within 10% accuracy for normal livers. So, if you have 15 milligrams per 10 grams weight liver per 12 hours, that is 30 milligrams in 24 hours. Now, normal haptoglobin in the rat is approximately 50 to 70 mg 100 ml^{-1}; you can do a quick calculation which would show that that is more than enough to accommodate the gross requirement for haptoglobin synthesis and turnover. I am quite confident of that. Incidentally, that is also true of the other plasma proteins that we have studied with perfused liver.

References

1. Salaman, D. F. and Korner, A. Rapid stimulation of nucleolar ribonucleic acid synthesis by growth hormone. *Biochem. J.*, **125** (1971), 72
2. Salaman, D. F., Betteridge, S. and Korner, A. Early effects of growth hormone on nucleolar and nucleoplasmic RNA synthesis and RNA polymerase activity in normal rat liver. *Biochem. Biophys. Acta.*, **272** (1972), 382

Index